49.95
70C

Graduate Texts in Mathematics **178**

Editorial Board
S. Axler   F.W. Gehring   K.A. Ribet

Springer
New York
Berlin
Heidelberg
Barcelona
Budapest
Hong Kong
London
Milan
Paris
Santa Clara
Singapore
Tokyo

Graduate Texts in Mathematics

1 TAKEUTI/ZARING. Introduction to Axiomatic Set Theory. 2nd ed.
2 OXTOBY. Measure and Category. 2nd ed.
3 SCHAEFER. Topological Vector Spaces.
4 HILTON/STAMMBACH. A Course in Homological Algebra. 2nd ed.
5 MAC LANE. Categories for the Working Mathematician.
6 HUGHES/PIPER. Projective Planes.
7 SERRE. A Course in Arithmetic.
8 TAKEUTI/ZARING. Axiomatic Set Theory.
9 HUMPHREYS. Introduction to Lie Algebras and Representation Theory.
10 COHEN. A Course in Simple Homotopy Theory.
11 CONWAY. Functions of One Complex Variable I. 2nd ed.
12 BEALS. Advanced Mathematical Analysis.
13 ANDERSON/FULLER. Rings and Categories of Modules. 2nd ed.
14 GOLUBITSKY/GUILLEMIN. Stable Mappings and Their Singularities.
15 BERBERIAN. Lectures in Functional Analysis and Operator Theory.
16 WINTER. The Structure of Fields.
17 ROSENBLATT. Random Processes. 2nd ed.
18 HALMOS. Measure Theory.
19 HALMOS. A Hilbert Space Problem Book. 2nd ed.
20 HUSEMOLLER. Fibre Bundles. 3rd ed.
21 HUMPHREYS. Linear Algebraic Groups.
22 BARNES/MACK. An Algebraic Introduction to Mathematical Logic.
23 GREUB. Linear Algebra. 4th ed.
24 HOLMES. Geometric Functional Analysis and Its Applications.
25 HEWITT/STROMBERG. Real and Abstract Analysis.
26 MANES. Algebraic Theories.
27 KELLEY. General Topology.
28 ZARISKI/SAMUEL. Commutative Algebra. Vol.I.
29 ZARISKI/SAMUEL. Commutative Algebra. Vol.II.
30 JACOBSON. Lectures in Abstract Algebra I. Basic Concepts.
31 JACOBSON. Lectures in Abstract Algebra II. Linear Algebra.
32 JACOBSON. Lectures in Abstract Algebra III. Theory of Fields and Galois Theory.
33 HIRSCH. Differential Topology.
34 SPITZER. Principles of Random Walk. 2nd ed.
35 ALEXANDER/WERMER. Several Complex Variables and Banach Algebras. 3rd ed.
36 KELLEY/NAMIOKA et al. Linear Topological Spaces.
37 MONK. Mathematical Logic.
38 GRAUERT/FRITZSCHE. Several Complex Variables.
39 ARVESON. An Invitation to C^*-Algebras.
40 KEMENY/SNELL/KNAPP. Denumerable Markov Chains. 2nd ed.
41 APOSTOL. Modular Functions and Dirichlet Series in Number Theory. 2nd ed.
42 SERRE. Linear Representations of Finite Groups.
43 GILLMAN/JERISON. Rings of Continuous Functions.
44 KENDIG. Elementary Algebraic Geometry.
45 LOÈVE. Probability Theory I. 4th ed.
46 LOÈVE. Probability Theory II. 4th ed.
47 MOISE. Geometric Topology in Dimensions 2 and 3.
48 SACHS/WU. General Relativity for Mathematicians.
49 GRUENBERG/WEIR. Linear Geometry. 2nd ed.
50 EDWARDS. Fermat's Last Theorem.
51 KLINGENBERG. A Course in Differential Geometry.
52 HARTSHORNE. Algebraic Geometry.
53 MANIN. A Course in Mathematical Logic.
54 GRAVER/WATKINS. Combinatorics with Emphasis on the Theory of Graphs.
55 BROWN/PEARCY. Introduction to Operator Theory I: Elements of Functional Analysis.
56 MASSEY. Algebraic Topology: An Introduction.
57 CROWELL/FOX. Introduction to Knot Theory.
58 KOBLITZ. p-adic Numbers, p-adic Analysis, and Zeta-Functions. 2nd ed.
59 LANG. Cyclotomic Fields.
60 ARNOLD. Mathematical Methods in Classical Mechanics. 2nd ed.

continued after index

F.H. Clarke
Yu.S. Ledyaev
R.J. Stern
P.R. Wolenski

Nonsmooth Analysis and Control Theory

 Springer

F.H. Clarke
Institut Desargues
Université de Lyon I
Villeurbanne, 69622
France

Yu.S. Ledyaev
Steklov Mathematics Institute
Moscow, 117966
Russia

R.J. Stern
Department of Mathematics
Concordia University
7141 Sherbrooke St. West
Montreal, PQ H4B 1R6
Canada

P.R. Wolenski
Department of Mathematics
Louisiana State University
Baton Rouge, LA 70803-0001
USA

Editorial Board

S. Axler
Mathematics Department
San Francisco State
　University
San Francisco, CA 94132
USA

F. W. Gehring
Mathematics Department
East Hall
University of Michigan
Ann Arbor, MI 48109
USA

K.A. Ribet
Department of Mathematics
University of California
　at Berkeley
Berkeley, CA 94720-3840
USA

Mathematics Subject Classification (1991): 49J52, 58C20, 90C48

With 8 figures.

Library of Congress Cataloging-in-Publication Data
Nonsmooth analysis and control theory / F.H. Clarke . . . [et al.].
　　　p.　　cm. – (Graduate texts in mathematics ; 178)
　Includes bibliographical references and index.
　ISBN 0-387-98336-8 (hardcover : alk. paper)
　1. Control theory.　2. Nonsmooth optimization.　I. Clarke, Francis H.　II. Series.
　QA402.3.N66　　1998
　515'.64–dc21　　　　　　　　　　　　　　　　　　　　　　　　　97-34140

Printed on acid-free paper.

©1998 Springer-Verlag New York, Inc.
All rights reserved. This work may not be translated or copied in whole or in part without the written permission of the publisher (Springer-Verlag New York, Inc., 175 Fifth Avenue, New York, NY 10010, USA), except for brief excerpts in connection with reviews or scholarly analysis. Use in connection with any form of information storage and retrieval, electronic adaptation, computer software, or by similar or dissimilar methodology now known or hereafter developed is forbidden.
The use of general descriptive names, trade names, trademarks, etc., in this publication, even if the former are not especially identified, is not to be taken as a sign that such names, as understood by the Trade Marks and Merchandise Marks Act, may accordingly be used freely by anyone.

Production managed by Anthony K. Guardiola; manufacturing supervised by Jacqui Ashri.
Photocomposed pages prepared from the authors' LaTeX files.
Printed and bound by Edwards Brothers, Inc., Ann Arbor, MI.
Printed in the United States of America.

9 8 7 6 5 4 3 2 1

ISBN 0-387-98336-8　Springer-Verlag　New York　Berlin　Heidelberg　　SPIN 10557384

The authors dedicate this book:
  to Gail, Julia, and Danielle;
  to Sofia, Simeon, and Irina;
  to Judy, Adam, and Sach; and
  to Mary and Anna.

Preface

>   Pardon me for writing such a long letter; I had not the time to write a short one.
>
>   —Lord Chesterfield

Nonsmooth analysis refers to differential analysis in the absence of differentiability. It can be regarded as a subfield of that vast subject known as nonlinear analysis. While nonsmooth analysis has classical roots (we claim to have traced its lineage back to Dini), it is only in the last decades that the subject has grown rapidly. To the point, in fact, that further development has sometimes appeared in danger of being stymied, due to the plethora of definitions and unclearly related theories.

One reason for the growth of the subject has been, without a doubt, the recognition that nondifferentiable phenomena are more widespread, and play a more important role, than had been thought. Philosophically at least, this is in keeping with the coming to the fore of several other types of irregular and nonlinear behavior: catastrophes, fractals, and chaos.

In recent years, nonsmooth analysis has come to play a role in functional analysis, optimization, optimal design, mechanics and plasticity, differential equations (as in the theory of viscosity solutions), control theory, and, increasingly, in analysis generally (critical point theory, inequalities, fixed point theory, variational methods ...). In the long run, we expect its methods and basic constructs to be viewed as a natural part of differential analysis.

We have found that it would be relatively easy to write a very long book on nonsmooth analysis and its applications; several times, we did. We have now managed not to do so, and in fact our principal claim for this work is that it presents the essentials of the subject clearly and succinctly, together with some of its applications and a generous supply of interesting exercises. We have also incorporated in the text a number of new results which clarify the relationships between the different schools of thought in the subject. We hope that this will help make nonsmooth analysis accessible to a wider audience. In this spirit, the book is written so as to be used by anyone who has taken a course in functional analysis.

We now proceed to discuss the contents. Chapter 0 is an Introduction in which we allow ourselves a certain amount of hand-waving. The intent is to give the reader an *avant-goût* of what is to come, and to indicate at an early stage why the subject is of interest.

There are many exercises in Chapters 1 to 4, and we recommend (to the active reader) that they be done. Our experience in teaching this material has had a great influence on the writing of this book, and indicates that comprehension is proportional to the exercises done. The end-of-chapter problems also offer scope for deeper understanding. We feel no guilt in calling upon the results of exercises later as needed.

Chapter 1, on proximal analysis, should be done carefully by every reader of this book. We have chosen to work here in a Hilbert space, although the greater generality of certain Banach spaces having smooth norms would be another suitable context. We believe the Hilbert space setting makes for a more accessible theory on first exposure, while being quite adequate for later applications.

Chapter 2 is devoted to the theory of generalized gradients, which constitutes the other main approach (other than proximal) to developing nonsmooth analysis. The natural habitat of this theory is Banach space, which is the choice made. The relationship between these two principal approaches is now well understood, and is clearly delineated here. As for the preceding chapter, the treatment is not encyclopedic, but covers the important ideas.

In Chapter 3 we develop certain special topics, the first of which is value function analysis for constrained optimization. This topic is previewed in Chapter 0, and §3.1 is helpful, though not essential, in understanding certain proofs in the latter part of Chapter 4. The next topic, mean value inequalities, offers a glimpse of more advanced calculus. It also serves as a basis for the solvability results of the next section, which features the Graves–Lyusternik Theorem and the Lipschitz Inverse Function Theorem. Section 3.4 is a brief look at a *third* route to nonsmooth calculus, one that bases itself upon directional subderivates. It is shown that the salient points of this theory can be derived from the earlier results. We also present here a self-contained proof of Rademacher's Theorem. In §3.5 we develop some machinery that is used in the following chapter, notably measurable selection. We take a quick look at variational functionals, but by-and-large, the calculus of variations has been omitted. The final section of the chapter examines in more detail some questions related to tangency.

Chapter 4, as its title implies, is a self-contained introduction to the theory of control of ordinary differential equations. This is a biased introduction, since one of its avowed goals is to demonstrate virtually all of the preceding theory in action. It makes no attempt to address issues of modeling or of implementation. Nonetheless, most of the central issues in control are studied, and we believe that any serious student of mathematical control theory will find it essential to have a grasp of the tools that are developed here via nonsmooth analysis: invariance, viability, trajectory monotonicity, viscosity solutions, discontinuous feedback, and Hamiltonian inclusions. We believe that the unified and geometrically motivated approach presented here for the first time has merits that will continue to make themselves felt in the subject.

We now make some suggestions for the reader who does not have the time to cover all of the material in this book. If control theory is of less interest, then Chapters 1 and 2, together with as much of Chapter 3 as time allows, constitutes a good introduction to nonsmooth analysis. At the other extreme is the reader who wishes to do Chapter 4 virtually in its entirety. In that case, a jump to Chapter 4 directly after Chapter 1 is feasible; only occasional references to material in Chapters 2 and 3 is made, up to §4.8, and in such a way that the reader can refer back without difficulty. The two final sections of Chapter 4 have a greater dependence on Chapter 2, but can still be covered if the reader will admit the proofs of the theorems.

A word on numbering. All items are numbered in sequence within a section; thus Exercise 7.2 precedes Theorem 7.3, which is followed by Corollary 7.4. For references between two chapters, an extra initial digit refers to the chapter number. Thus a result that would be referred to as Theorem 7.3 within Chapter 1 would be invoked as Theorem 1.7.3 from within Chapter 4. All equation numbers are simple, as in (3), and start again at (1) at the beginning of each section (thus their effect is only local). A reference to §3 is to the third section of the current chapter, while §2.3 refers to the third section of Chapter 2.

A summary of our notational conventions is given in §0.5, and a Symbol Glossary appears in the Notes and Comments at the end of the book.

We would like to express our gratitude to the personnel of the Centre de Recherches Mathématiques (CRM) of l'Université de Montréal, and in particular to Louise Letendre, for their invaluable help in producing this book.

Finally, we learned as the book was going to press, of the death of our friend and colleague Andrei Subbotin. We wish to express our sadness at his passing, and our appreciation of his many contributions to our subject.

Francis Clarke, Lyon
Yuri Ledyaev, Moscow
Ron Stern, Montréal
Peter Wolenski, Baton Rouge

May 1997

Contents

Preface   vii

List of Figures   xiii

0 Introduction   1
    1   Analysis Without Linearization   1
    2   Flow-Invariant Sets   7
    3   Optimization   10
    4   Control Theory   15
    5   Notation   18

1 Proximal Calculus in Hilbert Space   21
    1   Closest Points and Proximal Normals   21
    2   Proximal Subgradients   27
    3   The Density Theorem   39
    4   Minimization Principles   43
    5   Quadratic Inf-Convolutions   44
    6   The Distance Function   47
    7   Lipschitz Functions   51
    8   The Sum Rule   54
    9   The Chain Rule   58
    10 Limiting Calculus   61
    11 Problems on Chapter 1   63

xii   Contents

2 Generalized Gradients in Banach Space — 69
1. Definition and Basic Properties 69
2. Basic Calculus 74
3. Relation to Derivatives 78
4. Convex and Regular Functions 80
5. Tangents and Normals 83
6. Relationship to Proximal Analysis 88
7. The Bouligand Tangent Cone and Regular Sets 90
8. The Gradient Formula in Finite Dimensions 93
9. Problems on Chapter 2 96

3 Special Topics — 103
1. Constrained Optimization and Value Functions 103
2. The Mean Value Inequality 111
3. Solving Equations 125
4. Derivate Calculus and Rademacher's Theorem 136
5. Sets in L^2 and Integral Functionals 148
6. Tangents and Interiors 165
7. Problems on Chapter 3 170

4 A Short Course in Control Theory — 177
1. Trajectories of Differential Inclusions 177
2. Weak Invariance 188
3. Lipschitz Dependence and Strong Invariance 195
4. Equilibria 202
5. Lyapounov Theory and Stabilization 208
6. Monotonicity and Attainability 215
7. The Hamilton–Jacobi Equation and Viscosity Solutions . . 222
8. Feedback Synthesis from Semisolutions 228
9. Necessary Conditions for Optimal Control 230
10. Normality and Controllability 244
11. Problems on Chapter 4 247

Notes and Comments   **257**

List of Notation   **263**

Bibliography   **265**

Index   **273**

List of Figures

0.1	Torricelli's table.	12
0.2	Discontinuity of the local projection.	13
1.1	A set S and some of its boundary points.	22
1.2	A point x_1 and its five projections.	24
1.3	The epigraph of a function.	30
1.4	ζ belongs to $\partial_P f(x)$.	35
4.1	The set S of Exercise 2.12.	195
4.2	The set S of Exercise 4.3.	204

0 Introduction

> Experts are not supposed to read this book at all.
> —R.P. Boas, *A Primer of Real Functions*

We begin with a motivational essay that previews a few issues and several techniques that will arise later in this book.

1 Analysis Without Linearization

Among the issues that routinely arise in mathematical analysis are the following three:

- to minimize a function $f(x)$;

- to solve an equation $F(x) = y$ for x as a function of y; and

- to derive the stability of an equilibrium point x^* of a differential equation $\dot{x} = \varphi(x)$.

None of these issues imposes by its nature that the function involved (f, F, or φ) be smooth (differentiable); for example, we can reasonably aim to minimize a function which is merely continuous, if growth or compactness is postulated.

Nonetheless, the role of derivatives in questions such as these has been central, due to the classical technique of *linearization*. This term refers to the construction of a linear local approximation of a function by means of its derivative at a point. Of course, this approach requires that the derivative exists. When applied to the three scenarios listed above, linearization gives rise to familiar and useful criteria:

- at a minimum x, we have $f'(x) = 0$ (Fermat's Rule);

- if the $n \times n$ Jacobian matrix $F'(x)$ is nonsingular, then $F(x) = y$ is locally invertible (the Inverse Function Theorem); and

- if the eigenvalues of $\varphi'(x^*)$ have negative real parts, the equilibrium is locally stable.

The main purpose of this book is to introduce and motivate a set of tools and methods that can be used to address these types of issues, as well as others in analysis, optimization, and control, when the underlying data are not (necessarily) smooth.

In order to illustrate in a simple setting how this might be accomplished, and in order to make contact with what could be viewed as the first theorem in what has become known as *nonsmooth analysis*, let us consider the following question: to characterize in differential, thus local terms, the global property that a given continuous function $f \colon \mathbb{R} \to \mathbb{R}$ is decreasing (i.e., $x \leq y \implies f(y) \leq f(x)$).

If the function f admits a continuous derivative f', then the integration formula

$$f(y) = f(x) + \int_x^y f'(t)\,dt$$

leads to a sufficient condition for f to be decreasing: that $f'(t)$ be nonpositive for each t. It is easy to see that this is necessary as well, so a satisfying characterization via f' is obtained.

If we go beyond the class of continuously differentiable functions, the situation becomes much more complex. It is known, for example, that there exists a *strictly* decreasing continuous f for which we have $f'(t) = 0$ almost everywhere. For such a function, the derivative appears to fail us, insofar as characterizing decrease is concerned.

In 1878, Ulysse Dini introduced certain constructs, one of which is the following (lower, right) *derivate*:

$$Df(x) := \liminf_{t \downarrow 0} \frac{f(x+t) - f(x)}{t}.$$

Note that $Df(x)$ can equal $+\infty$ or $-\infty$. It turns out that Df will serve our purpose, as we now see.

1 Analysis Without Linearization

1.1. Theorem. *The continuous function* $f\colon \mathbb{R} \to \mathbb{R}$ *is decreasing iff*

$$Df(x) \leq 0 \ \forall x \in \mathbb{R}.$$

Although this result is well known, and in any case greatly generalized in a later chapter, let us indicate a nonstandard proof of it now, in order to bring out two themes that are central to this book: optimization and nonsmooth calculus.

Note first that $Df(x) \leq 0$ is an evident necessary condition for f to be decreasing, so it is the sufficiency of this property that we must prove.

Let x, y be any two numbers with $x < y$. We will prove that for any $\delta > 0$, we have

$$\min\{f(t)\colon y \leq t \leq y+\delta\} \leq f(x). \tag{1}$$

This implies $f(y) \leq f(x)$, as required.

As a first step in the proof of (1), let g be a function defined on $(x-\delta, y+\delta)$ with the following properties:

(a) g is continuously differentiable, $g(t) \geq 0$, $g(t) = 0$ iff $t = y$;

(b) $g'(t) < 0$ for $t \in (x-\delta, y)$ and $g'(t) \geq 0$ for $t \in [y, y+\delta)$; and

(c) $g(t) \to \infty$ as $t \downarrow x - \delta$, and also as $t \uparrow y + \delta$.

It is easy enough to give an explicit formula for such a function; we will not do so.

Now consider the minimization over $(x-\delta, y+\delta)$ of the function $f+g$; by continuity and growth, the minimum is attained at a point z. A necessary condition for a local minimum of a function is that its Dini derivate be nonnegative there, as is easily seen. This gives

$$D(f+g)(z) \geq 0.$$

Because g is smooth, we have the following fact (in nonsmooth calculus!):

$$D(f+g)(z) = Df(z) + g'(z).$$

Since $Df(z) \leq 0$ by assumption, we derive $g'(z) \geq 0$, which implies that z lies in the interval $[y, y+\delta)$. We can now estimate the left side of (1) as follows:

$$\min\{f(t)\colon y \leq t \leq y+\delta\} \leq f(z)$$
$$\leq f(z) + g(z) \text{ (since } g \geq 0\text{)}$$
$$\leq f(x) + g(x) \text{ (since } z \text{ minimizes } f+g\text{)}.$$

We now observe that the entire argument to this point will hold if g is replaced by εg, for any positive number ε (since εg continues to satisfy the listed properties for g). This observation implies (1) and completes the proof. □

We remark that the proof of Theorem 1.1 will work just as well if f, instead of being continuous, is assumed to be *lower semicontinuous*, which is the underlying hypothesis made on the functions that appear in Chapter 1.

An evident corollary of Theorem 1.1 is that a continuous *everywhere differentiable* function f is decreasing iff its derivative $f'(x)$ is always nonpositive, since when $f'(x)$ exists it coincides with $Df(x)$. This could also be proved directly from the Mean Value Theorem, which asserts that when f is differentiable we have

$$f(y) - f(x) = f'(z)(y - x)$$

for some z between x and y.

Proximal Subgradients

We will now consider monotonicity for functions of several variables. When x, y are points in $\mathbb{R}^n$, the inequality $x \leq y$ will be understood in the component-wise sense: $x_i \leq y_i$ for $i = 1, 2, \ldots, n$. We say that a given function $f \colon \mathbb{R}^n \to \mathbb{R}$ is decreasing provided that $f(y) \leq f(x)$ whenever $x \leq y$.

Experience indicates that the best way to extend Dini's derivates to functions of several variables is as follows: for a given direction v in $\mathbb{R}^n$ we define

$$Df(x; v) := \liminf_{\substack{t \downarrow 0 \\ w \to v}} \frac{f(x + tw) - f(x)}{t}.$$

We call $Df(x; v)$ a *directional subderivate*. Let $\mathbb{R}^n_+$ denote the positive orthant in $\mathbb{R}^n$:

$$\mathbb{R}^n_+ := \{x \in \mathbb{R}^n : x \geq 0\}.$$

We omit the proof of the following extension of Theorem 1.1, which can be given along the lines of that of Theorem 1.1.

1.2. Theorem. *The continuous function $f \colon \mathbb{R}^n \to \mathbb{R}$ is decreasing iff $Df(x; v) \leq 0 \ \forall x$ in $\mathbb{R}^n$, $\forall v \in \mathbb{R}^n_+$.*

When f is continuously differentiable, it is the case that $Df(x; v)$ agrees with $\langle \nabla f(x), v \rangle$, an observation that leads to the following consequence of the theorem:

1.3. Corollary. *A continuously differentiable function $f \colon \mathbb{R}^n \to \mathbb{R}$ is decreasing iff $\nabla f(x) \leq 0 \ \forall x \in \mathbb{R}^n$.*

Since it is easier in principle to examine one gradient vector than an infinite number of directional subderivates, we are led to seek an object that could replace $\nabla f(x)$ in a result such as Corollary 1.3, when f is nondifferentiable.

A concept that turns out to be a powerful tool in characterizing a variety of functional properties is that of the *proximal subgradient*. A vector ζ in $\mathbb{R}^n$ is said to be a proximal subgradient of f at x provided that there exist a neighborhood U of x and a number $\sigma > 0$ such that

$$f(y) \geq f(x) + \langle \zeta, y - x \rangle - \sigma \|y - x\|^2 \ \forall y \in U.$$

The set of such ζ, if any, is denoted $\partial_P f(x)$ and is referred to as the *proximal subdifferential*. The existence of a proximal subgradient ζ at x corresponds to the possibility of approximating f from below (thus in a *one-sided* manner) by a function whose graph is a parabola. The point $(x, f(x))$ is a contact point between the graph of f and the parabola, and ζ is the slope of the parabola at that point. Compare this with the usual derivative, in which the graph of f is approximated by an affine function.

Among the many properties of $\partial_P f$ developed later will be a Mean Value Theorem asserting that for given points x and y, for any $\varepsilon > 0$, we have

$$f(y) - f(x) \leq \langle \zeta, y - x \rangle + \varepsilon,$$

where ζ belongs to $\partial_P f(z)$ for some point z which lies within ε of the line segment joining x and y. This theorem requires of f merely lower semicontinuity. A consequence of this is the following.

1.4. Theorem. *A lower semicontinuous function* $f\colon \mathbb{R}^n \to \mathbb{R}$ *is decreasing iff* $\zeta \leq 0 \ \forall \zeta$ *in* $\partial_P f(x)$, $\forall x$ *in* $\mathbb{R}^n$.

We remark that Theorem 1.4 subsumes Theorem 1.2, as a consequence of the following implication, which the reader may readily confirm:

$$\zeta \in \partial_P f(x) \implies Df(x; v) \geq \langle \zeta, v \rangle \ \forall v.$$

While characterizations such as the one given by Theorem 1.4 are of intrinsic interest, it is reassuring to know that they can be and have been of actual use in practice. For example, in developing an existence theory in the calculus of variations, one approach leads to the following function f:

$$f(t) := \max\left\{ \int_0^1 L\big(s, x(s), \dot{x}(s)\big)\, ds \colon \|\dot{x}\|_2 \leq t \right\},$$

where the maximum is taken over a certain class of functions $x\colon [0,1] \to \mathbb{R}^n$, and where the function L is given. In the presence of the constraint $\|\dot{x}\|_2 \leq t$, the maximum is attained, but the object is to show that the maximum is attained even in the absence of that constraint. The approach hinges upon showing that for t sufficiently large, the function f becomes constant. Since f is increasing by definition, this amounts to showing that f is (eventually) decreasing, a task that is accomplished in part by Theorem 1.4, since there is no a priori reason for f to be smooth.

This example illustrates how nonsmooth analysis can play a partial but useful role as a tool in the analysis of apparently unrelated issues; detailed examples will be given later in connection with control theory.

It is a fact that $\partial_P f(x)$ can in general be empty almost everywhere (a.e.), even when f is a continuously differentiable function on the real line. Nonetheless, as illustrated by Theorem 1.4, and as we will see in much more complex settings, the proximal subdifferential determines the presence or otherwise of certain basic functional properties. As in the case of the derivative, the utility of $\partial_P f$ is based upon the existence of a calculus allowing us to obtain estimates (as in the proximal version of the Mean Value Theorem cited above), or to express the subdifferentials of complicated functionals in terms of the simpler components used to build them. Proximal calculus (among other things) is developed in Chapters 1 and 3, in a Hilbert space setting.

Generalized Gradients

We continue to explore the decrease properties of a given function $f\colon \mathbb{R}^n \to \mathbb{R}$, but now we introduce, for the first time, an element of volition: we wish to *find a direction* in which f decreases.

If f is smooth, linearization provides an answer: Provided that $\nabla f(x) \neq 0$, the direction $v := -\nabla f(x)$ will do, in the sense that

$$f(x+tv) < f(x) \quad \text{for } t > 0 \text{ sufficiently small.} \tag{2}$$

What if f is nondifferentiable? In that case, the proximal subdifferential $\partial_P f(x)$ may not be of any help, as when it is empty, for example.

If f is locally Lipschitz continuous, there is another nonsmooth calculus available, that which is based upon the *generalized gradient* $\partial f(x)$. A locally Lipschitz function is differentiable almost everywhere; this is Rademacher's Theorem, which is proved in Chapter 3. Its derivative f' generates $\partial f(x)$ as follows ("co" means "convex hull"):

$$\partial f(x) = \mathrm{co}\Big\{\lim_{i \to \infty} \nabla f(x_i) \colon x_i \to x,\, f'(x_i) \text{ exists}\Big\}.$$

Then we have the following result on decrease directions:

1.5. Theorem. *The generalized gradient $\partial f(x)$ is a nonempty compact convex set. If $0 \notin \partial f(x)$, and if ζ is the element of $\partial f(x)$ having minimal norm, then $v := -\zeta$ satisfies* (2).

The calculus of generalized gradients (Chapter 2) will be developed in an arbitrary Banach space, in contrast to proximal calculus.

Lest our discussion of decrease become too monotonous, we turn now to another topic, one which will allow us to preview certain *geometric* concepts that lie at the heart of future developments. For we have learned, since Dini's time, that a better theory results if functions and sets are put on an equal footing.

2 Flow-Invariant Sets

Let S be a given closed subset of $\mathbb{R}^n$ and let $\varphi\colon \mathbb{R}^n \to \mathbb{R}^n$ be locally Lipschitz. The question that concerns us here is whether the trajectories $x(t)$ of the differential equation with initial condition

$$\dot{x}(t) = \varphi\bigl(x(t)\bigr), \quad x(0) = x_0, \tag{1}$$

leave S invariant, in the sense that if x_0 lies in S, then $x(t)$ also belongs to S for $t > 0$. If this is the case, we say that the system (S, φ) is *flow-invariant*.

As in the previous section (but now for a set rather than a function), linearization provides an answer when the set S lends itself to it; that is, it is sufficiently smooth. Suppose that S is a *smooth manifold*, which means that locally it admits a representation of the form

$$S = \bigl\{x \in \mathbb{R}^n : h(x) = 0\bigr\},$$

where $h\colon \mathbb{R}^n \to \mathbb{R}^m$ is a continuously differentiable function with a nonvanishing derivative on S. Then if the trajectories of (1) remain in S, we have $h\bigl(x(t)\bigr) = 0$ for $t \geq 0$. Differentiating this for $t > 0$ gives $h'\bigl(x(t)\bigr)\dot{x}(t) = 0$. Substituting $\dot{x}(t) = \varphi(x(t))$, and letting t decrease to 0, leads to

$$\bigl\langle \nabla h_i(x_0), \varphi(x_0) \bigr\rangle = 0 \quad (i = 1, 2, \ldots, m).$$

The *tangent space* to the manifold S at x_0 is by definition the set

$$\bigl\{v \in \mathbb{R}^n : \langle \nabla h_i(x_0), v \rangle = 0, i = 1, 2, \ldots, m \bigr\},$$

so we have proven the necessity part of the following:

2.1. Theorem. *Let S be a smooth manifold. For (S, φ) to be flow-invariant, it is necessary and sufficient that, for every $x \in S$, $\varphi(x)$ belong to the tangent space to S at x.*

There are situations in which we are interested in the flow invariance of a set which is not a smooth manifold, for example, $S = \mathbb{R}^n_+$, which corresponds to $x(t) \geq 0$. It will turn out that it is just as simple to prove the sufficiency part of the above theorem in a nonsmooth setting, once we have decided upon how to define the notion of tangency when S is an arbitrary closed set. To this end, consider the *distance function* d_S associated with S:

$$d_S(x) := \min\{\|x - s\| : s \in S\},$$

a globally Lipschitz, nondifferentiable function that turns out to be very useful. Then, if $x(\cdot)$ is a solution of (1), where $x_0 \in S$, we have $f(0) = 0$, $f(t) \geq 0$ for $t \geq 0$, where f is the function defined by

$$f(t) := d_S(x(t)).$$

What property would ensure that $f(t) = 0$ for $t \geq 0$; that is, that $x(t) \in S$? Clearly, that f be decreasing: monotonicity comes again to the fore! In the light of Theorem 1.1, f is decreasing iff $Df(t) \leq 0$, a condition which at $t = 0$ says

$$\liminf_{t \downarrow 0} \frac{d_S(x(t))}{t} \leq 0.$$

Since d_S is Lipschitz, and since we have

$$x(t) = x_0 + t\varphi(x_0) + o(t),$$

the lower limit in question is equal to

$$\liminf_{t \downarrow 0} \frac{d_S(x_0 + t\varphi(x_0))}{t}.$$

This observation suggests the following definition and essentially proves the ensuing theorem, which extends Theorem 2.1 to arbitrary closed sets.

2.2. Definition. *A vector v is tangent to a closed set S at a point x if*

$$\liminf_{t \downarrow 0} \frac{d_S(x + tv)}{t} = 0.$$

The set of such vectors is a cone, and is referred to as the *Bouligand tangent cone* to S at x, denoted $T_S^B(x)$. It coincides with the tangent space when S is a smooth manifold.

2.3. Theorem. *Let S be a closed set. Then (S, φ) is flow-invariant iff*

$$\varphi(x) \in T_S^B(x) \ \forall x \in S.$$

When S is a smooth manifold, its *normal space* at x is defined as the space orthogonal to its tangent space, namely

$$\operatorname{span}\{\nabla h_i(x) : i = 1, 2, \ldots, m\},$$

and a restatement of Theorem 2.1 in terms of normality goes as follows: (S,φ) is flow-invariant iff $\langle \zeta, \varphi(x) \rangle \leq 0$ whenever $x \in S$ and ζ is a normal vector to S at x.

We now consider how to develop in the nonsmooth setting the concept of an outward normal to an arbitrary closed subset S of $\mathbb{R}^n$. The key is *projection*: Given a point u not in S, and let x be a point in S that is closest to u; we say that x lies in the projection of u onto S. Then the vector $u - x$ (and all its nonnegative multiples) defines a *proximal normal direction* to S at x. The set of all vectors constructed this way (for fixed x, by varying u) is called the *proximal normal cone* to S at x, and denoted $N_S^P(x)$. It coincides with the normal space when S is a smooth manifold.

It is possible to characterize flow-invariance in terms of proximal normals as follows:

2.4. Theorem. *Let S be a closed set. Then (S,φ) is flow-invariant iff*

$$\langle \zeta, \varphi(x) \rangle \leq 0 \ \forall \zeta \in N_S^P(x), \ \forall x \in S.$$

We can observe a certain *duality* between Theorems 2.3 and 2.4. The former characterizes flow-invariance in terms *internal* to the set S, via tangency, while the latter speaks of normals generated by looking *outside* the set. In the case of a smooth manifold, the duality is exact: the tangential and normal conditions are restatements of one another. In the general nonsmooth case, this is no longer true (pointwise, the sets T_S^B and N_S^P are not obtainable one from the other).

While the word "duality" may have to be interpreted somewhat loosely, this element is an important one in our overall approach to developing nonsmooth analysis. The dual objects often work well in tandem. For example, while tangents are often convenient to verify flow-invariance, proximal normals lie at the heart of the "proximal aiming method" used in Chapter 4 to define stabilizing feedbacks.

Another type of duality that we seek involves coherence between the various analytical and geometrical constructs that we define. To illustrate this, consider yet another approach to studying the flow-invariance of (S,φ), that which seeks to characterize the property (cited above) that the function $f(t) = d_S(x(t))$ be decreasing in terms of the proximal subdifferential of f (rather than subderivates). If an appropriate "chain rule" is available, then we could hope to use it in conjunction with Theorem 1.4 in order to reduce the question to an inequality:

$$\langle \partial_P d_S(x), \varphi(x) \rangle \leq 0 \ \forall x \in S.$$

Modulo some technicalities that will interest us later, this is feasible. In the light of Theorem 2.4, we are led to suspect (or hope for) the following fact:

$$N_S^P(x) = \text{the cone generated by } \partial_P d_S(x).$$

This type of formula illustrates what we mean by coherence between constructs, in this case between the proximal normal cone to a set and the proximal subdifferential of its distance function.

3 Optimization

As a first illustration of how nonsmoothness arises in the subject of optimization, we consider *minimax problems*. Let a smooth function f depend on two variables x and u, where the first is thought of as being a choice variable, while the second cannot be specified; it is known only that u varies in a set M. We seek to minimize f.

Corresponding to a choice of x, the worst possibility over the values of u that may occur corresponds to the following value of f: $\max_{u \in M} f(x, u)$. Accordingly, we consider the problem

$$\underset{x}{\text{minimize}}\, g(x), \text{ where } g(x) := \max_{u \in M} f(x, u).$$

The function g so defined will not generally be smooth, even if f is a nice function and the maximum defining g is attained. To see this in a simple setting, consider the upper envelope g of two smooth functions f_1, f_2. (We suggest that the reader make a sketch at this point.) Then g will have a corner at a point x where $f_1(x) = f_2(x)$, provided that

$$f_1'(x) \neq f_2'(x).$$

Returning to the general case, we remark that under mild hypotheses, the generalized gradient $\partial g(x)$ can be calculated; we find

$$\partial g(x) = \text{co}\{f_x'(x, u) \colon u \in M(x)\},$$

where

$$M(x) := \{u \in M \colon f(x, u) = g(x)\}.$$

This characterization can then serve as the initial step in approaching the problem, either analytically or numerically. There may then be explicit constraints on x to consider.

A problem having a very specific structure, and one which is of considerable importance in engineering and optimal design, is the following *eigenvalue* problem. Let the $n \times n$ symmetric matrix A depend on a parameter x in some way, so that we write $A(x)$. A familiar criterion in designing the underlying system which is represented by $A(x)$ is that the maximal eigenvalue Λ of $A(x)$ be made as small as possible. This could correspond to a question of stability, for example.

It turns out that this problem is of minimax type, for by Rayleigh's formula for the maximal eigenvalue we have

$$\Lambda(x) = \max\{\langle u, A(x)u\rangle : \|u\| = 1\}.$$

The function $\Lambda(\cdot)$ will generally be nonsmooth, even if the dependence $x \mapsto A(x)$ is itself smooth. For example, the reader may verify that the maximal eigenvalue $\Lambda(x,y)$ of the matrix

$$A(x,y) := \begin{bmatrix} 1+x & y \\ y & 1-x \end{bmatrix}$$

is given by $1 + \|(x,y)\|$. Note that the minimum of this function occurs at $(0,0)$, precisely its point of nondifferentiability. This is not a coincidence, and it is now understood that nondifferentiability is to be expected as an intrinsic feature of design problems generally, in problems as varied as designing an optimal control or finding the shape of the strongest column.

Another class of problems in which nondifferentiability plays a role is that of L^1-*optimization*. In its discrete version, the problem consists of minimizing a function f of the form

$$f(x) := \sum_{i=1}^{p} m_i \|x - s_i\|. \tag{1}$$

Such problems arise, for example, in approximation and statistics, where L^1-approximation possesses certain features that can make it preferable to the more familiar (and smooth) L^2-approximation.

Let us examine such a problem in the context of a simple physical system.

Torricelli's Table

A table has holes in it at points whose coordinates are $s_1, s_2, \ldots, s_p$. Strings are attached to masses $m_1, m_2, \ldots, m_p$, passed through the corresponding hole, and then are all tied to a point mass m whose position is denoted x (see Figure 0.1). If friction and the weight of the strings are negligible, the equilibrium position x of the nexus is precisely the one that minimizes the function f given by (1), since $f(x)$ can be recognized as the potential energy of the system.

The proximal subdifferential of the function $x \mapsto \|x - s\|$ is the closed unit ball if $x = s$, and otherwise is the singleton set consisting of its derivative, the point $(x-s)/\|x-s\|$. Using this fact, and some further calculus, we can derive the following necessary condition for a point x to minimize f;

$$0 \in \sum_{i=1}^{p} m_i \partial_P \|(\cdot) - s_i\|(x). \tag{2}$$

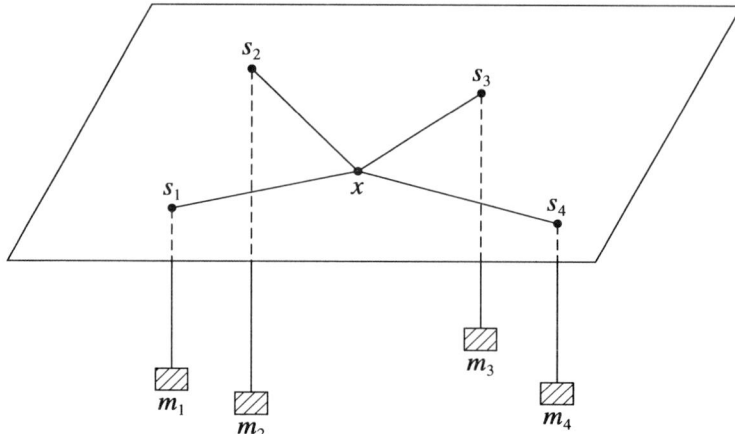

FIGURE 0.1. Torricelli's table.

Of course, (2) is simply Fermat's rule in subdifferential terms, interpreted for the particular function f that we are dealing with.

There is not necessarily a unique point x that satisfies relation (2), but it is the case that any point satisfying (2) globally minimizes f. This is because f is *convex*, another functional class that plays an important role in the subject. A consequence of convexity is that there are no purely local minima in this problem.

When $p = 3$, each $m_i = 1$, and the three points are the vertices of a triangle, the problem becomes that of finding a point such that the sum of its distances from the vertices is minimal. The solution is called the *Torricelli point*, after the seventeenth-century mathematician.

The fact that (2) is necessary and sufficient for a minimum allows us to recover easily certain classical conclusions regarding this problem. As an example, the reader is invited to establish that the Torricelli point coincides with a vertex of the triangle iff the angle at that vertex is 120° or more.

Returning now to the general case of our table, it is possible to make the system far more complex by the addition of one more string and one more mass m_0, if we allow that mass to hang over the outside edge of the table. Then the extra string will automatically trace a line segment from x to a point $s(x)$ on the edge of the table that is closest to x (locally at least, in the sense that $s(x)$ is the closest point to x on the edge, relative to a neighborhood of $s(x)$.) If S is the set defined as the closure of the complement of the table, the potential energy (up to a constant) of the

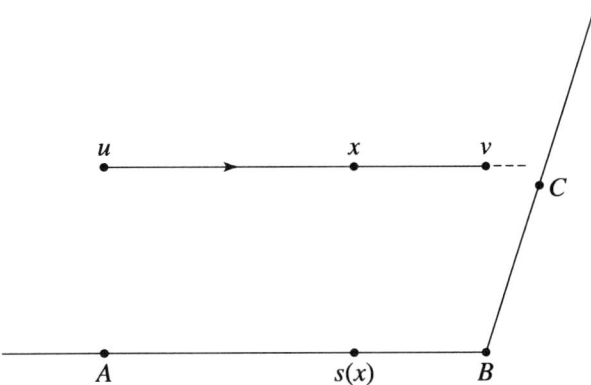

FIGURE 0.2. Discontinuity of the local projection.

system is now, at its lowest level,

$$\tilde{f}(x) := m_0 d_S(x) + \sum_{i=1}^{p} m_i \|x - s_i\|.$$

The function $\tilde{f}$ is not only nonsmooth, as was f, but also nonconvex, and will admit *local minima* at different energy levels. The points s on the boundary of S which are feasible as points through which would pass the over-the-table string (at equilibrium) are precisely those for which the proximal normal cone $N_S^P(s)$ is nonzero. Such points can be rather sparse, though they are always dense in the boundary of S. For a rectangular table, there are exactly four points at which N_S^P is $\{0\}$.

If $x(t)$ represents a displacement undergone by the nexus over time, Newton's Law implies

$$M\ddot{x} = \sum_{i=1}^{p} m_i \frac{(s_i - x)}{\|s_i - x\|} + m_0 \frac{s(x) - x}{\|s(x) - x\|} \qquad (3)$$

on any time interval during which $x \neq s_i$, $x \neq s(x)$, where M is the total mass of the system, namely $m + m_0 + \sum m_i$. The local projection $x \mapsto s(x)$ will be discontinuous in general, so in solving (3), there arises the issue of a differential equation incorporating a *discontinuous* function of the state.

Figure 0.2 illustrates the discontinuity of $s(x)$ in a particular case. As x traces the line segment from u toward v, the corresponding $s(x)$ traces the segment joining A and B. When x goes beyond v, $s(x)$ abruptly moves to the vicinity of the point C. (The figure omits all the strings acting upon x.)

We will treat the issue of discontinuous differential equations in Chapter 4, where it arises in connection with feedback control design.

Constrained Optimization

In minimizing a function $f(x)$, it is often necessary to take account of explicit constraints on the point x, for example, that x lie in a given set S. There are two methods for dealing with such problems that figure prominently in this book.

The first of these, called *exact penalization*, seeks to replace the constrained optimization problem

$$\text{minimize } f(x) \text{ subject to } x \in S$$

by the unconstrained problem

$$\text{minimize } f(x) + K d_S(x),$$

where d_S is the distance function introduced in §2. Under mild conditions, this constraint-removal technique is justified, for K sufficiently large.

Since the distance function is *never* differentiable at all boundary points of S, however, and since that is precisely where the solutions of the new problem are likely to lie, we are subsequently obliged to deal with a nonsmooth minimization problem, even if the original problem has smooth data f, S.

The second general technique for dealing with constrained optimization, called *value function analysis*, is applied when the constraint set S has an explicit functional representation, notably in terms of equalities and inequalities. A simple case to illustrate: we seek to minimize $f(x)$ subject to $h(x) = 0$. Let us embed the problem in a family of similar ones, parametrized by a perturbation term in the equality constraint. Specifically, the problem $P(\alpha)$ is the following:

$$P(\alpha): \text{ minimize } f(x) \text{ over } x \text{ subject to } h(x) + \alpha = 0.$$

Let $V(\alpha)$, the associated value function of this perturbation scheme, designate the minimum value of the problem $P(\alpha)$.

Our original problem is simply $P(0)$. If x_0 is a solution of $P(0)$, then of course $h(x_0) = 0$ (since x_0 must be *feasible* for $P(0)$), and we have $V(0) = f(x_0)$. This last observation implies that

$$f(x_0) - V(-h(x_0)) = 0,$$

whereas it follows from the very definition of V that, for any x whatsoever, we have

$$f(x) - V(-h(x)) \geq 0.$$

(We ask our readers to convince themselves of this.) Put another way, these observations amount to saying that the function

$$x \mapsto f(x) - V(-h(x))$$

attains a minimum at $x = x_0$, whence
$$f'(x_0) + V'(0)h'(x_0) = 0,$$
a conclusion that we recognize as the Lagrange Multiplier Rule (with, as a bonus, a sensitivity interpretation of the multiplier, $V'(0)$).

If our readers are dubious about this simple proof of the Multiplier Rule, they are justified in being so. Still, the only fallacy involved is the implicit assumption that V is differentiable. Nonsmooth analysis will allow us to develop a rigorous argument along the lines of the above, in Chapter 3.

4 Control Theory

In the control theory of ordinary differential equations, the standard model revolves around the system
$$\dot{x}(t) = f(x(t), u(t)) \text{ a.e.}, \quad 0 \leq t \leq T, \tag{1}$$
where the (measurable) *control function* $u(\cdot)$ is chosen subject to the constraint
$$u(t) \in U \text{ a.e.}, \tag{2}$$
and where the ensuing *state* $x(\cdot)$ is subject to an initial condition $x(0) = x_0$ and perhaps other constraints. This indirect control of $x(\cdot)$ via the choice of $u(\cdot)$ is to be exercised for a purpose, of which there are two principal sorts: *positional* ($x(t)$ is to remain in a given set in $\mathbb{R}^n$, or approach that set) and *optimal* ($x(\cdot)$, together with $u(\cdot)$, is to minimize a given functional).

As is the case in optimization, certain problems arise in which the underlying data are nonsmooth; minimax criteria are an example. In this section, however, we wish to convey to the reader how considerations of nondifferentiability arise from the very way in which we might hope to solve the problem. Our illustrative example will be one that combines positional and optimal considerations, namely the *minimal time* problem.

It consists of finding the least $T \geq 0$ together with a control function $u(\cdot)$ on $[0, T]$ having the property that the resulting state x satisfies $x(T) = 0$. Informally, it is required to steer the initial state x_0 to the origin in least time.

Let us introduce the following set-valued mapping F:
$$F(x) := f(x, U).$$
A *trajectory* of F on an interval $[0, T]$ is an absolutely continuous function $x(\cdot)$ on $[0, T]$ which satisfies
$$\dot{x}(t) \in F(x(t)) \text{ a.e.}, \quad 0 \leq t \leq T. \tag{3}$$

Under mild hypotheses, it is a fact that $x(\cdot)$ is a trajectory (i.e., satisfies (3)) iff there is a control function $u(\cdot)$ (i.e., a measurable function $u(\cdot)$ satisfying (2)) for which the differential equation (1) linking x and u holds. (See Chapter 3 for this; here, we are not even going to state hypotheses at all.)

In terms of trajectories, then, the problem is to find one which is optimal from x_0; that is, one which reaches the origin as quickly as possible. Let us undertake the quest.

We begin by introducing the *minimal time function* $T(\cdot)$, defined on $\mathbb{R}^n$ as follows:

$$T(\alpha) := \min\{T \geq 0 \colon \text{ some trajectory } x(\cdot) \text{ satisfies } x(0) = \alpha,\ x(T) = 0.\}$$

An issue of *controllability* arises here: Is it always possible to steer α to 0 in finite time? We will study this question in Chapter 4; for now, let us assume that such is the case.

The *principle of optimality* is the dual observation that if $x(\cdot)$ is any trajectory, the function

$$t \mapsto T(x(t)) + t$$

is increasing, and that if x is optimal, then the same function is constant. In other terms, if $x(\cdot)$ is an optimal trajectory joining α to 0, then

$$T(x(t)) = T(\alpha) - t \quad \text{for } 0 \leq t \leq T(\alpha),$$

since an optimal trajectory from the point $x(t)$ is furnished by the truncation of $x(\cdot)$ to the interval $[t, T(\alpha)]$. If $x(\cdot)$ is any trajectory, then the inequality

$$T(x(t)) \geq T(\alpha) - t$$

is a reflection of the fact that in going to the point $x(t)$ from α (in time t), we may have acted optimally (in which case equality holds) or not (then inequality holds).

Since $t \mapsto T(x(t)) + t$ is increasing, we expect to have

$$\langle \nabla T(x(t)), \dot{x}(t)\rangle + 1 \geq 0, \tag{4}$$

with equality when $x(\cdot)$ is an optimal trajectory. The possible values of $\dot{x}(t)$ for a trajectory being precisely the elements of the set $F(x(t))$, we arrive at

$$\min_{v \in F(x)} \{\langle \nabla T(x), v\rangle\} + 1 = 0.$$

We define the (lower) *Hamiltonian function* h as follows:

$$h(x, p) := \min_{v \in F(x)} \langle p, v\rangle.$$

In terms of h, the partial differential equation obtained above reads

$$h(x, \nabla T(x)) + 1 = 0, \tag{5}$$

a special case of the *Hamilton–Jacobi equation*.

Here is the first step in our quest: use the Hamilton–Jacobi equation (5), together with the boundary condition $T(0) = 0$, to find $T(\cdot)$. How will this help us find the optimal trajectory?

To answer this question, we recall that an optimal trajectory is such that equality holds in (4). This suggests the following procedure: for each x, let $\hat{v}(x)$ be a point in $F(x)$ satisfying

$$\min_{v \in F(x)} \langle \nabla T(x), v \rangle = \langle \nabla T(x), \hat{v}(x) \rangle = -1. \tag{6}$$

Then, if we construct $x(\cdot)$ via the initial-value problem

$$\dot{x}(t) = \hat{v}(x(t)), \quad x(0) = \alpha, \tag{7}$$

we will have a trajectory that is optimal (from α)!

Here is why: Let $x(\cdot)$ satisfy (7); then $x(\cdot)$ is a trajectory, since $\hat{v}(x)$ belongs to $F(x)$. Furthermore,

$$\frac{d}{dt} T(x(t)) = \langle \nabla T(x(t)), \dot{x}(t) \rangle$$
$$= \langle \nabla T(x(t)), \hat{v}(x(t)) \rangle = -1.$$

In consequence, we find

$$T(x(t)) = T(\alpha) - t,$$

which implies that at $t = T(\alpha)$, we must have $x = 0$. Therefore $x(\cdot)$ is an optimal trajectory.

Let us stress the important point that $\hat{v}(\cdot)$ generates the optimal trajectory from *any* initial value α (via (7)), and so constitutes what can be considered the Holy Grail for this problem: an *optimal feedback synthesis*. There can be no more satisfying answer to the problem: If you find yourself at x, just choose $\dot{x} = \hat{v}(x)$ to approach the origin as fast as possible.

Unfortunately, there are serious obstacles to following the route that we have just outlined, beginning with the fact that T is nondifferentiable, as simple examples show. (T is a value function, analogous to the one we met in §3.)

We will therefore have to examine anew the argument that led to the Hamilton–Jacobi equation (5), which in any case, will have to be recast in some way to accommodate nonsmooth solutions. Having done so, will the generalized Hamilton–Jacobi equation admit T as the unique solution?

The next step (after characterizing T) offers fresh difficulties of its own. Even if T were smooth, there would be in general no *continuous* function $\hat{v}(\cdot)$ satisfying (6) for each x. The meaning and existence of a trajectory $x(\cdot)$ generated by $\hat{v}(\cdot)$ via (7) is therefore problematic in itself.

The intrinsic difficulties of the "dynamic programming" approach to the minimal-time problem, which is what we have outlined above, have been an historical focal point of activity in differential equations and control, and it is only recently that fully satisfying answers to all the questions raised above have been found. We will present them in Chapter 4, together with results bearing on other basic topics in mathematical control theory: invariance, equilibria, stability, and necessary and sufficient conditions for optimality.

Let us begin now to be more precise.

5 Notation

We expect our readers to have taken a course in functional analysis, and we hope that the following notation appears natural to them.

X is a real Hilbert space or Banach space with norm $\|\cdot\|$. The *open ball* in X (of radius 1, centered at 0) is denoted by B, its closure by $\bar{B}$. We also write B_X if X is to be distinguished from other spaces.

The *inner product* of ζ and x is denoted $\langle \zeta, x \rangle$, a notation which is also employed when X is a Banach space for the evaluation, at $x \in X$, of the linear functional $\zeta \in X^*$ (the space of continuous linear functionals defined on X).

The open unit ball in X^* is written B_*. The notation

$$x = \underset{i \to \infty}{\text{w-lim}}\, x_i$$

means that the sequence $\{x_i\}$ converges weakly to x. Similarly, w^* refers to the weak* topology on the space X^*. $L_n^p[a, b]$ refers to the set of p-integrable functions from $[a, b]$ to $\mathbb{R}^n$.

For the two subsets S_1 and S_2 of X, the set $S_1 + S_2$ is given by

$$\{s = s_1 + s_2 \colon s_1 \in S_1, s_2 \in S_2\}.$$

The open ball of radius $r > 0$, centered at x, is denoted by either $B(x; r)$ or $x + rB$, where (strictly speaking) the latter should be written $\{x\} + rB$. The closure of $B(x; r)$ is written as either $\bar{B}(x; r)$ or $x + r\bar{B}$.

We confess to writing "iff" for "if and only if." The symbol := means "equal by definition."

We lean toward mnemonic notation in general. For a given set S, the expressions
$$\text{int } S, \quad \text{cl } S, \quad \text{bdry } S, \quad \text{co } S, \quad \overline{\text{co}}\, S,$$
signify the interior, closure, boundary, convex hull, and closed convex hull of S, respectively.

A list of the principal notational constructs used in the book is given in the Notes and Comments at the end. A reference such as Theorem 1.2.3 refers to Theorem 2.3 of Chapter 1, which will be found in §2.3. From within Chapter 1, it is referred to simply as Theorem 2.3.

1
Proximal Calculus in Hilbert Space

> Shall we begin with a few Latin terms?
>
> —*Dangerous Liaisons*, the Film.

We introduce in this chapter two basic constructs of nonsmooth analysis: proximal normals (to a set) and proximal subgradients (of a function). Proximal normals are direction vectors pointing outward from a set, generated by projecting a point onto the set. Proximal subgradients have a certain local support property to the epigraph of a function. It is a familiar device to view a function as a set (through its graph), but we develop the duality between functions and sets to a much greater extent, extending it to include the calculus of these normals and subgradients. The very existence of a proximal subgradient often says something of interest about a function at a point; the Density Theorem of §3 is a deep result affirming existence on a substantial set. From it we deduce two minimization principles. These are theorems bearing upon situations where a minimum is "almost attained," and which assert that a small perturbation leads to actual attainment. We will meet some useful classes of functions along the way: convex, Lipschitz, indicator, and distance functions. Finally, we will see some elements of proximal calculus, notably the sum and chain rules.

1 Closest Points and Proximal Normals

Let X be a real Hilbert space, and let S be a nonempty subset of X. Suppose that x is a point not lying in S. Suppose further that there exists

1. Proximal Calculus in Hilbert Space

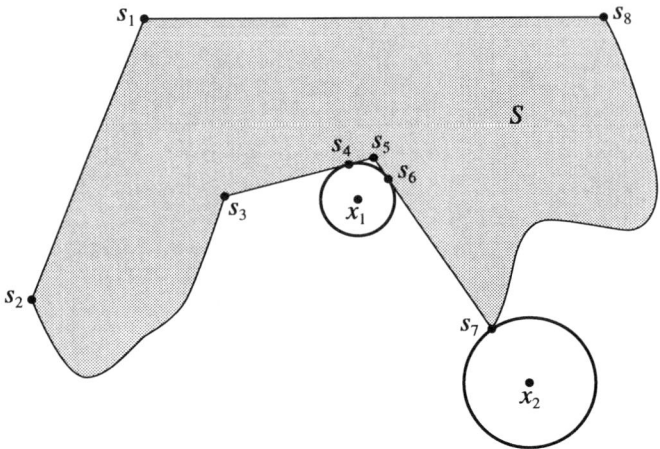

FIGURE 1.1. A set S and some of its boundary points.

a point s in S whose distance to x is minimal. Then s is called a *closest point* or a *projection* of x onto S. The set of all such closest points is denoted by $\mathrm{proj}_S(x)$. It is clear that $s \in \mathrm{proj}_S(x)$ iff $\{s\} \subset S \cap \overline{B}(x; \|x - s\|)$ and $S \cap B(x; \|x - s\|) = \emptyset$. See Figure 1.1.

The vector $x - s$ determines what we will call a *proximal normal direction* to S at s; any nonnegative multiple $\zeta = t(x - s)$, $t \geq 0$, of such a vector will be called a proximal normal (or a P-normal) to S at s. The set of all ζ obtainable in this manner is termed the *proximal normal cone* to S at s, and is denoted by $N_S^P(s)$; it is clear that $N_S^P(s)$ is in fact a cone; i.e., a set closed under forming nonnegative scalar multiples. Intuitively, a proximal normal vector at a given point defines a direction of perpendicular departure from the set.

Suppose $s \in S$ is such that $s \notin \mathrm{proj}_S(x)$ for all x not in S (which is certainly the case if s lies in int S). Then we set $N_S^P(s) = \{0\}$. When $s \notin S$, then $N_S^P(s)$ remains undefined. In Figure 1.1, the points s_3 and s_5 have P-normal cones equal to $\{0\}$, and the points s_1, s_2, s_7, and s_8 have at least two independent vectors in their P-normal cones. The remaining boundary points of S have their P-normal cone generated by a single nonzero vector.

Notice that we have not asserted above that the point x must admit a closest point s in S. In finite dimensions, there is little difficulty in assuring that projections exist, for it suffices that S be closed. We will in fact only focus on closed sets S, but nonetheless, the issue of the existence of closest points in infinite dimensions is far more subtle, and will be an important point later.

1.1. Exercise. Let X admit a countable orthonormal basis $\{e_i\}_{i=1}^{\infty}$, and set
$$S := \left\{\frac{i+1}{i} e_i : i \geq 1\right\}.$$
Prove that S is closed, and that $\text{proj}_S(0) = \emptyset$.

The above concepts can be described in terms of the *distance function* $d_S \colon X \to \mathbb{R}$, which is given by
$$d_S(x) := \inf\{\|x - s\| : s \in S\}.$$

Occasionally it is more convenient to write $d(x; S)$ for $d_S(x)$. The set $\text{proj}_S(x)$ consists of those points (if any) at which the infimum defining $d_S(x)$ is attained. We also have the formula
$$N_S^P(s) = \{\zeta \colon \exists t > 0 \text{ so that } d_S(s + t\zeta) = t\|\zeta\|\}.$$

Some further basic properties of d_S are listed in the following exercise:

1.2. Exercise.

(a) Show that x belongs to $\text{cl}\, S$ iff $d_S(x) = 0$.

(b) Suppose that S and S' are two subsets of X. Show that $d_S = d_{S'}$ iff $\text{cl}\, S = \text{cl}\, S'$.

(c) Show that d_S satisfies
$$|d_S(x) - d_S(y)| \leq \|x - y\| \ \forall x, y \in X,$$
which says that d_S is *Lipschitz* of rank 1, on X.

(d) If S is a closed subset of $\mathbb{R}^n$, show that $\text{proj}_S(x) \neq \emptyset$ for all x, and that the set $\{s \in \text{proj}_S(x) \colon x \in \mathbb{R}^n \backslash S\}$ is dense in $\text{bdry}\, S$. (*Hint.* Let $s \in \text{bdry}\, S$, and let $\{x_i\}$ be a sequence not in S that converges to s. Show that any sequence $\{s_i\}$ chosen with $s_i \in \text{proj}\, x_i$, converges to s.)

Suppose now that $s \in \text{proj}_S(x)$. This is equivalent to the condition
$$\|x - s'\| \geq \|x - s\| \ \forall s' \in S.$$

If we square both sides of this inequality and expand in terms of the inner product, we thus obtain the conclusion that $s \in \text{proj}_S(x)$ iff
$$\langle x - s, s' - s \rangle \leq \tfrac{1}{2}\|s' - s\|^2 \ \forall s' \in S.$$

This in turn is clearly equivalent to
$$\langle [s + t(x - s)] - s, s' - s \rangle \leq \tfrac{1}{2}\|s' - s\|^2 \ \forall t \in [0, 1], \ \forall s' \in S,$$

24   1. Proximal Calculus in Hilbert Space

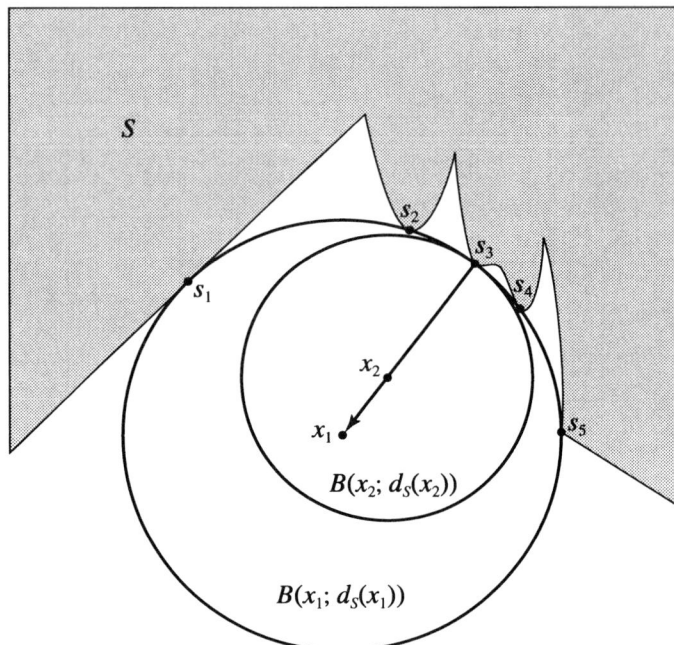

FIGURE 1.2. A point x_1 and its five projections.

which (by the preceding characterization) holds iff for all $t \in [0,1]$, we have $s \in \text{proj}_S(s + t(x-s))$. These remarks are summarized in the following:

1.3. Proposition. *Let S be a nonempty subset of X, and let $x \in X$, $s \in S$. The following are equivalent:*

(a) $s \in \text{proj}_S(x)$;

(b) $s \in \text{proj}_S(s + t(x-s))$ $\forall t \in [0,1]$;

(c) $d_S(s + t(x-s)) = t\|x - s\|$ $\forall t \in [0,1]$; *and*

(d) $\langle x - s, s' - s \rangle \leq \frac{1}{2}\|s' - s\|^2$ $\forall s' \in S$.

1.4. Exercise. For $0 < t < 1$ in Proposition 1.3(b), we have
$$\text{proj}_S(s + t(x - s)) = \{s\};$$
that is, if x has a closest point s in S, then $s + t(x - s)$ has a *unique* closest point in S. (See Figure 1.2, taking $x = x_1$, $s = s_3$, and $s + t(x - s) = x_2$.)

The first part of the following result follows readily from the cone property of $N_S^P(s)$ and the characterization (d) of Proposition 1.3; the second part demonstrates that P-normality is essentially a local property: the proximal normal cones $N^P_{S_1}(s)$ and $N^P_{S_2}(s)$ are the same if the two sets S_1 and S_2 are the same in a neighborhood of s. The inequality in Proposition 1.5(a) is called the *proximal normal inequality*.

1.5. Proposition.

(a) *A vector ζ belongs to $N^P_S(s)$ iff there exists $\sigma = \sigma(\zeta,s) \geq 0$ such that*

$$\langle \zeta, s' - s \rangle \leq \sigma \|s' - s\|^2 \ \forall s' \in S.$$

(b) *Furthermore, for any given $\delta > 0$, we have $\zeta \in N^P_S(s)$ iff there exists $\sigma = \sigma(\zeta, s) \geq 0$ such that*

$$\langle \zeta, s' - s \rangle \leq \sigma \|s' - s\|^2 \ \forall s' \in S \cap B(s;\delta).$$

The only item requiring proof is the following:

1.6. Exercise. Prove that if the inequality of (b) holds for some σ and δ, then that of (a) holds for some possibly larger σ.

The previous proposition makes it evident that $N^P_S(s)$ is convex; however, it need be neither open nor closed. That $N^P_S(s)$ can be trivial (i.e., reduce to $\{0\}$) even when S is a closed subset of $\mathbb{R}^n$ and s lies in bdry S, can easily be seen by considering the set

$$S := \{(x,y) \in \mathbb{R}^2 : y \geq -|x|\}.$$

There are no points outside S whose closest point in S is $(0,0)$ (to put this another way: no ball whose interior fails to intersect S can have $(0,0)$ on its boundary). Thus $N^P_S(0,0) = \{0\}$. A slightly more complicated but smoother example is the following:

1.7. Exercise. Consider S defined as

$$S := \{(x,y) \in \mathbb{R}^2 : y \geq -|x|^{3/2}\}.$$

Show that for $(x,y) \in$ bdry S, $N^P_S(x,y) = (0,0)$ iff $(x,y) = (0,0)$.

1.8. Exercise. Let $X = X_1 \oplus X_2$ be an orthogonal decomposition, and suppose $S \subseteq X$ is closed, $s \in S$, and $\zeta \in N^P_S(s)$. Write $s = (s_1, s_2)$ and $\zeta = (\zeta_1, \zeta_2)$ according to the given decomposition, and define $S_1 = \{s'_1 : (s'_1, s_2) \in S\}$, and similarly define S_2. Show that $\zeta_i \in N^P_{S_i}(s_i)$, $i = 1, 2$.

26   1. Proximal Calculus in Hilbert Space

The next two propositions illustrate that the concept of a proximal normal generalizes two classical definitions, that of a normal direction to a C^2 manifold as defined in differential geometry, and that of a normal vector in the context of convex analysis.

Consider a closed subset S of $\mathbb{R}^n$ that admits a representation of the form

$$S = \{x \in \mathbb{R}^n : h_i(x) = 0, i = 1, 2, \ldots, k\}, \tag{1}$$

where each $h_i : \mathbb{R}^n \to \mathbb{R}$ is C^1. If the vectors $\{\nabla h_i(s)\}$ ($i = 1, 2, \ldots, k$) are linearly independent at each $s \in S$, then S is a C^1 manifold of dimension $n - k$.

1.9. Proposition. *Let $s \in S$, where S is given by (1), and assume that the vectors $\{\nabla h_i(s)\}$ ($i = 1, 2, \ldots, k$) are linearly independent. Then:*

(a) $N_S^P(s) \subseteq \mathrm{span}\{\nabla h_i(s)\}$ ($i = 1, 2, \ldots, k$).

(b) *If in addition each h_i is C^2, then equality holds in* (a).

Proof. Let ζ belong to $N_S^P(s)$. By Proposition 1.5, there exists a constant $\sigma > 0$ so that

$$\langle \zeta, s' - s \rangle \leq \sigma \|s' - s\|^2,$$

whenever s' belongs to S. Put another way, this is equivalent to saying that the point s minimizes the function $s' \mapsto \langle -\zeta, s' \rangle + \sigma \|s' - s\|^2$ over all points s' satisfying $h_i(s') = 0$ ($i = 1, 2, \ldots, k$). The Lagrange multiplier rule of classical calculus provides a set of scalars $\{\mu_i\}_{i=1}^k$ such that $\zeta = \sum_i \mu_i \nabla h_i(s)$, which establishes (a).

Now let ζ have the form $\sum_i \mu_i \nabla h_i(s)$, where each h_i is C^2. Consider the C^2 function

$$g(x) := \langle -\zeta, x \rangle + \sum_i \mu_i h_i(x) + \sigma \|x - s\|^2,$$

where $\sigma > 0$. Then $g'(s) = 0$, and for σ sufficiently large we have $g''(s) > 0$ (positive definite), from which it follows that g admits a local minimum at s. Consequently, if s' is near enough to s and satisfies $h_i(s') = 0$ for each i, we have

$$g(s') = \langle -\zeta, s' \rangle + \sigma \|s' - s\|^2 \geq g(s) = \langle -\zeta, s \rangle.$$

This confirms the proximal normal inequality and completes the proof. □

The special case in which S is convex is an important one.

1.10. Proposition. *Let S be closed and convex. Then*

(a) $\zeta \in N_S^P(s)$ *iff*

$$\langle \zeta, s' - s \rangle \leq 0 \ \forall s' \in S.$$

(b) *If X is finite-dimensional and $s \in \text{bdry}(S)$, then $N_S^P(s) \neq \{0\}$.*

Proof. The inequality in (a) holds iff the proximal normal inequality holds with $\sigma = 0$. Hence the "if" statement is immediate from Proposition 1.5(a). To see the converse, let $\zeta \in N_S^P(s)$ and $\sigma > 0$ be chosen as in the proximal normal inequality. Let s' be any point in S. Since S is convex, the point $\tilde{s} := s + t(s' - s) = ts' + (1-t)s$ also belongs to S for each $t \in (0,1)$. The proximal normal inequality applied to $\tilde{s}$ gives

$$\langle \zeta, t(s' - s) \rangle \leq \sigma t^2 |s' - s|^2.$$

Dividing across by t and letting $t \downarrow 0$ yields the desired inequality.

To prove (b), let $\{s_i\}$ be a sequence in S converging to s so that $N_S^P(s_i) \neq \{0\}$ for all i. Such a sequence exists by Exercise 1.2(d). Let $\zeta_i \in N_S^P(s_i)$ satisfy $\|\zeta_i\| = 1$, and passing to a subsequence if necessary, assume that $\zeta_i \to \zeta$ as $i \to \infty$, and note that $\|\zeta\| = 1$. By part (a), we have

$$\langle \zeta_i, s' - s_i \rangle \leq 0 \ \forall s' \in S.$$

Letting $i \to \infty$ yields

$$\langle \zeta, s' - s \rangle \leq 0 \ \forall s' \in S,$$

which, again by part (a), says that $\zeta \in N_S^P(s)$. □

Let $0 \neq \zeta \in X$ and $r \in \mathbb{R}$. A *hyperplane* (with associated normal vector ζ) is any set of the form $\{x \in X : \langle \zeta, x \rangle = r\}$, and a *half-space* is a set of the form $\{x \in X : \langle \zeta, x \rangle \leq r\}$. Proposition 1.10(b) is a *separation theorem*, for it says that each point in the boundary of a convex set lies in some hyperplane, with the set itself lying in one of the associated half-spaces. An example given in the end-of-chapter problems shows that this fact fails in general when X is infinite dimensional, although separation does hold under additional hypotheses.

We now turn our attention from sets to functions.

2 Proximal Subgradients

We begin by establishing some notation and recalling some facts about functions.

A quite useful convention prevalent in the theories of integration and optimization, which we will also adopt, is to allow for functions $f \colon X \to (-\infty, +\infty]$; that is, functions which are *extended real-valued*. As we will see, there are many advantages in allowing f to actually attain the value $+\infty$ at a given point. To single out those points at which f is not $+\infty$, we define the (effective) *domain* as the set

$$\text{dom}\, f := \{x \in X : f(x) < \infty\}.$$

The *graph* and *epigraph* of f are given, respectively, by

$$\operatorname{gr} f := \{(x, f(x)) : x \in \operatorname{dom} f\},$$
$$\operatorname{epi} f := \{(x, r) \in \operatorname{dom} f \times \mathbb{R} : r \geq f(x)\}.$$

Just as sets are customarily assumed to be closed, the usual background assumption on f is that of *lower semicontinuity*. A function $f \colon X \to (-\infty, +\infty]$ is lower semicontinuous at x provided that

$$\liminf_{x' \to x} f(x') \geq f(x).$$

This condition is clearly equivalent to saying that for all $\varepsilon > 0$, there exists $\delta > 0$ so that $y \in B(x; \delta)$ implies $f(y) \geq f(x) - \varepsilon$, where as usual, $\infty - r$ is interpreted as ∞ when $r \in \mathbb{R}$.

Complementary to lower semicontinuity is *upper semicontinuity*: f is upper semicontinuous at x if $-f$ is lower semicontinuous at x. Lower semicontinuous functions are featured prominently in our development, but of course our results have upper semicontinuous analogues, although we will rarely state them. This preference for lower semicontinuity explains why $+\infty$ is allowed as a function value and not $-\infty$.

As is customary, we say that a function f is *continuous* at $x \in X$ provided it is finite-valued near x and for all $\varepsilon > 0$, there exists $\delta > 0$ so that $y \in B(x; \delta)$ implies $|f(x) - f(y)| \leq \varepsilon$. For finite-valued f, this is equivalent to saying that f is both lower and upper semicontinuous at x. If f is lower semicontinuous (respectively, upper semicontinuous, continuous) at *each* point x in an open set $U \subset X$, then f is called lower semicontinuous (respectively, upper semicontinuous, continuous) on U.

To restrict certain pathological functions from entering the discussion, we designate by $\mathcal{F}(U)$, where $U \subseteq X$ is open, the class of all functions $f \colon X \to (-\infty, \infty]$ which are lower semicontinuous on U and such that $\operatorname{dom} f \cap U \neq \emptyset$. If $U = X$, then we simply write $\mathcal{F}$ for $\mathcal{F}(X)$.

Let S be a subset of X. The *indicator function* of S, denoted either by $I_S(\cdot)$ or $I(\cdot; S)$, is the extended-valued function defined by

$$I_S(x) := \begin{cases} 0 & \text{if } x \in S, \\ +\infty & \text{otherwise.} \end{cases}$$

Let $U \subset X$ be an open convex set. A function $f \colon X \to (-\infty, \infty]$ is said to be *convex* on U provided

$$f(tx + (1-t)y) \leq tf(x) + (1-t)f(y) \ \forall x, y \in U, \ 0 < t < 1.$$

A function f which is convex on X is simply said to be convex. Note that $\operatorname{dom} f$ is necessarily a convex set if f is convex.

The following exercise contains some elementary properties of lower semicontinuous and convex functions. Parts (a) and (b) in particular help to demonstrate why the epigraph, rather than the graph, of a function plays the fundamental role in the analysis of lower semicontinuous functions. Note that $X \times \mathbb{R}$, the space in which epi f lives, is always viewed as a Hilbert space with inner product $\langle (x,r), (x',r') \rangle := \langle x, x' \rangle + rr'$.

2.1. Exercise. Suppose $f\colon X \to (-\infty, +\infty]$.

(a) Show that f is lower semicontinuous on X iff epi f is closed in $X \times \mathbb{R}$, and this is true iff each r-level set $\{x\colon f(x) \leq r\}$ is closed, $r \in \mathbb{R}$. Note that gr f need not be closed when f is lower semicontinuous.

(b) Show that f is convex on X iff epi f is a convex subset of $X \times \mathbb{R}$.

(c) When f is an indicator function, $f = I_S$, then $f \in \mathcal{F}(X)$ iff S is nonempty and closed, and f is convex iff S is convex.

(d) Suppose that $(\zeta, -\lambda) \in X \times \mathbb{R}$ belongs to $N^P_{\text{epi } f}(x, r)$ for some $(x, r) \in \text{epi } f$, where $f \in \mathcal{F}$. Prove that $\lambda \geq 0$, that $r = f(x)$ if $\lambda > 0$, and that $\lambda = 0$ if $r > f(x)$. In this last case, show that $(\zeta, 0) \in N^P_{\text{epi } f}(x, f(x))$.

(e) Give an example of a continuous $f \in \mathcal{F}(\mathbb{R})$ such that at some point x we have $(1, 0) \in N^P_{\text{epi } f}(x, f(x))$.

(f) If $S = \text{epi } f$, where $f \in \mathcal{F}$, prove that for all x, $d_S(x, r)$ is nonincreasing as a function of r.

A vector $\zeta \in X$ is called a *proximal subgradient* (or *P-subgradient*) of a lower semicontinuous function f at $x \in \text{dom } f$ provided that

$$(\zeta, -1) \in N^P_{\text{epi } f}(x, f(x)).$$

The set of all such ζ is denoted $\partial_P f(x)$, and is referred to as the *proximal subdifferential*, or *P-subdifferential*. Note that because a cone is involved, if $\alpha > 0$ and $(\zeta, -\alpha) \in N^P_{\text{epi } f}(x, f(x))$, then $\zeta/\alpha \in \partial_P f(x)$. It also follows immediately from our study of the proximal normal cone that $\partial_P f(x)$ is convex, however it is not necessarily open, closed, or nonempty. The function $f\colon \mathbb{R} \to \mathbb{R}$ defined by $f(x) = -|x|$ is a simple example of a continuous function having $\partial_P f(0) = \emptyset$.

Figure 1.3 illustrates the epigraph of a function f together with some vectors of the form $(\zeta, -1)$, $\zeta \in \partial_P f(x)$. There exists a single P-subgradient at x_1, as well as at all the unlabeled points. At x_2, there are no P-subgradients, and there are multiple P-subgradients at the three remaining labeled points. At x_4, the proximal subdifferential is an unbounded set; the (horizontal) dashed arrow here is not associated with a P-subgradient, although it does represent a P-normal to epi f.

The indicator function is one of several ways in which we pass between sets and functions. It is also useful in optimization: note that minimizing f over a set S is equivalent to minimizing the function $f + I_S$ globally.

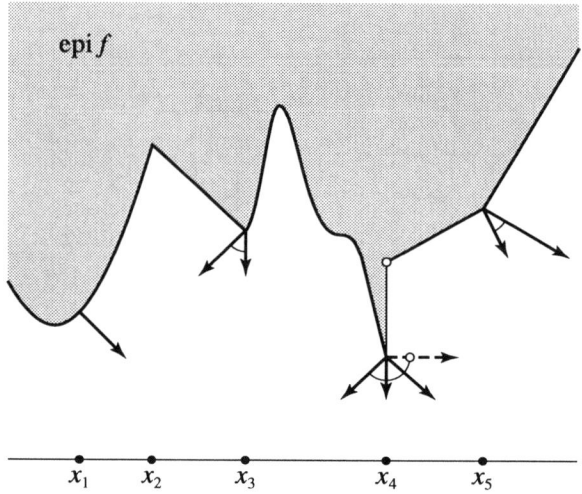

FIGURE 1.3. The epigraph of a function.

2.2. Exercise. Let $f = I_S$. Prove that for $x \in S$ we have

$$\partial_P f(x) = \partial_P I_S(x) = N_S^P(x).$$

The main theme of this chapter is to develop the calculus rules governing the proximal subgradient. We will see that to a surprising degree, many of the usual properties enjoyed by the classical derivative carry over to the proximal subgradient $\partial_P f(x)$. As a first illustration of this, we give an exercise which echos the vanishing of the derivative at a local minimum. A point $x \in X$ is said to attain the minimum of f on S provided $x \in S \cap \mathrm{dom}\, f$ and

$$f(x) \leq f(y) \; \forall y \in S.$$

If there exists an open neighborhood U of $x \in X$ on which x attains the minimum of f, then x is said to be a local minimum of f. If x is a minimum of f on $U = X$, then x is called a global minimum.

2.3. Exercise. Suppose $f \in \mathcal{F}$.

(a) Show that if f attains a local minimum at x, then $0 \in \partial_P f(x)$.

(b) Suppose $S \subset X$ is compact and satisfies $S \cap \mathrm{dom}\, f \neq \emptyset$. Show that f is bounded below on S, and attains its minimum over S.

Classical Derivatives

Before developing further properties of P-subgradients, we need to recall some facts about classical derivatives. We will do so rather quickly. The *directional derivative* of f at $x \in \text{dom}\, f$ in the direction $v \in X$ is defined as

$$f'(x;v) := \lim_{t \downarrow 0} \frac{f(x+tv) - f(x)}{t}, \qquad (1)$$

when the limit exists. We say that f is *Gâteaux* differentiable at x provided the limit in (2.1) exists for all $v \in X$, and there exists a (necessarily unique) element $f'_G(x) \in X$ (called the *Gâteaux derivative*) that satisfies

$$f'(x;v) = \langle f'_G(x), v \rangle \ \forall v \in X. \qquad (2)$$

A function may possess a directional derivative at x in every direction and yet fail to possess a Gâteaux derivative, as is evidenced by $f(x) = \|x\|$ at $x = 0$. In this case, we have $f'(0;v) = \|v\|$. Also, a lower semicontinuous function may have a Gâteaux derivative at a point x but not be continuous there.

Suppose that (2) holds at a point x, and in addition that the convergence in (1) is uniform with respect to v in bounded subsets of X. We then say that f is *Fréchet differentiable* at x, and in this case write $f'(x)$ (the *Fréchet derivative*) in place of $f'_G(x)$. Equivalently this means that for all $r > 0$ and $\varepsilon > 0$, there exists $\delta > 0$ so that

$$\left| \frac{f(x+tv) - f(x)}{t} - \langle f'(x), v \rangle \right| < \varepsilon$$

holds for all $|t| < \delta$ and $\|v\| \leq r$.

The two notions of differentiability are not equivalent, even in finite dimensions. We can easily show that Fréchet differentiability at x implies continuity at x, which is not the case for Gâteaux differentiability.

Many of the elementary properties of the derivative encountered in the multivariate calculus (i.e., when $X = \mathbb{R}^n$) have exact analogues using either Fréchet or Gâteaux derivatives, where f' or f'_G take the place of the usual gradient ∇f. To illustrate in some detail, suppose $f, g \colon X \to \mathbb{R}$ have Fréchet derivatives at $x \in X$. Then $f \pm g$, fg, and f/g (with $g(x) \neq 0$) all have Fréchet derivatives at x obeying the classical rules:

$$(f \pm g)'(x) = f'(x) \pm g'(x),$$
$$(fg)'(x) = f'(x)g(x) + f(x)g'(x),$$
$$\left(\frac{f}{g}\right)'(x) = \left(\frac{f'(x)g(x) - f(x)g'(x)}{g^2(x)}\right).$$

The proofs of these facts are the same as in the classical case.

The *Mean Value Theorem* can be stated as follows: suppose $f \in \mathcal{F}(X)$ is Gâteaux differentiable on an open neighborhood that contains the line

segment $[x, y] := \{tx + (1-t)y \colon 0 \le t \le 1\}$, where $x, y \in X$. That is, there exists an open set U containing the line segment $[x, y]$ such that f is differentiable at every point of U. Then there exists a point $z := tx+(1-t)y$, $0 < t < 1$, so that
$$f(y) - f(x) = \langle f'_G(z), y - x \rangle.$$
A proof of the Mean Value Theorem can be given by applying the classical one-dimensional mean value theorem to the function $g \colon [0, 1] \to \mathbb{R}$ defined by $g(t) = f(x + t(y - x))$.

Another useful result is the *Chain Rule*. In order to state it, we first need to extend the above notions of differentiability to maps between two Hilbert spaces. Suppose X_1 and X_2 are Hilbert spaces with norms $\|\cdot\|_1$ and $\|\cdot\|_2$, respectively, and suppose $F \colon X_1 \to X_2$ is a mapping between these spaces. We write $\mathcal{L}(X_1, X_2)$ for the space of bounded linear transformations from X_1 to X_2 endowed with the usual operator norm. The scalar case $X_2 = \mathbb{R}$ was discussed above, in which case $\mathcal{L}(X_1, \mathbb{R})$ was identified with X_1 in the usual way.

Let $x \in X_1$. The Gâteaux derivative, should it exist, of F at x is an element $F'_G(x) \in \mathcal{L}(X_1, X_2)$ that satisfies
$$\lim_{t \downarrow 0} \left\| \frac{F(x+tv) - F(x)}{t} - F'_G(x)(v) \right\|_2 = 0,$$
for all $v \in X_1$. Should in addition the above limit hold uniformly over v in bounded sets of X_1, then F is Fréchet differentiable and we write $F'(x)$ in place of $F'_G(x)$.

As in the scalar case, the derivative of the sum of two functions mapping X_1 to X_2 is the sum of the derivatives. Let us now consider the Chain Rule. Suppose X_1, X_2, and X_3 are all Hilbert spaces, and $F \colon X_1 \to X_2$, $G \colon X_2 \to X_3$. Assume that F is Fréchet differentiable at $x \in X_1$, and G is Fréchet differentiable at $F(x) \in X_2$. Then the composition $G \circ F \colon X_1 \to X_3$ is Fréchet differentiable at x and
$$(G \circ F)'(x) = G'(F(x)) F'(x),$$
where $G'(F(x)) F'(x) \in \mathcal{L}(X_1, X_3)$ signifies the composition of $F'(x)$ with $G'(F(x))$.

Suppose $U \subseteq X$ is open and $f \colon U \to \mathbb{R}$ is Fréchet differentiable on U. If $f'(\cdot) \colon U \to X$ is continuous on U, then we say that f is C^1 on U, and write $f \in C^1(U)$. It turns out that if f is Gâteaux differentiable on U with a continuous derivative there, then $f \in C^1(U)$. Now suppose further that the map $f'(\cdot) \colon U \to X$ is itself Fréchet differentiable on U with its derivative at $x \in U$ denoted by $f''(x) \in \mathcal{L}(X, X)$ (in the multivariate calculus, $f''(x)$ is the Hessian). For each $x \in U$, f then admits a local second-order Taylor expansion with remainder, which means there exists a neighborhood $B(x;\eta)$ of x so that for every $y \in B(x;\eta)$ we have

$$f(y) = f(x) + \langle f'(x), y-x \rangle + \tfrac{1}{2}\langle f''(z)(y-x), y-x \rangle,$$

where z is some element on the line segment connecting x and y. We note that if the norms of $f''(y)$ are bounded over $y \in B(x;\eta)$ by the constant $2\sigma > 0$, then this implies

$$f(y) \geq f(x) + \langle f'(x), y-x \rangle - \sigma \|y-x\|^2 \tag{3}$$

for all $y \in B(x;\eta)$.

If it should also happen that $f'' \colon X \to \mathcal{L}(X,X)$ is continuous on U, then f is said to be twice continuously differentiable on U, and we write $f \in C^2(U)$, or simply $f \in C^2$ if $U = X$. We note that if $f \in C^2(U)$, then for each $x \in U$ there exists a neighborhood $B(x;\eta)$ and a constant σ so that (3) holds, since the continuity of f'' at x implies that the norms of f'' are bounded in a neighborhood of x.

2.4. Exercise.

(a) Let $x \in X$ and define $f \colon X \to \mathbb{R}$ by $f(y) = \|y-x\|^2$. Show that $f \in C^2$, and that for each $y \in X$, we have $f'(y) = 2(y-x)$ and $f''(y) = 2\mathcal{I}$, where $\mathcal{I} \in \mathcal{L}(X,X)$ is the identity transformation.

(b) Suppose $c > 0$ is a constant, and x and ζ are fixed elements in X. Define $g \colon X \to \mathbb{R}$ by

$$g(y) = \left[c^2 + 2c\langle \zeta, y-x \rangle - \|y-x\|^2 \right]^{1/2}.$$

Show that $g \in C^2(U)$ for some neighborhood U of x, and that $g'(x) = \zeta$.

(c) Let $f(x) = \|x\|$. Then $f'(x)$ exists for $x \neq 0$, and equals $x/\|x\|$.

We now return to developing properties of the proximal subgradient. The following characterization is the most widely used description of the proximal subgradient, and we give it a name: by the *proximal subgradient inequality*, we mean the inequality appearing in the following result:

2.5. Theorem. *Let $f \in \mathcal{F}$ and let $x \in \mathrm{dom}(f)$. Then $\zeta \in \partial_P f(x)$ if and only if there exist positive numbers σ and η such that*

$$f(y) \geq f(x) + \langle \zeta, y-x \rangle - \sigma \|y-x\|^2 \quad \forall y \in B(x;\eta). \tag{4}$$

Proof. Let us first prove the "if" part of the theorem's statement. The inequality (4) implies that

$$\alpha - f(x) + \sigma\left[\|y-x\|^2 + (\alpha - f(x))^2\right] \geq \langle \zeta, y-x \rangle$$

for all $y \in B(x;\eta)$ and for all $\alpha \geq f(y)$. This in turn implies

$$\langle (\zeta, -1), [(y, \alpha) - (x, f(x))] \rangle \leq \sigma \|(y, \alpha) - (x, f(x))\|^2$$

for all points $(y, \alpha) \in \text{epi}(f)$ near $(x, f(x))$. In view of Proposition 1.5, this implies that $(\zeta, -1) \in N^P_{\text{epi } f}(x, f(x))$.

Let us now turn to the "only if" part. To this end, suppose that $(\zeta, -1) \in N^P_{\text{epi } f}(x, f(x))$. Then by Proposition 1.3 there exists $\delta > 0$ such that

$$(x, f(x)) \in \text{proj}_{\text{epi } f}\big((x, f(x)) + \delta(\zeta, -1)\big).$$

This evidently implies

$$\|\delta(\zeta, -1)\|^2 \leq \|[(x, f(x)) + \delta(\zeta, -1)] - (y, \alpha)\|^2$$

for all $(y, \alpha) \in \text{epi } f$; see Figure 1.4. Upon taking $\alpha = f(y)$, the last inequality yields

$$\delta^2 \|\zeta\|^2 + \delta^2 \leq \|x - y + \delta\zeta\|^2 + \big(f(x) - f(y) - \delta\big)^2,$$

which can be rewritten as

$$\big(f(y) - f(x) + \delta\big)^2 \geq \delta^2 + 2\delta \langle \zeta, y - x \rangle - \|x - y\|^2. \tag{5}$$

It is clear that the right-hand side of (5) is positive for all y sufficiently near x, say for $y \in B(x;\eta)$. By shrinking $\eta > 0$ if necessary, we can also ensure (by the lower semicontinuity of f) that $y \in B(x;\eta)$ implies

$$f(y) - f(x) + \delta > 0.$$

Hence taking square roots of (5) gives us that

$$f(y) \geq g(y) := f(x) - \delta + \big\{\delta^2 + 2\delta \langle \zeta, y - x \rangle - \|x - y\|^2\big\}^{1/2} \tag{6}$$

for all $y \in B(x;\eta)$. Direct calculations show that $g'(x) = \zeta$ and that g'' exists and is bounded, say by $2\sigma > 0$, on a neighborhood of x (Exercise 2.4). Again if η is shrunk further if necessary, we have (as noted above in connection with the inequality (3))

$$g(y) \geq g(x) + \langle \zeta, y - x \rangle - \sigma \|y - x\|^2 \; \forall y \in B(x;\eta).$$

But then by (6), and since $f(x) = g(x)$, we see that

$$f(y) \geq f(x) + \langle \zeta, y - x \rangle - \sigma \|y - x\|^2 \; \forall y \in B(x;\eta),$$

which is (4) as required. $\square$

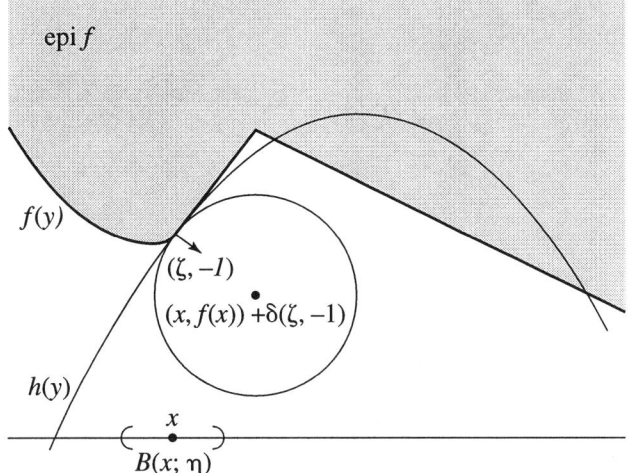

FIGURE 1.4. ζ belongs to $\partial_P f(x)$.

The definition of proximal subgradients via proximal normals to an epigraph is a geometric approach, and the characterization in Theorem 2.5 can also be interpreted geometrically. The proximal subgradient inequality (4) asserts that near x, $f(\cdot)$ majorizes the quadratic function

$$h(y) := f(x) + \langle \zeta, y - x \rangle - \sigma \|y - x\|^2,$$

with equality at $y = x$ (since obviously $h(x) = f(x)$). It is worth noting that this is equivalent to saying that $y \mapsto f(y) - h(y)$ has a local minimum at $y = x$ with min value equal to 0. Put into purely heuristic terms, the content of Theorem 2.5 is that the existence of such a parabola h which "locally fits under" the epigraph of f at $(x, f(x))$ is equivalent to the existence of a ball in $X \times \mathbb{R}$ touching the epigraph nonhorizontally at that point; this is, in essence, what the proof of the theorem shows. See Figure 1.4.

The description of proximal subgradients contained in Theorem 2.5 is generally more useful in analyzing lower semicontinuous functions than is a direct appeal to the definition. The first corollary below illustrates this, and relates $\partial_P f$ to classical differentiability. It also states that for convex functions, the inequality (4) holds globally in an even simpler form; this is the functional analogue of the simplified proximal normal inequality for convex sets (Proposition 1.10).

2.6. Corollary. *Let $f \in \mathcal{F}$ and $U \subset X$ be open.*

(a) *Assume that f is Gâteaux differentiable at $x \in U$. Then*

$$\partial_P f(x) \subseteq \{f'_G(x)\}.$$

(b) *If $f \in C^2(U)$, then*
$$\partial_P f(x) = \{f'(x)\}$$
for all $x \in U$.

(c) *If f is convex, then $\zeta \in \partial_P f(x)$ iff*
$$f(y) \geq f(x) + \langle \zeta, y - x \rangle \ \forall y \in X. \tag{7}$$

Proof.

(a) Suppose f has a Gâteaux derivative at x and that $\zeta \in \partial_P f(x)$. For any $v \in X$, if we write $y = x + tv$, the proximal subgradient inequality (4) implies that there exists $\sigma > 0$ such that
$$\frac{f(x + tv) - f(x)}{t} - \langle \zeta, v \rangle \geq -t\sigma \|v\|^2$$
for all sufficiently small positive t. Upon letting $t \downarrow 0$ we obtain
$$\langle f'_G(x) - \zeta, v \rangle \geq 0.$$
Since v was arbitrary, the conclusion $\zeta = f'_G(x)$ follows.

(b) If $f \in C^2(U)$ and $x \in U$, then we have $f'(x) \in \partial_P f(x)$ by Theorem 2.5, since (3) implies (4) if ζ is set equal to $f'(x)$. That $\partial_P f(x)$ contains only $f'(x)$ follows from part (a).

(c) Obviously if ζ satisfies (7), then (4) holds with $\sigma = 0$ and any $\eta > 0$, so that $\zeta \in \partial_P f(x)$. Conversely, suppose $\zeta \in \partial_P f(x)$, and σ and η are chosen as in (4). Let $y \in X$. Then for any t in $(0, 1)$ sufficiently small so that $(1 - t)x + ty \in B(x; \eta)$, we have by the convexity of f and (4) (where we substitute $(1 - t)x + ty$ for y) that
$$(1-t)f(x) + tf(y) \geq f\big((1-t)x + ty\big)$$
$$\geq f(x) + t\langle \zeta, y - x \rangle - t^2 \sigma \|y - x\|^2.$$
Simplifying and dividing by t, we conclude
$$f(y) \geq f(x) + \langle \zeta, y - x \rangle - t\sigma \|y - x\|^2.$$
Letting $t \downarrow 0$ yields (7). $\square$

The containment in Corollary 2.6(a) is the best possible conclusion under the stated assumptions, since even when $X = \mathbb{R}$ and f is continuously differentiable, the nonemptiness of the proximal subdifferential is not assured. The already familiar C^1 function $f(x) = -|x|^{3/2}$ admits no proximal subgradient at $x = 0$ (see Exercise 1.7).

The first part of the following corollary has already been observed (Exercise 2.3). Despite its simplicity, it is the fundamental fact that generates proximal subgradients on many occasions. The second part says that the "first-order" necessary condition for a minimum is also sufficient in the case of convex functions, which is a principal reason for their importance.

2.7. Corollary. *Suppose $f \in \mathcal{F}$.*

(a) *If f has a local minimum at x, then $0 \in \partial_P f(x)$.*

(b) *Conversely, if f is convex and $0 \in \partial_P f(x)$, then x is a global minimum of f.*

Proof.

(a) The definition of a local minimum says there exists $\eta > 0$ so that
$$f(y) \geq f(x) \ \forall y \in B(x; \eta),$$
which is the proximal subgradient inequality with $\zeta = 0$ and $\sigma = 0$. Thus Theorem 2.5 implies that $0 \in \partial_P f(x)$.

(b) Under these hypotheses, (7) holds with $\zeta = 0$. Thus $f(y) \geq f(x)$ for all $y \in X$, which says that x is a global minimum of f. □

The proximal subdifferential is a "one-sided" object suitable to the analysis of lower semicontinuous functions. For a theory applicable to upper semicontinuous functions f, the *proximal superdifferential* $\partial^P f(x)$ is the appropriate object, and can be defined simply as $-\partial_P(-f)(x)$. In the subsequent development, analogues for upper semicontinuous functions will usually not be stated because they require only evident modifications, such as replacing "sub" by "super," "$\leq$" by "$\geq$," "minimum" by "maximum," and "convex" by "concave." Nonetheless, we will have occasional use for supergradients.

2.8. Exercise.

(a) Suppose $-f \in \mathcal{F}$ and $x \in \operatorname{dom}(-f)$. Show that $\zeta \in \partial^P f(x)$ iff there exist positive numbers σ and η so that
$$f(y) - \langle \zeta, y - x \rangle - \sigma \|y - x\|^2 \leq f(x) \ \forall y \in B(x; \eta).$$

(b) Suppose $U \subset X$ is open, $x \in U$, $f : U \to \mathbb{R}$ is continuous on U, and that both $\partial_P f(x)$ and $\partial^P f(x)$ are nonempty. Prove that f is Fréchet differentiable at x, and that we have $\partial_P f(x) = \{f'(x)\} = \partial^P f(x)$.

(c) Suppose $f \in \mathcal{F}$ is convex and continuous at $x \in \operatorname{int} \operatorname{dom} f$. Show that $\partial_P f(x) \neq \emptyset$. (*Hint.* Apply the Separation Theorem (see, e.g., Rudin (1973)) to $(x, f(x)) \in \operatorname{bdry} \operatorname{epi} f)$. Deduce further that if $\partial^P f(x) \neq \emptyset$, then f is Fréchet differentiable at x.

There is a natural way to define *partial* proximal subgradients. Suppose X is decomposed into orthogonal subspaces $X = X_1 \oplus X_2$, and $f \in \mathcal{F}$. Let $x \in X$, and write $x = (x_1, x_2)$ according to the direct sum decomposition. The notation $\partial_P f(\cdot, x_2)(x_1)$ denotes the proximal subdifferential evaluated at x_1 of the function $x_1' \mapsto f(x_1', x_2)$ defined on X_1. (This is the *partial proximal subdifferential* taken with respect to the first coordinate.) Similar considerations apply to $\partial_P f(x_1, \cdot)(x_2)$. The functional analogue to Exercise 1.8 is the following:

2.9. Exercise. Using the notation of the previous paragraph, suppose there exists $\zeta \in \partial_P f(x)$, and write $\zeta = (\zeta_1, \zeta_2)$ according to the direct sum decomposition. Show that $\zeta_1 \in \partial_P f(\cdot, x_2)(x_1)$ and $\zeta_2 \in \partial_P f(x_1, \cdot)(x_2)$. Give an example where the converse fails.

We will develop proximal calculus in some detail later, but let us note right away that we cannot expect a calculus Sum Rule of the form

$$\partial_P f(x) + \partial_P g(x) = \partial_P (f + g)(x) \tag{8}$$

to hold in much generality. One inclusion between these sets can be established easily, but unfortunately, this inclusion is seldom the one we need.

2.10. Exercise.

(a) Show that $\partial_P f(x) + \partial_P g(x) \subseteq \partial_P (f + g)(x)$.

(b) Give an example for which $\partial_P (f + g)(x)$ is nonempty but for which one of $\partial_P f(x)$ or $\partial_P g(x)$ is empty.

(c) Show that for all $c > 0$, we have $\partial_P (cf)(x) = c \partial_P f(x)$.

The following proposition is a proximal Sum Rule, and says in essence that the Sum Rule (8) does hold whenever one of the functions is C^2.

2.11. Proposition. *Suppose that $f \in \mathcal{F}$, and let $x \in X$. Suppose further that g is C^2 in a neighborhood of x. Then*

$$\zeta \in \partial_P (f + g)(x) \text{ implies } \zeta - g'(x) \in \partial_P f(x).$$

Proof. The inequality (3) applied to $-g$ implies the existence of a constant $\sigma' > 0$ such that

$$-g(y) + g(x) + \sigma' \|y - x\|^2 \geq \langle -g'(x), y - x \rangle$$

for all y near x. Since $\zeta \in \partial_P (f + g)(x)$, we have

$$f(y) + g(y) - f(x) - g(x) + \sigma \|y - x\|^2 \geq \langle \zeta, y - x \rangle$$

for some $\sigma > 0$ and all y near x. Upon adding these inequalities, we arrive at

$$f(y) - f(x) + (\sigma' + \sigma)\|y - x\|^2 \geq \langle \zeta - g'(x), y - x \rangle,$$

which holds for all y near x, and says via Theorem 2.5 that $\zeta - g'(x) \in \partial_P f(x)$. $\square$

2.12. Exercise. Let $f \in C^2$, and suppose that f attains a minimum over S at x. Prove that $-f'(x) \in N_S^P(x)$. (*Hint.* Consider $f + I_S$.)

A function $f \colon X \to (-\infty, \infty]$ is said to satisfy a *Lipschitz condition* of rank K on a given set S provided that f is finite on S and satisfies

$$|f(x) - f(y)| \leq K\|x - y\| \ \forall x, y \in S.$$

2.13. Exercise. Let f satisfy a Lipschitz condition of rank K on some neighborhood of a given point x_0. Show that any $\zeta \in \partial_P f(x_0)$ satisfies $\|\zeta\| \leq K$.

The question of a statement converse to that of this exercise is a deeper one that we will address in §7. A function f is said to be *Lipschitz near* x if it satisfies the Lipschitz condition (of some rank) on a neighborhood of x. A function f is said to be *locally Lipschitz* on S if f is Lipschitz near x for every $x \in S$.

2.14. Exercise. Let $f \in C^1(U)$, U open. Prove that f is locally Lipschitz on U. (*Hint.* Mean Value Theorem.) If S is a compact convex subset of U, show that f is Lipschitz on S with rank $K := \max\{\|f'(x)\| \colon x \in S\}$.

3 The Density Theorem

We now establish an important fact: the set $\operatorname{dom}(\partial_P f)$ of points in $\operatorname{dom} f$ at which at least one proximal subgradient exists is dense in $\operatorname{dom} f$.

3.1. Theorem. *Suppose $f \in \mathcal{F}$. Let $x_0 \in \operatorname{dom} f$, and let $\varepsilon > 0$ be given. Then there exists a point $y \in x_0 + \varepsilon B$ satisfying $\partial_P f(y) \neq \emptyset$ and $f(x_0) - \varepsilon \leq f(y) \leq f(x_0)$. In particular, $\operatorname{dom}(\partial_P f)$ is dense in $\operatorname{dom} f$.*

Proof. By lower semicontinuity, there exists δ with $0 < \delta < \varepsilon$ so that

$$x \in x_0 + \delta \overline{B} \implies f(x) \geq f(x_0) - \varepsilon. \tag{1}$$

We first give a simple motivational proof in the case $X = \mathbb{R}^n$. We define

$$g(x) := \begin{cases} \left[\delta^2 - \|x - x_0\|^2\right]^{-1} & \text{if } \|x - x_0\| < \delta, \\ +\infty & \text{otherwise.} \end{cases}$$

Note that g belongs to $C^2(x_0 + \delta B)$, and that $g(x) \to \infty$ as x approaches the boundary of $B(x_0; \delta)$. Now consider the function $(f + g) \in \mathcal{F}$ which is bounded below on $x_0 + \delta \overline{B}$. It follows that $f + g$ attains a minimum y over $x_0 + \delta \overline{B}$. Obviously $y \in x_0 + \delta B$, and $f + g$ has a local minimum at y, thus by Exercise 2.3 we have that $0 \in \partial_P(f + g)(y)$. It now follows from

Proposition 2.11 that $-g'(y) \in \partial_P f(y)$, and in particular that $\partial_P f(y) \neq \emptyset$. Since δ can be made arbitrarily small, it follows that $\text{dom}(\partial_P f)$ is dense in $\text{dom } f$.

In view of (1), we are left only with showing that $f(y) \leq f(x_0)$. We deduce this by noting that y is a minimum of $f + g$ and $g(x_0) \leq g(y)$, and hence

$$f(y) \leq f(x_0) + (g(x_0) - g(y)) \leq f(x_0).$$

The proof is thus complete if X is finite dimensional.

The proof in the general case is more complicated, due to the possible nonexistence of minimizers. An iterative procedure will lead to an appropriate minimum's existence, however.

We begin again by choosing δ so that (1) holds, and we take $\sigma > 2\varepsilon/\delta^2$. We will show that there exists a point $z \in x_0 + \delta B$ so that the function $x \mapsto f(x) + \sigma \|x - z\|^2$ has a minimum over $x_0 + \delta B$ at some point $y \in x_0 + \delta B$ that satisfies $f(y) \leq f(x_0)$. Once the existence of such y and z are established, the proof is easily completed as follows. Since y is a minimum, we have $0 \in \partial_P\big(f + \|(\cdot) - z\|^2\big)(y)$, which by Proposition 2.11 and Exercise 2.4 gives the inclusion $-2\sigma(y - z) \in \partial_P f(y)$. Also, $f(x_0) - \varepsilon \leq f(y) \leq f(x_0)$ is then immediate in view of (1).

We proceed to demonstrate the existence of points y and z having the properties described above. We define

$$S_0 := \left\{ x \in x_0 + \delta \overline{B} : f(x) + \frac{\sigma}{2}\|x - x_0\|^2 \leq f(x_0) \right\}.$$

We claim that S_0 is closed and that

$$x_0 \in S_0 \subset x_0 + \delta B. \tag{2}$$

Indeed, if $x \in S_0$, then by (1) and the choice of σ, we have

$$\|x - x_0\|^2 \leq \frac{2}{\sigma}\big[f(x_0) - f(x)\big] < \delta^2$$

which gives (2). The closedness of S_0 follows from the lower semicontinuity of f (see Exercise 2.1(a)). If S_0 contains only a single point y, then $y = x_0$ and the proof is completed, since then x_0 is a minimum of

$$x \mapsto f(x) + \sigma\|x - x_0\|^2$$

over $x \in x_0 + \delta B$. Since this is generally not the case, we employ an iterative procedure producing successively smaller sets that will shrink to a point y.

Let $x_1 \in S_0$ be chosen so that

$$f(x_1) + \frac{\sigma}{2}\|x_1 - x_0\|^2 \leq \inf_{x \in S_0}\left\{ f(x) + \frac{\sigma}{2}\|x - x_0\|^2 \right\} + \frac{\sigma}{4},$$

and define another closed set by

$$S_1 := \left\{ x \in S_0 \colon f(x) + \sigma\left[\frac{\|x - x_0\|^2}{2} + \frac{\|x - x_1\|^2}{4}\right] \right.$$
$$\left. \leq f(x_1) + \frac{\sigma}{2}\|x_1 - x_0\|^2 \right\}.$$

Note that $x_1 \in S_1$, so that S_1 is nonempty. Again, if x_1 were the only point in S_1 the proof would be complete (why?).

Inductively, if x_j and S_j are chosen for $j \geq 0$, we choose $x_{j+1} \in S_j$ so that

$$f(x_{j+1}) + \frac{\sigma}{2}\sum_{i=0}^{j} \frac{\|x_{j+1} - x_i\|^2}{2^i}$$
$$\leq \inf_{s \in S_j}\left\{ f(x) + \frac{\sigma}{2}\sum_{i=0}^{j} \frac{\|x - x_i\|^2}{2^i}\right\} + \frac{\sigma}{4^{j+1}}, \quad (3)$$

and we define

$$S_{j+1} := \left\{ x \in S_j \colon f(x) + \frac{\sigma}{2}\sum_{i=0}^{j+1} \frac{\|x - x_i\|^2}{2^i} \right.$$
$$\left. \leq f(x_{j+1}) + \frac{\sigma}{2}\sum_{i=0}^{j} \frac{\|x_{j+1} - x_i\|^2}{2^i}\right\}. \quad (4)$$

We obviously have $x_{j+1} \in S_{j+1} \subset S_j$ for each $j \geq 0$, and thus $\{S_j\}$ is a nested sequence of nonempty closed sets. We need only show that $\sup\{\|x - x'\| \colon x, x' \in S_j\} =: \mathrm{diam}(S_j) \to 0$ as $j \to \infty$ to conclude that the sequence $\{S_j\}$ shrinks to a single point. (This is a fact about complete metric spaces known as Cantor's Theorem.) Toward this end, let $x \in S_{j+1}$. Then for each $j \geq 0$, we have by (4) and (3) that

$$\frac{\sigma}{2}\frac{\|x - x_{j+1}\|^2}{2^{j+1}} \leq f(x_{j+1}) + \frac{\sigma}{2}\sum_{i=0}^{j} \frac{\|x_{j+1} - x_i\|^2}{2^i}$$
$$- \left\{ f(x) + \frac{\sigma}{2}\sum_{i=0}^{j} \frac{\|x - x_i\|^2}{2^i}\right\} \leq \frac{\sigma}{4^{j+1}}.$$

It follows that
$$\sup_{x \in S_{j+1}} \|x - x_{j+1}\| \leq 2^{-j/2},$$

and thus $\lim_{j \to \infty} \mathrm{diam}(S_j) = 0$. Consequently, since the Hilbert space X is complete, there is a point y such that $\bigcap_{j=1}^{\infty} S_j = \{y\}$. Of course, y lies in S_0, whence $\|y - x_0\| < \delta < \varepsilon$ (in view of (2)) and

$$f(y) \leq f(y) + \frac{\sigma}{2}\|y - x_0\|^2 \leq f(x_0)$$

(by the definition of S_0).

Let z be the point $(1/2)\sum_{j=0}^{\infty} x_j/2^j$ (the sum is uniformly convergent), and observe that the identity

$$\|x-z\|^2 = \frac{1}{2}\left(\sum_{i=0}^{\infty} \frac{\|x-x_i\|^2}{2^i}\right) - c \tag{5}$$

holds for any x, where $c := (1/2)(\sum_{i=0}^{\infty} \|x_i\|^2/2^i) - \|z\|^2$.

We now show that y minimizes $x \mapsto f(x) + \sigma\|x-z\|^2$ over $x_0 + \delta B$. Note first that since $x_{j+1} \in S_j$ for all j, it follows that

$$\left\{f(x_j) + \frac{\sigma}{2}\sum_{i=0}^{j-1}\frac{\|x_j - x_i\|^2}{2^i}\right\} \text{ is nonincreasing} \tag{6}$$

as j increases. Now let $x \in x_0 + \delta B$ be different from y, and let $k \geq 0$ be the least integer such that $x \notin S_k$. Let j be any index larger than $k-1$. Since $y \in S_{j+1}$, the definition (4) of S_{j+1} and (6) imply

$$f(y) + \frac{\sigma}{2}\sum_{i=0}^{j+1}\frac{\|y-x_i\|^2}{2^i} \leq \left\{f(x_{j+1}) + \frac{\sigma}{2}\sum_{i=0}^{j}\frac{\|x_{j+1}-x_i\|^2}{2^i}\right\}$$

$$\leq \left\{f(x_k) + \frac{\sigma}{2}\sum_{i=0}^{k-1}\frac{\|x_k-x_i\|^2}{2^i}\right\}. \tag{7}$$

Now k has been chosen so that $x \in S_{k-1}\backslash S_k$ (if $k = 0$, this should be interpreted with $S_{-1} := x_0 + \delta B$), and hence the definition of S_k yields that

$$f(x_k) + \frac{\sigma}{2}\sum_{i=0}^{k-1}\frac{\|x_k-x_i\|^2}{2^i} < f(x) + \frac{\sigma}{2}\sum_{i=0}^{k}\frac{\|x-x_i\|^2}{2^i}$$

$$\leq f(x) + \frac{\sigma}{2}\sum_{i=0}^{\infty}\frac{\|x-x_i\|^2}{2^i}. \tag{8}$$

We now combine (8) with (7) and let $j \to \infty$ to deduce that

$$f(y) + \frac{\sigma}{2}\sum_{i=0}^{\infty}\frac{\|y-x_i\|^2}{2^i} \leq f(x) + \frac{\sigma}{2}\sum_{i=0}^{\infty}\frac{\|x-x_i\|^2}{2^i}.$$

We now can simply add $-\sigma c$ to both sides of this inequality and use the representation (5) to conclude that

$$f(y) + \sigma\|y-z\|^2 \leq f(x) + \sigma\|x-z\|^2.$$

Since x was any point in $x_0 + \delta B$, we deduce that y is a minimum of $x \mapsto f(x) + \|x-z\|^2$ over $x_0 + \delta B$, as required. $\square$

4 Minimization Principles

As was pointed out at the time, a simple and direct proof of the Density Theorem 3.1 is possible in finite dimensions. The proof fails in infinite dimensions for a reason that is a persistent and thorny issue in analysis: the possible nonexistence of a minimizer. Specifically, a lower semicontinuous function on a closed bounded set may not attain a minimum, or even be bounded below. An important type of result affirms that in certain situations an arbitrarily small perturbation of the function in question *will* attain a minimum. Here is an example, known as Stegall's minimization principle.

4.1. Theorem. *Let $f \in \mathcal{F}$, and suppose that f is bounded below on the bounded closed set $S \subset X$, with $S \cap \mathrm{dom}\, f \neq \emptyset$. Then there exists a dense set of points x in X having the property that the function $y \mapsto f(y) - \langle x, y \rangle$ attains a unique minimum over S.*

The main use of this theorem involves taking x close to 0, so that a small linear perturbation of f attains a minimum. The next minimization principle we present is due to Borwein and Preiss, and is more complicated to state. It features a quadratic perturbation and two parameters, and can be used in such a way that the conclusion is strongly related to a *pregiven* point x_0 of interest.

4.2. Theorem. *Let $f \in \mathcal{F}$ be bounded below, and let $\varepsilon > 0$. Suppose that x_0 is a point satisfying $f(x_0) < \inf_{x \in X} f(x) + \varepsilon$. Then, for any $\lambda > 0$ there exist points y and z with*

$$\|z - x_0\| < \lambda, \quad \|y - z\| < \lambda, \quad f(y) \leq f(x_0),$$

and having the property that the function

$$x \mapsto f(x) + \frac{\varepsilon}{\lambda^2} \|x - z\|^2$$

has a unique minimum at $x = y$.

The proofs of both these minimization principles appear in the next section, for they are simple consequences of the proximal analysis of "infconvolutions" (together with the Density Theorem). The following exercise is helpful in understanding the content of Theorem 4.2.

4.3. Exercise. Let f, x_0, and ε be as in the statement of Theorem 4.2, where in addition f is Fréchet differentiable. Prove that there is a point y in $x_0 + 2\sqrt{\varepsilon} B$ such that $\|f'(y)\| \leq 2\sqrt{\varepsilon}$ and $f(y) \leq f(x_0)$. Proceed to deduce the existence of a minimizing sequence $\{y_i\}$ for f (i.e., one that satisfies $\lim_{i \to \infty} f(y_i) = \inf_{x \in X} f(x)$) such that $\lim_{i \to \infty} \|f'(y_i)\| = 0$.

5 Quadratic Inf-Convolutions

The *inf-convolution* of two functions f, g is another function h defined as follows:
$$h(x) := \inf_{y \in X} \{f(y) + g(x - y)\}.$$

The term "convolution" is suggested by the visual resemblance of this formula to the classical integral convolution formula. Our interest here involves only such inf-convolutions formed between a function $f \in \mathcal{F}$ and the quadratic function $x \mapsto \alpha\|x\|^2$, where $\alpha > 0$. Such functions have surprisingly far-reaching properties.

Given $f \in \mathcal{F}$ that is bounded below and $\alpha > 0$, we define $f_\alpha \colon X \to \mathbb{R}$ by
$$f_\alpha(x) := \inf_{y \in X} \{f(y) + \alpha\|x - y\|^2\}. \tag{1}$$

We recall some terminology: $\{x_i\}$ is said to be a minimizing sequence for an infimum of the type $\inf_{x \in S} g(x)$ provided that the points x_i all lie in S and satisfy $\lim_{i \to \infty} g(x_i) = \inf_{x \in S} g(x)$.

5.1. Theorem. *Suppose that $f \in \mathcal{F}$ is bounded below by some constant c, and f_α is defined as above with $\alpha > 0$. Then f_α is bounded below by c, and is Lipschitz on each bounded subset of X (and in particular is finite-valued). Furthermore, suppose $x \in X$ is such that $\partial_P f_\alpha(x)$ is nonempty. Then there exists a point $\bar{y} \in X$ satisfying the following:*

(a) *If $\{y_i\} \subset X$ is a minimizing sequence for the infimum in (1), then $\lim_{i \to \infty} y_i = \bar{y}$.*

(b) *The infimum in (1) is attained uniquely at $\bar{y}$.*

(c) *The Fréchet derivative $f'_\alpha(x)$ exists and equals $2\alpha(x - \bar{y})$. Thus the proximal subgradient $\partial_P f_\alpha(x)$ is the singleton $\{2\alpha(x - \bar{y})\}$.*

(d) $2\alpha(x - \bar{y}) \in \partial_P f(\bar{y})$.

Proof. Suppose we are given f and $\alpha > 0$ as above. It is clear from the definition that f_α is bounded below by c. We now show that f_α is Lipschitz on any bounded set $S \subset X$.

For any fixed $x_0 \in \text{dom } f \neq \emptyset$, note that $f_\alpha(x) \leq f(x_0) + \alpha\|x - x_0\|^2$ for all $x \in X$, and thus in particular $m := \sup\{f_\alpha(x) \colon x \in S\} < \infty$. Since $\alpha > 0$, and f is bounded below, we have that for any $\varepsilon > 0$ the set
$$C := \{z \colon \exists y \in S \text{ so that } f(z) + \alpha\|y - z\|^2 \leq m + \varepsilon\}$$
is bounded in X.

Now let x and y belong to S and $\varepsilon > 0$. Since $f_\alpha(\cdot)$ is given as an infimum, there exists $z \in C$ so that

$$f_\alpha(y) \geq f(z) + \alpha \|y - z\|^2 - \varepsilon.$$

Thus we have

$$\begin{aligned} f_\alpha(x) - f_\alpha(y) &\leq f_\alpha(x) - f(z) - \alpha\|y - z\|^2 + \varepsilon \\ &\leq f(z) + \alpha\|x - z\|^2 - f(z) - \alpha\|y - z\|^2 + \varepsilon \\ &= \alpha\|x - y\|^2 - 2\alpha\langle x - y, z - y\rangle + \varepsilon \\ &\leq \lambda\|x - y\| + \varepsilon, \end{aligned}$$

where $\lambda := \alpha \sup\{\|s' - s\| + 2\|z - s\| : s', s \in S, z \in C\} < \infty$. Reversing the roles of x and y, and then letting $\varepsilon \downarrow 0$, the above shows that f_α is Lipschitz of rank λ on S.

We now consider the other assertions in the theorem. Suppose $x \in X$ is such that there exists at least one $\zeta \in \partial_P f_\alpha(x)$. By the proximal subgradient inequality, there exist positive constants σ and η so that

$$\langle \zeta, y - x\rangle \leq f_\alpha(y) - f_\alpha(x) + \sigma\|y - x\|^2 \tag{2}$$

for all $y \in B(x; \eta)$. Now suppose $\{y_i\}$ is any minimizing sequence of (1) and thus there exists ε_i with $\varepsilon_i \downarrow 0$ as $i \to \infty$ such that

$$f_\alpha(x) \leq f(y_i) + \alpha\|y_i - x\|^2 = f_\alpha(x) + \varepsilon_i^2. \tag{3}$$

We observe that

$$f_\alpha(y) \leq f(y_i) + \alpha\|y_i - y\|^2, \tag{4}$$

since f_α is defined as an infimum over X. Inserting the inequalities (3) and (4) into (2) yields for each $y \in B(x; \eta)$ the conclusion

$$\begin{aligned} \langle \zeta, y - x\rangle &\leq \alpha\|y_i - y\|^2 - \alpha\|y_i - x\|^2 + \varepsilon_i^2 + \sigma\|y - x\|^2 \\ &= -2\alpha\langle y_i, y - x\rangle + \alpha\|y\|^2 - \alpha\|x\|^2 + \varepsilon_i^2 + \sigma\|y - x\|^2 \\ &= 2\alpha\langle x - y_i, y - x\rangle + \varepsilon_i^2 + (\alpha + \sigma)\|y - x\|^2, \end{aligned}$$

which rewritten says that

$$\langle \zeta - 2\alpha(x - y_i), y - x\rangle \leq \varepsilon_i^2 + (\alpha + \sigma)\|y - x\|^2 \tag{5}$$

for all $y \in B(x; \eta)$. Now let $v \in B$. Note that $y := x + \varepsilon_i v \in B(x; \eta)$ for large i since $\varepsilon_i \downarrow 0$ as $i \to \infty$. Hence for all large i, we can insert this value of y into (5) to deduce

$$\langle \zeta - 2\alpha(x - y_i), v\rangle \leq \varepsilon_i(1 + \alpha + \sigma).$$

Since $v \in B$ is arbitrary, it follows that

$$\|\zeta - 2\alpha(x - y_i)\| \leq \varepsilon_i(1 + \alpha + \sigma). \tag{6}$$

We now define $\bar{y} \in X$ by

$$\bar{y} := x - \frac{\zeta}{2\alpha},$$

and observe that (a) immediately follows from letting $i \to \infty$ in (6).

To see that $\bar{y}$ achieves the infimum in (1), it suffices to observe from (a) and the lower semicontinuity of f that

$$f_\alpha(x) \leq f(\bar{y}) + \alpha\|\bar{y} - x\|^2 \leq \liminf_{i \to \infty}\bigl[f(y_i) + \alpha\|y_i - x\|^2\bigr] = f_\alpha(x),$$

where the last equality comes from (3). It is also clear that $\bar{y}$ is unique, since if $\hat{y}$ is another minimizer of (1), the constant sequence $y_i := \hat{y}$ is minimizing, and therefore must converge to $\bar{y}$ by (a). Hence (b) holds.

The following observation about a supergradient will be useful in proving the Fréchet differentiability assertion:

$$2\alpha(x - \bar{y}) \in \partial^P f_\alpha(x). \tag{7}$$

To see this, let $y \in X$ and observe that

$$f_\alpha(y) \leq f(\bar{y}) + \alpha\|y - \bar{y}\|^2,$$

with equality holding if $y = x$. Then we see that

$$f_\alpha(x) - f_\alpha(y) \geq f(\bar{y}) + \alpha\|x - \bar{y}\|^2 - f(\bar{y}) - \alpha\|y - \bar{y}\|^2$$
$$= \langle 2\alpha(x - \bar{y}), y - x\rangle - \alpha\|x - y\|^2.$$

This confirms (7) (see Exercise 2.8(a)).

Part (c) of the theorem now follows from (7) combined with Exercise 2.8(b). As for part (d), observe that the function $x' \mapsto f(x') + \alpha\|x' - x\|^2$ attains a minimum at $x' = \bar{y}$, so that its proximal subdifferential there contains 0. With the help of Proposition 2.11 and Exercise 2.4(a), this translates to precisely statement (d). □

Two immediate consequences of Theorem 5.1 are the minimization principles presented in §4, as we now see.

5.2. Corollary. *Theorem 4.1 holds.*

Proof. Suppose $S \subset X$ is nonempty, closed, and bounded, and $f \in \mathcal{F}$ is bounded below on S with $\operatorname{dom} f \cap S \neq \emptyset$. Define

$$g(x) := \inf_{y \in X}\Bigl\{f(y) + I_S(y) - \tfrac{1}{2}\|y\|^2 + \tfrac{1}{2}\|x - y\|^2\Bigr\}, \tag{8}$$

which is easily seen to be a function of the form f_α as in (1) (where $f = f + I_S - (1/2)\|\cdot\|^2$ and $\alpha = 1/2$). Furthermore, expression (8) for $g(x)$ can be simplified to

$$g(x) = \inf_{y \in S} \{f(y) - \langle x, y \rangle\} + \tfrac{1}{2}\|x\|^2. \tag{9}$$

It is clear that for fixed $x \in X$, the sets of y attaining the infima in (8), (9), and in

$$\bar{g}(x) := \inf_{y \in S} \{f(y) - \langle x, y \rangle\} \tag{10}$$

all coincide. The Density Theorem 3.1 says that $\mathrm{dom}\,\partial_P g$ is dense in $\mathrm{dom}\,g = X$, and Theorem 5.1(b) says that for each $x \in \mathrm{dom}\,\partial_P g$, the infimum in (8) is uniquely attained. Hence for a dense set of $x \in X$, the infimum in (10) is attained at a unique $y \in S$, which is the assertion of the theorem. $\square$

5.3. Corollary. *Theorem 4.2 holds.*

Proof. Let f, x_0, and ε be as in the statement of Theorem 4.2, and let $\lambda > 0$. Consider the function f_α as given in (1) with $\alpha = \varepsilon/\lambda^2$:

$$f_\alpha(x) := \inf_{y' \in X} \left\{ f(y') + \frac{\varepsilon}{\lambda^2}\|y' - x\|^2 \right\}.$$

By the Density Theorem 3.1, there exists $z \in x_0 + \lambda B$ satisfying $f_\alpha(z) \leq f_\alpha(x_0) \leq f(x_0)$ and with $\partial_P f_\alpha(z) \neq \emptyset$. By Theorem 5.1(b), there is a unique point y at which the infimum defining $f_\alpha(z)$ is attained. All the assertions of the theorem are now immediate, except for the inequality $\|y - z\| < \lambda$, which we proceed to confirm.

We have

$$f(y) + \frac{\varepsilon}{\lambda^2}\|y - z\|^2 = f_\alpha(z) \leq f(x_0) < \inf_X(f) + \varepsilon,$$

and so

$$\frac{\varepsilon}{\lambda^2}\|y - z\|^2 < \inf_X(f) - f(y) + \varepsilon \leq \varepsilon,$$

which implies the inequality we seek. $\square$

5.4. Remark. The proof of Theorem 4.2 shows that we may in fact take z arbitrarily near x_0 (and not merely within distance λ of x_0).

6 The Distance Function

In this section we examine the proximal subgradients of the distance function d_S associated to a nonempty closed subset S of X. The results of the analysis will allow us to deduce geometric analogues of the minimization principles of §4.

6.1. Theorem. *Suppose $x \notin S$ and $\zeta \in \partial_P d_S(x)$. Then there exists a point $\bar{s} \in S$ so that the following holds:*

(a) *Every minimizing sequence $\{s_i\} \subset S$ of $\inf_{s \in S} \|s - x\|$ converges to $\bar{s}$.*

(b) *The set of closest points $\mathrm{proj}_S(x)$ in S to x is the singleton $\{\bar{s}\}$.*

(c) *The Fréchet derivative $d_S'(x)$ exists, and*

$$\{\zeta\} = \partial_P d_S(x) = \{d_S'(x)\} = \left\{\frac{x - \bar{s}}{\|x - \bar{s}\|}\right\}.$$

(d) $\zeta \in N_S^P(\bar{s})$.

Proof. Suppose $x \notin S$ and $\zeta \in \partial_P d_S(x)$. There exists $\sigma > 0$ and $\eta > 0$ so that

$$d_S(y) - d_S(x) \geq \langle \zeta, y - x \rangle - \sigma \|y - x\|^2 \quad \forall y \in B(x; \eta). \tag{1}$$

Note that for all $y \in X$ we have

$$d_S^2(y) = \inf_{z \in X} \left\{ I_S(z) + \|z - y\|^2 \right\}. \tag{2}$$

Hence for the choices $\alpha = 1$ and $f = I_S$, we see that $d_S^2(\cdot)$ is a quadratic inf-convolution function. The assertions of the present theorem will be derived as consequences of Theorem 5.1, but in order to do so, we must first establish that subgradients of d_S lead to subgradients of d_S^2. In particular, we show that $\zeta \in \partial_P d_S(x)$ implies $2d_S(x)\zeta \in \partial d_S^2(x)$, which can be viewed as a special case of the Chain Rule.

Observe that the elementary identity

$$d_S^2(y) - d_S^2(x) = 2d_S(x)\big(d_S(y) - d_S(x)\big) + \big(d_S(y) - d_S(x)\big)^2$$

holds for any $y \in X$. Using it and (1) we see for each $y \in B(x; \eta)$ that

$$d_S^2(y) - d_S^2(x) \geq 2d_S(x)\big(d_S(y) - d_S(x)\big)$$
$$\geq \langle 2d_S(x)\zeta, y - x \rangle - 2d_S(x)\sigma\|y - x\|^2,$$

which implies $2d_S(x)\zeta \in \partial_P d_S^2(x)$, as claimed.

Hence parts (a) and (b) of the present theorem follow from parts (a) and (b) of Theorem 5.1, respectively, since minimizing sequences and minima of (2) are precisely those for the infimum defining $d_S(x)$. We also have from Theorem 5.1(c) that $\partial_P d_S(x) = \{\zeta\}$ and

$$\zeta = \frac{x - \bar{s}}{d_S(x)} = \frac{x - \bar{s}}{\|x - \bar{s}\|}.$$

We turn now to differentiability. Theorem 5.1 implies that $d_S^2(\cdot)$ is Fréchet differentiable at x, whence for any $v \in X$, the limit as $t \downarrow 0$ of

$$\frac{d_S^2(x+tv) - d_S^2(x)}{t} = \left[\frac{d_S(x+tv) - d_S(x)}{t}\right][d_S(x+tv) + d_S(x)]$$

exists, with the convergence being uniform for v in bounded sets. We also know that the limit is $2d_S(x)\langle \zeta, v\rangle$. Since $d_S(x+tv) + d_S(x)$ evidently converges to $2d_S(x) > 0$ uniformly for v in bounded sets as $t \downarrow 0$ (as a consequence of the fact that d_S is globally Lipschitz), it follows that the difference quotient

$$\frac{d_S(x+tv) - d_S(x)}{t}$$

converges to $\langle \zeta, v\rangle$ uniformly on bounded sets as $t \downarrow 0$; i.e., $d_S'(x) = \zeta$.

Finally, we have that $\bar{s}$ minimizes the function $s \mapsto \|s - x\|^2$ over $s \in S$. It follows from Exercise 2.12 that $2(x - \bar{s}) \in N_S^P(\bar{s})$, which implies $\zeta = (x - \bar{s})/\|x - \bar{s}\| \in N_S^P(\bar{s})$, since this set is a cone. The theorem is proved. $\square$

The following corollary asserts that the points that admit unique closest points to a given closed set are dense (cf. Exercise 1.2(d)).

6.2. Corollary. *Suppose $S \subset X$ is closed.*

(a) *There is a dense set of points in $X \backslash S$ which admit unique closest points in S.*

(b) *The set of points $s \in \mathrm{bdry}\, S$ for which $N_S^P(s) \neq \{0\}$ is dense in $\mathrm{bdry}\, S$.*

Proof. By the Density Theorem 3.1, $\mathrm{dom}\,\partial_P d_S(\cdot)$ is dense in X, and by Theorem 6.1 each point x with $\partial_P d_S(x) \neq \emptyset$ has a unique closest point in S. This proves (a). To see that (b) holds, let $s \in \mathrm{bdry}\, S$ and $\varepsilon > 0$. By (a), there exists $x \notin S$ with $\|x - s\| < \varepsilon$ and so that $\mathrm{proj}_S(x) = \{\bar{s}\}$. Then $N_S^P(\bar{s})$ contains $x - \bar{s} \neq 0$. Since we have

$$\|\bar{s} - s\| \leq \|x - \bar{s}\| + \|x - s\| \leq 2\|x - s\| \leq 2\varepsilon,$$

assertion (b) follows. $\square$

The next proposition illustrates a mechanism for using the distance function as a tool in solving *constrained optimization* problems. Such a problem is of the form

$$\inf f(s) \text{ subject to } s \in S, \qquad (3)$$

where $f \in \mathcal{F}$ and $S \subset X$ is closed. We will study constrained optimization problems in detail in Chapter 3. The technique introduced below is called *exact penalization*, since it adds a penalty to the function to be minimized so as to obtain an equivalence between the original constrained problem and the new penalized and unconstrained problem. Exact penalization has many theoretical and numerical applications in optimization.

6.3. Proposition. *Suppose S is a closed subset of X, and that f is Lipschitz of rank K on an open set U that contains S. Assume that $s \in S$ solves (3). Then the function $x \mapsto f(x) + K d_S(x)$ attains its minimum over U at $x = s$. Conversely, if $K' > K$ and $x \mapsto f(x) + K' d_S(x)$ attains its minimum over U at $x = s$, then s belongs to S and solves (3).*

Proof. Suppose $x \in U$ and $\varepsilon > 0$. Let $s' \in S$ be such that $\|x - s'\| \leq d_S(x) + \varepsilon$. Since f is minimized over S at s, and using the Lipschitz property, we have

$$\begin{aligned} f(s) &\leq f(s') \\ &\leq f(x) + K\|s' - x\| \\ &\leq f(x) + K d_S(x) + \varepsilon K. \end{aligned}$$

Letting $\varepsilon \downarrow 0$ shows that $f + K d_S$ attains its minimum over U at $x = s$.

To prove the converse, suppose $K' > K$ and that a point $s \notin S$ minimizes $f + K' d_S$ over U. Pick $s' \in S$ so that $\|s' - s\| < (K'/K) d_S(s)$. Since f is Lipschitz of rank K, we have

$$f(s') \leq f(s) + K\|s' - s\|.$$

We are assuming that s minimizes $x \mapsto f(x) + K' d_S(x)$ over U, and since $s' \in S \subset U$, we conclude that

$$f(s) + K' d_S(s) \leq f(s') \leq f(s) + K\|s' - s\| < f(s) + K' d_S(s),$$

which is a contradiction. Thus $s \in S$. That s solves (3) is then immediate. $\square$

The next result forges another link between the geometric and functional points of view.

6.4. Theorem. *Suppose S is closed and $s \in S$. Then*

$$N_S^P(s) = \{t\zeta : t \geq 0, \zeta \in \partial_P d_S(s)\}.$$

Proof. Suppose $\zeta \in N_S^P(s)$. By the proximal normal inequality, there exists $\sigma > 0$ so that

$$\langle \zeta, s' - s \rangle \leq \sigma \|s' - s\|^2 \qquad (4)$$

for all $s' \in S$. It immediately follows that the C^2 function $x \mapsto -\langle \zeta, x \rangle + \sigma \|x - s\|^2$ has a minimum over S at $x = s$. Since this function is locally

Lipschitz, we conclude from Proposition 6.3 that for some constant K, the function
$$x \mapsto -\langle \zeta, x \rangle + \sigma \|x - s\|^2 + K d_S(x)$$
has a local minimum at $x = s$. It follows that $\zeta/K \in \partial_P d_S(s)$, which proves one of the desired inclusions.

To prove the reverse inclusion, suppose $\zeta \in \partial_P d_S(s)$. The proximal subgradient inequality provides a constant σ so that
$$d_S(x) - \langle \zeta, x - s \rangle + \sigma \|x - s\|^2 \geq d_S(s) = 0$$
for all x near s. In particular, there exists $\delta > 0$ so that (4) holds for all $s' \in S \cap \{s + \delta B\}$, which is equivalent to $\zeta \in N_S^P(s)$ by Proposition 1.5(b). Since $N_S^P(s)$ is a cone, the proof is now complete. □

7 Lipschitz Functions

Among the functions that have the Lipschitz property are the distance function d_S, the quadratic inf-convolution functions of §5, C^1 functions and, as we now see, all finite (nonpathological) convex functions. A function f is said to be *bounded above* near x if there exists $\eta > 0$ and $r \in \mathbb{R}$ such that
$$f(y) \leq r \ \forall y \in B(x; \eta).$$

7.1. Proposition. *Let $U \subset X$ be open and convex, and let $f \colon U \to \mathbb{R}$ be convex and finite on U. Suppose that f is bounded above at some point $\bar{x} \in U$. Then f is locally Lipschitz on U.*

Proof. Let $x \in U$, and let us suppose for the moment that f is bounded on a neighborhood of x. Choose $\eta > 0$ and $r \in \mathbb{R}$ so that $y \in B(x; 2\eta) \subset U$ implies that $|f(y)| \leq r$. Let y_1, y_2 be distinct points in $B(x; \eta)$, and set $\delta = \|y_1 - y_2\|$. Let $y_3 = y_2 + (\eta/\delta)(y_2 - y_1) \in B(x; 2\eta)$ and note then that
$$y_2 = \frac{\eta}{\eta + \delta} y_1 + \frac{\delta}{\eta + \delta} y_3.$$

By convexity, we have that
$$f(y_2) \leq \frac{\eta}{\eta + \delta} f(y_1) + \frac{\delta}{\eta + \delta} f(y_3),$$
which can be rearranged to
$$f(y_2) - f(y_1) \leq \frac{\delta}{\eta + \delta} [f(y_3) - f(y_1)] \leq \frac{\delta}{\eta} |f(y_3) - f(y_1)|.$$

Since $|f(y)| \leq r$ on $B(x; 2\eta)$ and $\delta = \|y_1 - y_2\|$, we conclude that

$$f(y_2) - f(y_1) \leq \frac{2r}{\eta} \|y_1 - y_2\|.$$

Switching the roles of y_1 and y_2 yields that f is Lipschitz on $B(x; \eta)$.

We now show f is locally bounded near any $x \in U$. Without loss of generality, we assume $\bar{x} = 0$. Suppose that $f(y) \leq r$ for all y satisfying $\|y\| < \eta$, where $B(0; \eta) \subset U$; let $x \in U$. There exists $t \in (0, 1)$ so that $z := x/t \in U$, and we claim that f is bounded above on $B(x; (1-t)\eta)$. Indeed, let $v \in B(x; (1-t)\eta)$, and choose $y \in \eta B$ so that $v = x + (1-t)y = tz + (1-t)y$. By convexity, we have

$$f(v) \leq tf(z) + (1-t)f(y) \leq tf(z) + r,$$

which shows that f is bounded above on $B(x; (1-t)\eta)$. We complete the proof by showing that f is bounded below on the same ball. Let $\bar{r}$ be an upper bound for f on $B(x; (1-t)\eta)$, and pick any point u in this set. Choose $u' \in B(x; (1-t)\eta)$ so that $(u + u')/2 = x$. Then $f(x) \leq (f(u) + f(u'))/2$ implies $f(u) \geq 2f(x) - f(u') \geq 2f(x) - \bar{r}$, establishing the required lower bound. □

7.2. Exercise. Show that the boundedness hypothesis of Proposition 7.1 is automatically satisfied if $X = \mathbb{R}^n$. (*Hint.* There exist finitely many points p_i such that $x \in \text{int co}\{p_i\} \subset U$.)

The class of Lipschitz functions constitutes an extremely interesting one in its own right. We now show in the general setting that the Lipschitz property can be characterized in proximal terms. It is interesting to compare the nature of this proof to the very different proofs in the smooth case, which use integration or the Mean Value Theorem. Here, optimization is the principal tool, and the criterion need only hold at certain points (where $\partial_P f(x)$ is nonempty).

7.3. Theorem. *Let $U \subset X$ be open and convex, and let $f \in \mathcal{F}(U)$. Then f is Lipschitz on U of rank $K \geq 0$ iff*

$$\|\zeta\| \leq K \; \forall \zeta \in \partial_P f(x), \; \forall x \in U. \tag{1}$$

Proof. If f has the stated Lipschitz property, then (1) follows, as already noted in Exercise 2.13. So let us now posit (1) and prove that the Lipschitz property holds. We claim that it suffices to prove the following local property: for each $x_0 \in U$ there is an open ball centered at x_0 on which f is Lipschitz of rank K. For suppose this local condition holds, and let x and y be any two points in U. Each point z of the line segment $[x, y]$ admits a ball $B(z; r)$ on which the Lipschitz condition holds. Since $[x, y]$ is compact, a finite number of balls $B(z_j; r_j)$ covers $[x, y]$. This allows us to find points

$x + t_i(y - x)$ in $[x, y]$ ($i = 0, 1\ldots, N$), with $0 = t_0 < t_1 < \cdots < t_N = 1$, such that successive points always lie in one of the sets $B(z_j; r_j)$. Then

$$f(y) - f(x) = \sum_{i=0}^{N-1} \left[f\big(x + t_{i+1}(y - x)\big) - f\big(x + t_i(y - x)\big) \right]$$
$$\leq \sum_{i=0}^{N-1} K(t_{i+1} - t_i) \|y - x\| = K\|y - x\|,$$

which confirms the global Lipschitz condition on U.

We turn now to verifying the local property. The first step will consist of showing that each point $x_0 \in \text{cl}(\text{dom}_U f) \cap U$ admits a neighborhood in which f is finite and Lipschitz of rank K. Here, $\text{dom}_U f$ means the set of those points in U at which f is finite, a set which is nonempty by definition of $\mathcal{F}(U)$.

Let $\varepsilon > 0$ be such that f is bounded below on $\overline{B}(x_0; 4\varepsilon) \subset U$; pick any $K' > K$. We denote by φ a function mapping the interval $[0, 3\varepsilon)$ in $\mathbb{R}$ to $[0, \infty)$ and having the following properties: $\varphi(\cdot)$ is strictly increasing, $\varphi(t) = K't$ for $0 \leq t \leq 2\varepsilon$, $\varphi'(t) \geq K'$ for $t \geq 2\varepsilon$, $\varphi(\cdot)$ is C^2 on $(0, 3\varepsilon)$, $\varphi(t) \to \infty$ as $t \to 3\varepsilon$. (A sketch will show the reader what is involved.)

Now fix any two points $y, z \in B(x_0; \varepsilon)$. The function

$$x \mapsto f(y + x) + \varphi(\|x\|)$$

is not identically $+\infty$ on $B(0; 3\varepsilon)$, since $x_0 \in \text{cl}(\text{dom}_U f)$. It is lower semicontinuous and bounded below on $\overline{B}(0; 3\varepsilon)$, and is equal to $+\infty$ on the boundary of that closed ball. We invoke the minimization principle embodied in Theorem 4.1 to deduce that for some $\beta \in X$ with $\|\beta\| < K' - K$, the function g defined as

$$g(x) := f(y + x) + \varphi(\|x\|) - \langle \beta, x \rangle$$

attains a minimum over $\overline{B}(0; 3\varepsilon)$ at a point u. Necessarily we have $\|u\| < 3\varepsilon$, so that $0 \in \partial_P g(u)$.

If $u \neq 0$, then the function $x \mapsto \varphi(\|x\|)$ is C^2 in a neighborhood of u, and so Proposition 2.11 applies to give

$$\beta - \varphi'(\|u\|) u/\|u\| \in \partial_P f(y + u).$$

But

$$\|\beta - \varphi'(\|u\|) u/\|u\|\| \geq \varphi'(\|u\|) - \|\beta\|$$
$$> K' - (K' - K) = K,$$

and $y+u$ is a point in $B(x_0; 4\varepsilon) \subset U$. This contradicts the given bound on $\partial_P f$ and implies that $u = 0$ necessarily.

Since $u = 0$ minimizes g over $B(0; 3\varepsilon)$, we deduce that $f(y) < \infty$, and

$$\begin{aligned} f(y) = g(0) &\leq g(z - y) \\ &= f(z) + \varphi(\|z - y\|) - \langle \beta, z - y \rangle \\ &\leq f(z) + K'\|z - y\| + (K' - K)\|z - y\|, \end{aligned}$$

where we have used the fact that $\|z - y\| < 2\varepsilon$ and the fact that $\varphi(t) = K't$ for $t \in [0, 2\varepsilon]$. Since y and z are arbitrary points in $B(x_0; \varepsilon)$ and $K' > K$ is arbitrary too, we have shown that f is Lipschitz of rank K on $B(x_0; \varepsilon)$. There remains to deal with the restriction that x_0 belongs to cl(dom$_U$ f). But the argument just given evidently implies that any such x_0 belongs to int(dom$_U$ f), so that cl(dom$_U$ f) is seen to be open (as well as closed) relative to U; consequently cl(dom$_U$ f) $\cap U =$ dom$_U$ $f = U$. Therefore the local property holds at every point of U and the proof is complete. □

7.4. Corollary. *Let $U \subset X$ be open and convex, and let $f \in \mathcal{F}(U)$. Then f is constant on U iff*

$$\partial_P f(x) \subset \{0\} \ \forall x \in U.$$

7.5. Exercise.

(a) Show that Corollary 7.4 is false if the hypothesis that U be convex is deleted.

(b) Show that if U is open (not necessarily convex), then $f \in \mathcal{F}(U)$ is locally Lipschitz on U iff $\partial_P f$ is locally bounded on U.

(c) Prove that a function f is locally Lipschitz on a compact set S iff f is globally Lipschitz on S.

8 The Sum Rule

Suppose that the sum of two functions $f_1 + f_2$ is minimized at a point x_0:

$$f_1(x) + f_2(x) \geq f_1(x_0) + f_2(x_0) \ \forall x \in X.$$

Then of course we have $0 \in \partial_P(f_1 + f_2)(x_0)$. Do we also have the "separated conclusion" $0 \in \partial_P f_1(x_0) + \partial_P f_2(x_0)$? That the answer in general is "no" is easy to see: take, for example, $f_1(x) := |x|$ and $f_2(x) := -|x|$ on $X = \mathbb{R}$, with $x_0 = 0$. Then $\partial_P f_2(x_0) = \emptyset$, so that the separated conclusion cannot hold. More generally, an exact Sum Rule of the form $\partial_P(f_1 + f_2)(x) = \partial_P f_1(x) + \partial_P f_2(x)$ is not available.

On the other hand, if f_1 and f_2 are functions of two different unrelated arguments, that is, if we have $f_1(x) + f_2(y)$, then at a minimizing pair (x_0, y_0)

of $f_1(x) + f_2(y)$ we evidently have the separate conclusions $0 \in \partial_P f_1(x_0)$ and $0 \in \partial_P f_2(y_0)$. There is a technique, sometimes called "decoupling," to approximate the first (coupled) situation above by the second (uncoupled) one. To apply this technique, we penalize the pairs (x, y) with $x \neq y$ by adding a large positive multiple of $\|x - y\|^2$; that is, we consider minimizing

$$f_1(x) + f_2(y) + r\|x - y\|^2$$

over the pairs (x, y) in $X \times X$. To make the decoupling technique work, we need to know that the minimum approaches that of the coupled case as $r \to \infty$. We will give a precise result of this type, whose relevance to the Sum Rule will be apparent shortly. Supplementary hypotheses are required, and we identify two alternatives, one of which involves weak lower semicontinuity. We say that a function f is *weakly lower semicontinuous* at x provided that

$$\liminf_{i \to \infty} f(x_i) \geq f(x)$$

whenever the sequence x_i converges weakly to x. (Note the sequential nature of this condition.)

8.1. Exercise.

(a) Let $f \in \mathcal{F}$ be convex. Prove that f is weakly lower semicontinuous at every point x. (*Hint.* Recall that a convex subset of a Hilbert space is closed iff it is weakly closed iff it is sequentially weakly closed.)

(b) Let S be a weakly closed subset of X. Then its indicator function I_S is weakly lower semicontinuous.

8.2. Proposition. *Let f_1 and f_2 belong to $\mathcal{F}(X)$, and let C be a closed convex bounded subset of X with $C \cap \mathrm{dom}\, f_1 \cap \mathrm{dom}\, f_2 \neq \emptyset$. Suppose that either*:

(i) *f_1 and f_2 are weakly lower semicontinuous on C (automatically the case if X is finite dimensional); or*

(ii) *one of the functions is Lipschitz on C and the other is bounded below on C.*

Then, for any positive sequence $\{r_n\}$ with $\lim_{n \to \infty} r_n = +\infty$, we have

$$\lim_{n \to \infty} \inf_{x, y \in C} \{f_1(x) + f_2(y) + r_n\|x - y\|^2\} = \inf_{x \in C} \{f_1(x) + f_2(x)\}. \tag{1}$$

Proof. The left side of (1) is no greater than the right, so only the opposite inequality need be proved. We address first the case in which hypothesis (i) holds. In that case, since C is weakly compact and since $(x, y) \mapsto \|x - y\|^2$ is convex on $X \times X$ (and hence, weakly lower semicontinuous), the function

$(x, y) \to f_1(x) + f_2(y) + r_n\|x - y\|^2$ admits a minimum over $C \times C$, at a point we will label (x_n, y_n). Let us extract a weakly convergent subsequence from $\{(x_n, y_n)\}$, without relabeling.

If x_0 is any point in $C \cap \operatorname{dom} f_1 \cap \operatorname{dom} f_2$, then we have

$$f_1(x_n) + f_2(y_n) + r_n\|x_n - y_n\|^2 \le f_1(x_0) + f_2(x_0),$$

which yields an a priori bound: $r_n\|x_n - y_n\|^2 \le m$. It follows that $\|x_n - y_n\| \to 0$ as $n \to \infty$, from which we deduce that the weakly convergent sequences $\{x_n\}$ and $\{y_n\}$ have the same weak limit $\bar{x} \in C$. Invoking weak lower semicontinuity now leads to

$$f_1(\bar{x}) + f_2(\bar{x}) \le \liminf_{n\to\infty} \{f_1(x_n) + f_2(y_n) + r_n\|x_n - y_n\|^2\},$$

which implies that the right side of (1) does not exceed the left.

We turn now to the case of hypothesis (ii), letting K be a Lipschitz constant for one of the functions on C, let us say f_1. For each n, let $(x_n, y_n) \in C \times C$ satisfy

$$f_1(x_n) + f_2(y_n) + r_n\|x_n - y_n\|^2 \le \inf_{x,y \in C} \{f_1(x) + f_2(y) + r_n\|x - y\|^2\} + \frac{1}{n}.$$

It follows essentially as in the first case above that $\|x_n - y_n\| \to 0$ as $n \to \infty$, since f_1 and f_2 are bounded below on C.

We argue now as follows:

$$\inf_{x,y\in C}\{f_1(x) + f_2(y) + r_n\|x - y\|^2\}$$

$$\ge f_1(x_n) + f_2(y_n) + r_n\|x_n - y_n\|^2 - \frac{1}{n}$$

$$\ge f_1(y_n) - K\|y_n - x_n\| + f_2(y_n) + r_n\|x_n - y_n\|^2 - \frac{1}{n}$$

$$\ge \inf_{x\in C}\{f_1(x) + f_2(x)\} - K\|y_n - x_n\| - \frac{1}{n}.$$

Passing to the limit as $n \to \infty$ yields the result. □

We are now ready to prove a result known as the "fuzzy Sum Rule."

8.3. Theorem. *Let $x_0 \in \operatorname{dom} f_1 \cap \operatorname{dom} f_2$, and let ζ belong to $\partial_P(f_1 + f_2)(x_0)$. Suppose that either:*

(i) *f_1 and f_2 are weakly lower semicontinuous (automatically the case if X is finite dimensional); or*

(ii) *one of the functions is Lipschitz near x_0.*

Then, for any $\varepsilon > 0$, there exist (for $i = 1, 2$) points $x_i \in B(x_0; \varepsilon)$ with $|f_i(x_0) - f_i(x_i)| < \varepsilon$ such that

$$\zeta \in \partial_P f_1(x_1) + \partial_P f_2(x_2) + \varepsilon B.$$

Proof. Since ζ belongs to $\partial_P(f_1 + f_2)(x_0)$, there exists a closed ball $C := \overline{B}(x_0; \eta)$ of positive radius η centered at x_0 and $\sigma \geq 0$ such that the function

$$x \mapsto f_1(x) + f_2(x) + \sigma\|x - x_0\|^2 - \langle \zeta, x - x_0 \rangle$$

attains a minimum over $\overline{B}(x_0; \eta)$ at $x = x_0$. We can take η small enough so that both f_1 and f_2 are bounded below on C and so that in case (ii), one of the functions is Lipschitz on that set. Now consider the problem of minimizing over $X \times X$ the function

$$\varphi_n(x, y) := f_1(x) + f_2(y) + \sigma\|y - x_0\|^2 - \langle \zeta, x - x_0 \rangle + I_C(x) + I_C(y) + r_n \|x - y\|^2,$$

where r_n is a positive sequence tending to $+\infty$ as $n \to \infty$. Applying Proposition 8.2 (with the role of $f_2(y)$ played here by

$$f_2(y) + \sigma \|y - x_0\|^2 - \langle \zeta, x - x_0 \rangle)$$

yields the conclusion that the nonnegative quantity

$$q_n := f_1(x_0) + f_2(x_0) - \inf_{X \times X} \varphi_n$$

tends to zero as $n \to \infty$. Put another way, we conclude that for n large, the point (x_0, x_0) "almost minimizes" φ_n.

We will invoke the Minimization Principle Theorem 4.2 with data $f := \varphi_n$, $\varepsilon := q_n + 1/n$, $\lambda := \sqrt{\varepsilon}$. We derive the existence of points (x_1^n, x_2^n) $(= y)$ and (z_1^n, z_2^n) $(= z)$ in $X \times X$ such that

$$\|(x_1^n, x_2^n) - (z_1^n, z_2^n)\| < \left(q_n + \frac{1}{n}\right)^{1/2}, \quad \|(z_1^n, z_2^n) - (x_0, x_0)\| < \left(q_n + \frac{1}{n}\right)^{1/2},$$

and such that the function

$$\varphi_n(x, y) + \|x - z_1^n\|^2 + \|y - z_2^n\|^2$$

is minimized over $X \times X$ at $(x, y) = (x_1^n, x_2^n)$. For n sufficiently large (since $q_n \to 0$), both x_1^n and x_2^n must lie in the interior of $C = \overline{B}(x_0; \eta)$, so that the fact of this minimization implies the separate necessary conditions

$$-2r_n(x_1^n - x_2^n) - 2(x_1^n - z_1^n) \in \partial_P f_1(x_1^n) \text{ (minimum in } x\text{)},$$
$$2r_n(x_1^n - x_2^n) - 2\sigma(x_2^n - x_0) - 2(x_2^n - z_2^n) \in \partial_P f_2(x_2^n) \text{ (minimum in } y\text{)}.$$

We derive from this

$$\zeta \in \partial_P f_1(x_1^n) + \partial_P f_2(x_2^n) + \gamma_n \overline{B},$$

where
$$\gamma_n := 2(\|x_1^n - z_1^n\| + \|x_2^n - z_2^n\| + \sigma\|x_2^n - x_0\|).$$

For all n large enough, the points x_1^n and x_2^n will lie within the pregiven ε of x_0, and γ_n will be less than ε too. So all that remains to check is the closeness of each of the function values $f_i(x_i^n)$ to $f_i(x_0)$ ($i = 1, 2$), for n sufficiently large. This will follow from the further assertion of Theorem 4.2 that we have
$$\varphi_n(x_1^n, x_2^n) \leq \varphi_n(x_0, x_0),$$
which implies that for all n large the following holds:
$$f_1(x_1^n) + f_2(x_2^n) \leq f_1(x_0) + f_2(x_0) + \|\zeta\| \, \|x_1^n - x_0\|.$$

We derive from this
$$\liminf_{n \to \infty} \{f_1(x_1^n) + f_2(x_2^n)\} \leq f_1(x_0) + f_2(x_0).$$

In view of the lower semicontinuity of f_1 and f_2, this yields the fact that for $i = 1, 2$ the sequence $f_i(x_i^n)$ converges to $f_i(x_0)$ as $n \to \infty$ (why?), completing the proof of the theorem. $\square$

8.4. Exercise.

(a) Let C_1 and C_2 be weakly closed subsets of X, and let $\zeta \in N_{C_1 \cap C_2}^P(x)$. Then for any $\varepsilon > 0$ there exist
$$x_1 \in C_1, \quad x_2 \in C_2, \quad \zeta_1 \in N_{C_1}^P(x_1), \quad \zeta_2 \in N_{C_2}^P(x_2),$$
such that
$$\|x_1 - x\| + \|x_2 - x\| < \varepsilon, \quad \|\zeta - \zeta_1 - \zeta_2\| < \varepsilon.$$

(b) Suppose that x minimizes the locally Lipschitz function f over the set C. Then for any $\varepsilon > 0$ there exist x_1 and x_2 in the ε-neighborhood of x such that
$$0 \in \partial_P f(x_1) + N_C^P(x_2) + \varepsilon B.$$

9 The Chain Rule

Suppose now that the function f is given as the composition $g \circ F$ of two functions $g: Y \to (-\infty, \infty]$ and $F: X \to Y$, where Y is another Hilbert space. Note that if g belongs to $\mathcal{F}(Y)$ and F is continuous (in particular, if F is locally Lipschitz, as we will assume), then it follows readily that $f = g \circ F$ belongs to $\mathcal{F}(X)$. In the classical smooth setting the Chain Rule asserts $f'(x) = g'(F(x)) \circ F'(x)$. In extending this to the nonsmooth setting, one of the apparent difficulties is that we have not defined a replacement for the derivative $F'(x)$ of a nondifferentiable *vector-valued* mapping such as F. We will get around this by a device which, in the smooth case, amounts to viewing $f'(x)$ as the derivative of the scalar-valued function $u \to g'(F(x)) \circ F(u)$. Also notable about our approach is that it reveals the composition of functions to be reducible to the case of the sum of two functions.

9.1. Theorem. *Let $g \in \mathcal{F}(Y)$, and let $F\colon X \to Y$ be locally Lipschitz. Set $f(x) := g(F(x))$, and let $\zeta \in \partial_P f(x_0)$. We assume that either:*

(i) *g is weakly lower semicontinuous and F is linear; or*

(ii) *g is Lipschitz near $F(x_0)$.*

Then for all $\varepsilon > 0$ there exist $\tilde{x}$ in $x_0 + \varepsilon B_X$, $\tilde{y}$ in $F(x_0) + \varepsilon B_Y$, and $\gamma \in \partial_P g(\tilde{y})$ such that $\|F(\tilde{x}) - F(x_0)\| < \varepsilon$ and

$$\zeta \in \partial_P \{\langle \gamma, F(\cdot) \rangle\}(\tilde{x}) + \varepsilon B_X.$$

Proof. Since ζ belongs to $\partial_P f(x_0)$, there exist $\sigma, \eta > 0$ such that

$$f(x) - \langle \zeta, x \rangle + \sigma \|x - x_0\|^2 \geq f(x_0) - \langle \zeta, x_0 \rangle$$

for all $x \in B(x_0; \eta)$. Let S denote the graph of F in the space $X \times Y$. Then another way of writing the preceding proximal subgradient inequality is the following:

$$g(y) - \langle \zeta, x \rangle + I_S(x,y) + \sigma \|x - x_0\|^2 \geq g(F(x_0)) - \langle \zeta, x_0 \rangle,$$

which implies that the left side, as a function of (x,y), attains a local minimum at $(x,y) = (x_0, F(x_0))$. We deduce

$$0 \in \partial_P \{g(y) - \langle \zeta, x \rangle + I_S(x,y) + \sigma \|x - x_0\|^2\}(x_0, F(x_0)).$$

In each of cases (i) and (ii) of the theorem, we are justified in applying the fuzzy Sum Rule, Theorem 8.3. We obtain, for any $\delta > 0$, the existence of points (x_1, y_1) and (x_2, y_2), both δ-near $(x_0, F(x_0))$, such that, for some $\gamma \in \partial_P g(y_1)$ we have

$$(\zeta - 2\sigma(x_1 - x_0), -\gamma) \in \partial_P I_S(x_2, y_2) + \delta B_{X \times Y}.$$

But $\partial_P I_S(x_2, y_2) = N_S^P(x_2, y_2)$ and it follows that $y_2 = F(x_2)$ and, from Exercise 9.2 below, that for some u, v with $\|u\| < \delta$, $\|v\| < \delta$, we have

$$\zeta - 2\sigma(x_1 - x_0) + u \in \partial_P \{\langle \gamma + v, F(\cdot) \rangle\}(x_2).$$

Now let us apply the fuzzy Sum Rule again to the functions $\langle \gamma, F(\cdot) \rangle$ and $\langle v, F(\cdot) \rangle$, which are locally Lipschitz. We deduce the existence of points x_3 and x_4 in the δ-neighborhood of x_2 such that

$$\zeta - 2\sigma(x_1 - x_0) + u \in \partial_P \{\langle \gamma, F(\cdot) \rangle\}(x_3) + \partial_P \{\langle v, F(\cdot) \rangle\}(x_4) + \delta B_X.$$

If K is a local Lipschitz constant for F, then the function $\langle v, F(\cdot)\rangle$ is locally Lipschitz of rank $K\|v\| < K\delta$, whence any element of $\partial_P\{\langle v, F(\cdot)\rangle\}(x_4)$ is of norm less than $K\delta$. We have now arrived at

$$\zeta - 2\sigma(x_1 - x_0) + u \in \partial_P\{\langle \gamma, F(\cdot)\rangle\}(x_3) + \delta(K+1)B_X.$$

It suffices now to take δ small enough to get the required result, with $\tilde{y} = y_1$ and $\tilde{x} = x_3$. The closeness of $F(\tilde{x}) = F(x_3)$ to $F(x_0)$ follows from the continuity of F. □

Here is the result needed in the proof above:

9.2. Exercise. Let $F\colon X \to Y$ be locally Lipschitz, and suppose that
$$(\alpha, -\gamma) \in N_S^P(x, F(x)),$$
where S is the graph of F. Prove that
$$\alpha \in \partial_P\{\langle \gamma, F(\cdot)\rangle\}(x).$$

The two following exercises illustrate the use of the Chain Rule, and will be used later.

9.3. Exercise. Let $g \in \mathcal{F}(Y)$ be weakly lower semicontinuous and let $A\colon X \to Y$ be a continuous linear operator. Suppose that ζ belongs to $\partial_P\{g(Ax)\}(x_0)$. Prove the existence, for any $\varepsilon > 0$, of $\tilde{y}$ within ε of Ax_0 such that
$$\zeta \in A^*\partial_P g(\tilde{y}) + \varepsilon B_X \quad \text{and} \quad |g(\tilde{y}) - g(Ax_0)| < \varepsilon.$$

(Here $A^*\colon Y \to X$ denotes the adjoint of A: $\langle A^*u, v\rangle = \langle u, Av\rangle$ for all $u \in Y$, $v \in X$.)

9.4. Exercise. Let $g \in \mathcal{F}(\mathbb{R}^n)$ be locally Lipschitz, and define f as follows:
$$f\colon L_n^2[a,b] \to \mathbb{R}, \quad f(v) := g\left(\int_a^b v(s)\,ds\right).$$
Let $\zeta \in L^2$ belong to $\partial_P f(v)$. Prove that for any $\varepsilon > 0$ there exist x and θ with $\theta \in \partial_P g(x)$ such that
$$\left\|x - \int_a^b v(s)\,ds\right\| < \varepsilon, \quad \|\zeta(t) - \theta\|_2 < \varepsilon.$$

Armed with the Chain Rule above and the Sum Rule of the preceding section, we can easily derive other basic results of "proximal fuzzy calculus." But we turn instead to the issue of passing to the limit in results such as the above.

10   Limiting Calculus

Consider once again the issue of a proximal Sum Rule. We have noted the easy fact

$$\partial_P f_1(x) + \partial_P f_2(x) \subseteq \partial_P(f_1 + f_2)(x) \tag{1}$$

and we have proved an approximate form of the reverse inclusion: if $\zeta \in \partial_P(f_1 + f_2)(x)$, then for any $\varepsilon > 0$ we have

$$\zeta \in \partial_P f_1(x_1) + \partial_P f_2(x_2) + \varepsilon B, \tag{2}$$

where x_1 and x_2 are ε-close to x, and where $f_i(x_i)$ and $f_i(x)$ are close as well ($i = 1, 2$). It is natural to consider passing to the limit in (2) as $\varepsilon \to 0$. In doing so, the outcome cannot necessarily be phrased in terms of $\partial_P f_1(x)$ and $\partial_P f_2(x)$ (which may be empty, for example). A construction that naturally recommends itself, and that we call the *limiting subdifferential* $\partial_L f(x)$ (or *L-subdifferential*), is the following:

$$\partial_L f(x) := \{\text{w-lim}\,\zeta_i : \zeta_i \in \partial_P f(x_i), x_i \xrightarrow{f} x\}. \tag{3}$$

That is, we consider the set of all vectors ζ that can be expressed as the weak limit (which is what "w-lim" signifies) of some sequence $\{\zeta_i\}$, where $\zeta_i \in \partial_P f(x_i)$ for each i, and where $x_i \to x$, $f(x_i) \to f(x)$. (Note: The notation

$$x_i \xrightarrow{f} x$$

encapsulates *both* of these convergences.)

Let us return now to the Sum Rule, but in limiting terms: let $\zeta \in \partial_L(f_1 + f_2)(x)$. By definition, we have $\zeta = \text{w-lim}\,\zeta_i$, where $\zeta_i \in \partial_P(f_1 + f_2)(x_i)$ and where $x_i \to x$, $(f_1 + f_2)(x_i) \to (f_1 + f_2)(x)$. If the fuzzy Sum Rule (Theorem 8.2) can be applied, for example if one of f_1, f_2 is Lipschitz locally, then we can write

$$\zeta_i = \theta_i + \xi_i + \varepsilon_i u_i, \tag{4}$$

where $\theta_i \in \partial_P f_1(x_i')$, $\xi_i \in \partial_P f_2(x_i'')$, $u_i \in B$, where x_i' and x_i'' lie in an ε_i-ball around x_i, and where $\varepsilon_i \downarrow 0$ as $i \to \infty$. We also have that $f_1(x_i')$ and $f_2(x_i'')$ are ε_i-close to $f_1(x_i)$ and $f_2(x_i)$, respectively.

We now wish to pass to the limit in (4); How can this be carried out? Again, assuming that one of the functions (let us say f_1) is Lipschitz locally would do the trick. For in that case $\partial_P f_1(\cdot)$ is bounded locally (by the Lipschitz rank) so that the sequence $\{\theta_i\}$ is bounded. But $\{\zeta_i\}$ is bounded too, since this sequence converges weakly (to ζ). It follows from (4) that $\{\xi_i\}$ is bounded. Extracting subsequences so that all these sequences converge weakly (to θ, ζ, ξ, respectively) gives $\zeta = \theta + \xi$. We also have that $\{f_1(x_i')\}$ and $\{f_2(x_i'')\}$ converge to $f_1(x)$ and $f_2(x)$, respectively (why?), so that $\theta \in \partial_L f_1(x)$, $\xi \in \partial_L f_2(x)$ by definition. The discussion has proved

10.1. Proposition. *If one of f_1, f_2 is Lipschitz near x, then*

$$\partial_L(f_1 + f_2)(x) \subseteq \partial_L f_1(x) + \partial_L f_2(x).$$

This is an appealing (nonfuzzy) Sum Rule. In view of the inclusion (1), we might be led to believe (briefly) that $\partial_L f_1(x) + \partial_L f_2(x) \subseteq \partial_L(f_1 + f_2)(x)$, and hence that equality actually holds in Proposition 10.1. But alas, the limiting form of (1) fails to hold, as shown in the next exercise. (We will find supplementary hypotheses in Chapter 2 under which equality does hold.)

10.2. Exercise. Set $X = \mathbb{R}$, $f_1(x) := |x|$, $f_2(x) := -|x|$. Show that $\partial_L f_1(0) = [-1,1]$, $\partial_L f_2(0) = \{-1,1\}$. Verify that the conclusion of Proposition 10.1 holds, but not with equality.

Let us consider now the limiting issue in geometric terms. The natural closure operation to apply to $N_S^P(\cdot)$ gives rise to $N_S^L(x)$, the *limiting normal* (or *L-normal*) *cone* to S at $x \in S$:

$$N_S^L(x) := \left\{ \text{w-lim}\, \zeta_i : \zeta_i \in N_S^P(x_i), x_i \xrightarrow{S} x \right\}.$$

Here, $x_i \xrightarrow{S} x$ signifies that $x_i \to x$ and that $x_i \in S$ $\forall i$. One motivation for defining a limiting normal cone is that $N_S^P(x)$ is potentially trivial (i.e., $= \{0\}$) for "many" x; in pointwise considerations it is $N_S^L(x)$ that may incorporate normality information. The following exercise vindicates this hope in $\mathbb{R}^n$, confirms a certain coherence between the functional and geometric closure operations, and derives some basic properties of the limiting constructs.

10.3. Exercise. Let S be closed and $f \in \mathcal{F}$.

(a) $(\zeta, -1) \in N_{\text{epi}\, f}^L(x, f(x))$ iff $\zeta \in \partial_L f(x)$, and $\partial_L I_S(x) = N_S^L(x)$.

(b) $N_S^L(x) = N_S^P(x)$ when S is convex.

(c) When $X = \mathbb{R}^n$, both $N_S^L(x)$ and $\partial_L f(x)$ are closed sets; if $x \in \text{bdry}\, S$, then $N_S^L(x) \neq \{0\}$, and if f is Lipschitz near x, then $\partial_L f(x) \neq \emptyset$.

Although a type of closure operation was used in defining $N_S^L(x)$, it is a fact that this set may fail to be closed when X is infinite dimensional; similarly, $\partial_L f(x)$ may not be closed if f fails to be Lipschitz. These facts make the limiting calculus most appealing in the presence of Lipschitz hypotheses, or in finite dimensions. Here is a limiting form of the Chain Rule in such a context, one whose use will be illustrated in the end-of-chapter problems.

10.4. Theorem. *Let $F: X \to \mathbb{R}^n$ be Lipschitz near x, and let $g: \mathbb{R}^n \to \mathbb{R}$ be Lipschitz near $F(x)$. Then*

$$\partial_L(g \circ F)(x) \subseteq \{\partial_L \langle \gamma, F(\cdot)\rangle(x): \gamma \in \partial_L g(F(x))\}.$$

11 Problems on Chapter 1

11.1. Give examples of each the following:

(a) A set S in $\mathbb{R}^2$ and a point $s \in S$ such that $N_S^P(s)$ is neither open nor closed, and a function $f\colon \mathbb{R} \to \mathbb{R}$ with a point $x \in \text{dom}\, f$ such that $\partial_P f(x)$ is neither open nor closed.

(b) A function $f \in \mathcal{F}$ that is not bounded below on $\overline{B}$.

(c) A function $f \in \mathcal{F}$ that is bounded below on $\overline{B}$ but does not attain its minimum there.

11.2. Suppose that $\{e_i\}_{i=1}^\infty$ is an orthonormal basis for X, and set $S := \overline{\text{co}}\{\pm e_i/i \colon i \geq 1\}$. Prove that $N_S^P(0) = \{0\}$.

11.3. Suppose $S \subset X$ is such that $\forall x \in S$, $\forall \zeta \in N_S^P(x)$, we have

$$\langle \zeta, x' - x \rangle \leq 0 \;\; \forall x' \in S.$$

Prove that S is convex. (This provides a converse to Proposition 1.10(a).)

11.4. Let S be bounded, closed, and nonempty. Prove that $\bigcup_{x \in S} N_S^P(x)$ is dense in X.

11.5. Let s belong to a closed set S, and suppose that d_S is differentiable at s. Show that $d_S'(s) = 0$ necessarily. Prove that if $N_S^P(s) \neq \{0\}$, then d_S cannot be differentiable at s.

11.6. Let $S \subseteq X$ be nonempty (but not necessarily closed), and suppose $f \in \mathcal{F}$ satisfies $f(x) \geq -c\|x\|$ for some $c > 0$.

(a) Let $K > c$, and define the function $g\colon X \to \mathbb{R}$ by

$$g(x) := \inf_{s \in S} \{f(s) + K\|s - x\|\}.$$

Show that g is Lipschitz on X of rank K.

(b) Suppose in addition that f is Lipschitz of rank K on S. Prove that g agrees with f on S, and deduce that Lipschitz functions defined on any subset of X can be extended to a Lipschitz function on all of X without increasing the Lipschitz rank.

11.7. Let $f \in \mathcal{F}$ and let $M \in \mathbb{R}$. Set

$$f^M(x) := \min\{f(x), M\}.$$

(a) Show that $f^M \in \mathcal{F}$.

(b) If $\zeta \in \partial_P f^M(x)$, prove that $\zeta \in \partial_P f(x)$ if $f(x) \leq M$, and that $\zeta = 0$ otherwise.

11.8. Suppose that $L: \mathbb{R}^n \to (-\infty, \infty]$ is lower semicontinuous, not identically $+\infty$, and bounded below. For any $v(\cdot) \in X := L_n^2[0, 1]$, define $f: X \to (-\infty, \infty]$ by

$$f(v(\cdot)) := \int_0^1 L(v(t))\, dt.$$

Prove that f is lower semicontinuous, and that for any $\alpha > 0$, the function

$$F(u(\cdot)) := \int_0^1 \left\{ L(u(t)) + \alpha \|v(t) - u(t)\|^2 \right\} dt$$

attains a unique minimum over X for a dense set of $v(\cdot) \in X$.

11.9. Suppose $f \in \mathcal{F}$ is bounded below, $\alpha > 0$, and f_α is defined as in §5.

(a) Let $x \in \operatorname{dom} f$. Prove that for each $\alpha > 0$ there exists $r_\alpha > 0$ so that each minimizer $\bar{y}$ in the infimum defining $f_\alpha(x)$ satisfies $\bar{y} \in B(x; r_\alpha)$, and $r_\alpha \to 0$ as $\alpha \to \infty$.

(b) Prove that $f_\alpha \uparrow f(x)$ as $\alpha \uparrow \infty$ for each $x \in X$, and that for $x \in \operatorname{dom} f$,

$$\liminf_{y \to x, \alpha \to \infty} f_\alpha(y) \geq f(x).$$

(Thus f_α is a locally Lipschitz lower approximation to f, one that improves as $\alpha \to \infty$; it is called the *Moreau–Yosida approximation*.)

11.10. Suppose $f \in \mathcal{F}$ is bounded below. Show that there exist sequences $\{y_i\}$ and $\{\zeta_i\}$ so that $\zeta_i \in \partial_P f(y_i)$ for each i, and satisfying $f(y_i) \to \inf_{x \in X} f(x)$ and $\|\zeta_i\| \to 0$ as $i \to \infty$.

11.11. Let $f \in \mathcal{F}$ and $x_0 \in \operatorname{dom} f$. Suppose that for some $\varepsilon > 0$, for all $x \in x_0 + \varepsilon B$ satisfying $|f(x) - f(x_0)| < \varepsilon$, for all $\zeta \in \partial_P f(x)$, we have $\|\zeta\| \leq K$. Prove that f is Lipschitz of rank K on a neighborhood of x_0. (*Hint.* Consider $\tilde{f}(x) := \min\{f(x), f(x_0) + \varepsilon/2\}$ in light of Problem 11.7.)

11.12. With $X = \mathbb{R}$, calculate $\partial_P f(0)$ and $\partial_L f(0)$ when $f(x) := x^2 \sin(1/x)$.

11.13.

(a) Prove the following Proximal Mean Value Theorem:

Theorem. *Let $f \in \mathcal{F}(X)$ be locally Lipschitz on a neighborhood of the line segment $[x, y]$. Then $\forall \varepsilon > 0$ there exists a point z in the ε-neighborhood of $[x, y]$ and $\zeta \in \partial_P f(z)$ such that*

$$f(y) - f(x) \leq \langle \zeta, y - x \rangle + \varepsilon.$$

(b) Show that the conclusion is false in the form

$$|f(y) - f(x) - \langle \zeta, y - x \rangle| < \varepsilon.$$

(c) Obtain a limiting form of the theorem in terms of $\partial_L f$ when $\varepsilon \to 0$. (*Hint.* Minimize $\varphi(t, u) := f(x + u) - tf(y) - (1 - t)f(x)$ over the set of (t, u) satisfying $u = t(y - x)$, $0 \le t < 1$.)

11.14. Let $f \colon \mathbb{R}^n \to \mathbb{R}$ be defined by

$$f(x) = f(x_1, x_2, \ldots, x_n) = \max_{1 \le i \le n} x_i.$$

Let $M(x)$ be the set of those indices i at which this maximum is attained. Prove that $\partial_P f(x)$ consists of all those vectors $(\zeta_1, \zeta_2, \ldots, \zeta_n)$ such that $\zeta_i \ge 0$, $\sum_{i=1}^n \zeta_i = 1$, $\zeta_i = 0$ if $i \notin M(x)$. (*Hint.* f is convex.)

11.15. (Monotonicity.) Let C be a cone in X, and let $f \in \mathcal{F}(X)$. We say that f is C-*decreasing* if $y - x \in C$ implies $f(y) \le f(x)$. Prove that if f is locally Lipschitz, then f is C-decreasing iff $\zeta \in C^\circ$ $\forall \zeta \in \partial_P f(x)$, $\forall x \in X$, where the *polar* C° of C is defined as

$$C^\circ := \{ z \in X \colon \langle z, c \rangle \le 0 \; \forall c \in C \}.$$

11.16. Prove the Limiting Chain Rule, Theorem 10.4.

11.17. (Upper envelopes.) Let $f_i \in \mathcal{F}(X)$ be locally Lipschitz, $i = 1, 2, \ldots, n$, and set $f(x) := \max_{1 \le i \le n} f_i(x)$. We denote by $M(x)$ the set of those indices i at which the maximum defining $f(x)$ is attained.

(a) Prove that f is locally Lipschitz, and that

$$\operatorname{co} \bigcup_{i \in M(x)} \partial_P f_i(x) \subset \partial_P f(x).$$

(b) If $\zeta \in \partial_L f(x)$, then there exist $\gamma_i \ge 0$ ($i = 1, 2, \ldots, n$) with $\sum_{i=1}^n \gamma_i = 1$ and $\gamma_i = 0$ for $i \notin M(x)$ such that $\zeta \in \partial_L(\sum_{i=1}^n \gamma_i f_i)(x)$.

11.18. (The Lagrange Multiplier Rule for inequality constraints.) Let x_0 solve the problem of minimizing $g_0(x)$ subject to $g_i(x) \le 0$ ($i = 1, 2, \ldots, n$), where all the functions involved are locally Lipschitz.

(a) Show that x_0 minimizes (subject to no constraints) the function

$$x \mapsto \max\{g_0(x) - g_0(x_0), g_1(x), \ldots, g_n(x)\}.$$

(b) Deduce the existence of $\gamma_i \ge 0$ ($i = 0, 1, \ldots, n$) with $\sum_{i=0}^n \gamma_i = 1$ and $\gamma_i g_i(x_0) = 0$ for $i \ge 1$, such that $0 \in \partial_L(\sum_{i=0}^n \gamma_i g_i)(x_0)$.

11.19. (Partial subgradients.) Let f be locally Lipschitz on $X_1 \times X_2$, and let $\bar{\zeta}_1 \in \partial_P f(\cdot, \bar{x}_2)(\bar{x}_1)$ (i.e., $\bar{\zeta}_1$ is a proximal subgradient at $\bar{x}_1$ of the function $x_1 \mapsto f(x_1, \bar{x}_2)$). Then for any $\varepsilon > 0$ there exist $(\tilde{x}_1, \tilde{x}_2)$ in $(\bar{x}_1, \bar{x}_2) + \varepsilon B$ and $(\tilde{\zeta}_1, \tilde{\zeta}_2)$ in $\partial_P f(\tilde{x}_1, \tilde{x}_2)$ such that $\|\tilde{\zeta}_1 - \bar{\zeta}_1\| < \varepsilon$. (*Hint.* Take $F(x) := (x, \bar{x}_2)$ in Theorem 9.1.)

11.20. This exercise is devoted to some relationships between Fréchet and Gâteaux differentiability.

(a) Suppose $f \colon X \to \mathbb{R}$ is Fréchet differentiable at x. Show that f is continuous at x.

(b) Suppose $f \colon X \to \mathbb{R}$ is Gâteaux differentiable at x. Show that for each $v \in X$, the function $g(t) := f(x + tv)$ defined for small values of $|t|$ is continuous at $t = 0$.

(c) Consider the following functions:

$$f_1(x,y) = \begin{cases} \dfrac{y^2}{x} & \text{if } x \neq 0, \\ 0 & \text{if } x = 0, \end{cases} \qquad f_2(x,y) = \begin{cases} \dfrac{xy}{x^2 + y^2} & \text{if } (x,y) \neq (0,0), \\ 0 & \text{if } (x,y) = (0,0). \end{cases}$$

Show that f_1 has a Gâteaux derivative at the origin, but that f_1 is not continuous there (and therefore not Fréchet differentiable). Show that the partial derivatives of f_2 exist at $(0,0)$, but the Gâteaux derivative $f_2'(0,0)$ does not.

(d) Suppose that $f \colon \mathbb{R}^n \to \mathbb{R}$ is Lipschitz near x. Then at x, Gâteaux differentiability implies Fréchet differentiability (this is false in general in infinite dimensional spaces).

(e) With X a Hilbert space, show that $f'(x)$ exists iff we have, for some $\zeta \in X$,
$$\lim_{i \to \infty} \frac{f(x + t_i v_i) - f(x)}{t_i} = \langle \zeta, v \rangle \quad \forall x,$$
whenever t_i decreases to 0 and v_i converges weakly to v (and then $f'(x) = \zeta$).

(f) Suppose f has a Gâteaux derivative at each point near $x \in \mathcal{F}$, and $y \mapsto f'(y)$ is continuous at x. Show that f is Fréchet differentiable at x.

11.21. Let X be separable, and let $f \in \mathcal{F}(X)$ be Lipschitz near a point $\bar{x}$, let $\{x_i\}$ be a sequence converging to $\bar{x}$, and let $\{\zeta_i\}$ be a sequence converging weakly to ζ, where $\zeta_i \in \partial_L f(x_i)$ for each i. Prove that $\zeta \in \partial_L f(\bar{x})$. (*Hint.* Recall that the weak topology restricted to a closed bounded subset of X is equivalent to a metric one.)

11.22. Let X admit an orthonormal basis $\{e_i\}_{i=1}^{\infty}$, and let $f \in \mathcal{F}(X)$ be such that at every $x \in \text{dom } f$ we have, for every index i, the conditions

$$\liminf_{t\downarrow 0} \frac{f(x+te_i)-f(x)}{t} \leq 0, \quad \liminf_{t\downarrow 0} \frac{f(x-te_i)-f(x)}{t} \leq 0.$$

Prove that f is a constant function.

11.23. (Horizontal Approximation Theorem.) This problem (needed in Chapter 4) will prove that horizontal proximal normals to epigraphs can be approximated by nonhorizontal ones (which then correspond to subgradients). Formally:

Theorem. *Let $f \in \mathcal{F}(\mathbb{R}^n)$, and let $(\theta, 0) \in N^P_{\text{epi } f}(x, f(x))$. Then for every $\varepsilon > 0$ there exist $x' \in x + \varepsilon B$ and $(\zeta, -\lambda) \in N^P_{\text{epi } f}(x', f(x'))$ such that*

$$\lambda > 0, \quad |f(x') - f(x)| < \varepsilon, \quad \|(\theta, 0) - (\zeta, -\lambda)\| < \varepsilon.$$

(a) Show how the case $\theta = 0$ is an immediate consequence of the Density Theorem 3.1.

We assume henceforth that $x = 0$, $f(0) = 0$, $\theta \neq 0$, and that the point $(\theta, 0)$ admits *unique* closest point $(0,0)$ in epi $f =: S$. (Why does this not constitute a loss of generality?)

(b) We have $d_S(\theta, t) > d_S(\theta, 0) = \|\theta\| \ \forall t < 0$.

(c) The function $t \mapsto d_S(\theta, t)$ has strictly negative proximal subgradients at points $t < 0$ arbitrarily near 0.

(d) There exist points (x, t) arbitrarily near $(\theta, 0)$ with $t < 0$ such that an (in fact, the only) element of $\partial_P d_S(x, t)$ has strictly negative final component. (*Hint.* Problem 11.19.)

(e) As the points (x, t) of part (d) converge to $(\theta, 0)$, the corresponding element of $\partial_P d_S(x, t)$ converges to $(\theta, 0)$, completing the proof.

11.24. Let $f: X \to \mathbb{R}$ be continuous and convex. Prove that f is Fréchet differentiable at a dense set of points. (*Hint.* Recall Exercise 2.8.)

11.25. Let f be locally Lipschitz on X.

(a) Prove that $\partial_L f(x)$ reducing to a singleton is a sufficient but not a necessary condition for $f'(x)$ to exist.

(b) When X is finite dimensional, prove that f is C^1 on an open set U iff $\partial_L f(x)$ reduces to a singleton for every $x \in U$.

11.26. Let $f \in \mathcal{F}(X)$, and set $S := \{x \in X : f(x) \leq 0\}$. Let $\bar{x}$ satisfy $f(\bar{x}) = 0$. Then

(a) $\bigcup_{\lambda \geq 0} \lambda \partial_P f(\bar{x}) \subset N_S^P(\bar{x})$.

(b) If f is Lipschitz near $\bar{x}$ and $0 \notin \partial_L f(\bar{x})$, then we have

$$N_S^L(\bar{x}) \subset \bigcup_{\lambda \geq 0} \lambda \partial_L f(\bar{x}).$$

(*Hint.* One approach uses Problems 11.18 and 11.25.)

(c) If f is C^2 near $\bar{x}$ and $f'(\bar{x}) \neq 0$, then

$$N_S^P(\bar{x}) = N_S^L(\bar{x}) = \{\lambda f'(\bar{x}) : \lambda \geq 0\}.$$

11.27. Let S be a nonempty closed subset of X, $\bar{x} \in S$. We will prove the formula
$$N_S^L(\bar{x}) = \bigcup_{\lambda \geq 0} \lambda \partial_L d_S(\bar{x}).$$

(a) Prove that if $\zeta \in N_S^P(x)$, then for any $\varepsilon > 0$ we have $\zeta/(\|\zeta\| + \varepsilon) \in \partial_P d_S(x)$. (*Hint.* Use exact penalization.)

(b) Deduce that $N_S^L(\bar{x}) \subset \bigcup_{\lambda \geq 0} \lambda \partial_L d_S(\bar{x})$.

(c) Complete the proof by confirming the opposite inclusion.

11.28. Let $f: \mathbb{R}^n \to \mathbb{R}$ be locally Lipschitz. Prove that $\partial_L f$ is countably generated, in the following sense. There exists a countable set Ω in $\mathbb{R}^n \times \mathbb{R}^n$ such that for every x, we have

$$\partial_L f(x) = \left\{ \lim_{i \to \infty} \zeta_i : (x_i, \zeta_i) \in \Omega \ \forall i, \ \lim_{i \to \infty} x_i = x \right\}.$$

(*Hint.* Consider a countable dense subset of $\operatorname{gr} \partial_P f$.)

11.29. (Proximal Gronwall Inequality.) Let $f: [0, T] \to \mathbb{R}$ be locally Lipschitz, where $T > 0$, and suppose there exists a constant $M \geq 0$ such that
$$\zeta \leq M f(t) \ \forall \zeta \in \partial_P f(t), \ \forall t \in (0, T).$$
Then
$$f(t) \leq e^{Mt} f(0) \ \forall t \in [0, T].$$

11.30. Let $f \in \mathcal{F}$ and $U \subset X$ be open. Show that f is C^1 on U iff for all $x \in U$ and $\varepsilon > 0$ there exists $\delta > 0$ such that

$$\|x_i - x\| < \delta, \quad \zeta_i \in \partial_P f(x_i) \quad (i = 1, 2)$$

implies that $\|\zeta_1 - \zeta_2\| < \varepsilon$.

# 2
Generalized Gradients in Banach Space

> I drew up the state of my affairs in writing; not so much to leave them to any that were to come after me, for I was like to have but few heirs, as to deliver my thoughts from daily poring upon them, and afflicting my mind.
>
> —Daniel Defoe, *Robinson Crusoe*

The calculus of generalized gradients is the best-known and most frequently invoked part of nonsmooth analysis. Unlike proximal calculus, it can be developed in an arbitrary Banach space X. In this chapter we make a fresh start in such a setting, but this time, in contrast to Chapter 1, we begin with functions and not sets. We present the basic results for the class of *locally Lipschitz* functions. Then the associated geometric concepts are introduced, including for the first time a look at *tangency*. In fact, we examine two notions of tangency; sets for which they coincide are termed *regular* and enjoy useful properties. We proceed to relate the generalized gradient to the constructs of the preceding chapter when X is a Hilbert space. Finally, we derive a useful limiting-gradient characterization when the underlying space is finite dimensional.

1 Definition and Basic Properties

Throughout this chapter, X is a real Banach space. Let $f\colon X \to \mathbb{R}$ be *Lipschitz of rank K* near a given point $x \in X$; that is, for some $\varepsilon > 0$, we have
$$|f(y) - f(z)| \leq K\|y - z\| \ \forall y, z \in B(x;\varepsilon).$$

2. Generalized Gradients in Banach Space

The *generalized directional derivative* of f at x in the direction v, denoted $f^\circ(x; v)$, is defined as follows:

$$f^\circ(x; v) := \limsup_{\substack{y \to x \\ t \downarrow 0}} \frac{f(y + tv) - f(y)}{t},$$

where of course y is a vector in X and t is a positive scalar. Note that this definition does not presuppose the existence of any limit (since it involves an upper limit only), that it involves only the behavior of f arbitrarily near x, and that it differs from the traditional definition of the directional derivative in that the base point (y) of the difference quotient varies. The utility of f° stems from the following basic properties. (A function g is *positively homogeneous* if $g(\lambda v) = \lambda g(v)$ for $\lambda \geq 0$, and *subadditive* if $g(v + w) \leq g(v) + g(w)$.)

1.1. Proposition. *Let f be Lipschitz of rank K near x. Then:*

(a) *The function $v \mapsto f^\circ(x; v)$ is finite, positively homogeneous, and subadditive on X, and satisfies*

$$|f^\circ(x; v)| \leq K\|v\|.$$

(b) $f^\circ(x; v)$ *is upper semicontinuous as a function of (x, v) and, as a function of v alone, is Lipschitz of rank K on X.*

(c) $f^\circ(x; -v) = (-f)^\circ(x; v)$.

Proof. In view of the Lipschitz condition, the absolute value of the difference quotient in the definition of $f^\circ(x; v)$ is bounded by $K\|v\|$ when y is sufficiently near x and t sufficiently near 0. It follows that $|f^\circ(x; v)|$ admits the same upper bound. The fact that $f^\circ(x; \lambda v) = \lambda f^\circ(x; v)$ for any $\lambda \geq 0$ is immediate, so let us turn now to the subadditivity. With all the upper limits below understood to be taken as $y \to x$ and $t \downarrow 0$, we calculate:

$$f^\circ(x; v + w) = \limsup \frac{f(y + tv + tw) - f(y)}{t}$$

$$\leq \limsup \frac{f(y + tv + tw) - f(y + tw)}{t} + \limsup \frac{f(y + tw) - f(y)}{t}$$

(since the upper limit of a sum is bounded above by the sum of the upper limits). The first upper limit in this last expression is $f^\circ(x; v)$, since the term $y + tw$ represents in essence just a dummy variable converging to x. We conclude

$$f^\circ(x; v + w) \leq f^\circ(x; v) + f^\circ(x; w).$$

which establishes (a).

Now let $\{x_i\}$ and $\{v_i\}$ be arbitrary sequences converging to x and v, respectively. For each i, by definition of the upper limit, there exist y_i in X and $t_i > 0$ such that

$$\|y_i - x_i\| + t_i < \frac{1}{i},$$

$$f°(x_i; v_i) - \frac{1}{i} \leq \frac{f(y_i + t_i v_i) - f(y_i)}{t_i}$$

$$= \frac{f(y_i + t_i v) - f(y_i)}{t_i} + \frac{f(y_i + t_i v_i) - f(y_i + t_i v)}{t_i}.$$

Note that the last term is bounded in magnitude by $K\|v_i - v\|$ (in view of the Lipschitz condition). Upon taking upper limits (as $i \to \infty$), we derive

$$\limsup_{i \to \infty} f°(x_i; v_i) \leq f°(x; v),$$

which establishes the upper semicontinuity.

Finally, let any v and w in X be given. We have

$$f(y + tv) - f(y) \leq f(y + tw) - f(y) + K\|v - w\|t$$

for y near x, t near 0. Dividing by t and taking upper limits as $y \to x$, $t \downarrow 0$, gives

$$f°(x; v) \leq f°(x; w) + K\|v - w\|.$$

Since this also holds with v and w switched, (b) follows. To prove (c), we calculate:

$$f°(x; -v) := \limsup_{\substack{x' \to x \\ t \downarrow 0}} \frac{f(x' - tv) - f(x')}{t}$$

$$= \limsup_{\substack{u \to x \\ t \downarrow 0}} \frac{(-f)(u + tv) - (-f)(u)}{t}, \text{ where } u := x' - tv$$

$$= (-f)°(x; v),$$

as stated. □

1.2. Exercise. Let f and g be Lipschitz near x. Prove that for any $v \in X$,

$$(f + g)°(x; v) \leq f°(x; v) + g°(x; v).$$

A function such as $v \mapsto f°(x; v)$ which is positively homogeneous and subadditive on X is the *support function* of a uniquely determined closed convex set in X^* (the dual space of continuous linear functionals on X).

Some terminology is in order. Given a nonempty subset Σ of X^*, its support function is the function $H_\Sigma \colon X \to (-\infty, \infty]$ defined as follows:

$$H_\Sigma(v) := \sup\{\langle \zeta, v \rangle \colon \zeta \in \Sigma\},$$

72   2. Generalized Gradients in Banach Space

where we have used the familiar convention of denoting the value of the linear functional ζ at v by $\langle \zeta, v \rangle$. We gather some useful facts about support functions in the next result.

1.3. Proposition.

(a) Let Σ be a nonempty subset of X^*. Then H_Σ is positively homogeneous, subadditive, and lower semicontinuous.

(b) If Σ is convex and w^*-closed, then a point ζ in X^* belongs to Σ iff we have $H_\Sigma(v) \geq \langle \zeta, v \rangle$ for all v in X.

(c) More generally, if Σ and Λ are two nonempty, convex, and w^*-closed subsets of X^*, then $\Sigma \supset \Lambda$ iff $H_\Sigma(v) \geq H_\Lambda(v)$ for all v in X.

(d) If $p: X \to \mathbb{R}$ is positively homogeneous and subadditive and bounded on the unit ball, then there is a uniquely defined nonempty, convex, and w^*-compact subset Σ of X^* such that $p = H_\Sigma$.

Proof. That H_Σ is positively homogeneous and subadditive follows immediately from its definition. As the upper envelope of continuous functions, H_Σ is automatically lower semicontinuous, whence (a). We turn now to (b), which is easily seen to amount to the following assertion: if $\zeta \notin \Sigma$, then for some $v \in X$ we have $H_\Sigma(v) < \langle \zeta, v \rangle$. This is proven by applying the Hahn–Banach Separation Theorem (see e.g., Rudin (1973)) to the topological vector space consisting of X^* with its weak*-topology, bearing in mind that the dual of that space is identified with X. The proof of (c) is immediate in light of (b); there remains (d).

Given p, we set

$$\Sigma := \{\zeta \in X^* : p(v) \geq \langle \zeta, v \rangle \ \forall v \in X\}.$$

Then Σ is seen to be convex as a consequence of the properties of p, and w^*-closed as the intersection of a family of w^*-closed subsets. If K is a bound for p on $B(0;1)$, then we have $\langle \zeta, v \rangle \leq K$ for all $v \in B(0;1)$, for any element ζ of Σ. It follows that Σ is bounded, and hence w^*-compact by Alaoglu's Theorem. Clearly we have $p \geq H_\Sigma$; let us prove equality. Let $v \in X$ be given. Then, by a standard form of the Hahn–Banach Theorem (Rudin (1973, Theorem 3.2)), there exists $\zeta \in X^*$ such that $\langle \zeta, w \rangle \leq p(w)$ $\forall w \in X$, with $\langle \zeta, v \rangle = p(v)$. Then $\zeta \in \Sigma$, so that $H_\Sigma(v) = p(v)$ as required. Finally, the uniqueness of Σ follows from (c).   □

Returning now to our function f, and taking for the function p of the proposition the function $f^\circ(x; \cdot)$, we define the *generalized gradient* of f at x, denoted $\partial f(x)$, to be the (nonempty) w^*-compact subset of X^* whose support function is $f^\circ(x; \cdot)$. Thus $\zeta \in \partial f(x)$ iff $f^\circ(x; v) \geq \langle \zeta, v \rangle$ for all v in X. Since $f^\circ(x; \cdot)$ does not depend on which one of two equivalent norms on X is chosen, it follows that $\partial f(x)$ too is independent of the particular norm on X.

Some immediate intuition about ∂f is available from the following exercise, where we see that the relationship between f° and ∂f generalizes the classical formula $f'(x;v) = \langle f'(x), v \rangle$ for the directional derivative $f'(x;v)$.

1.4. Exercise.

(a) Let $f: \mathbb{R}^n \to \mathbb{R}$ be C^1. Prove that $f^\circ(x;v) = \langle f'(x), v \rangle$ and that $\partial f(x) = \{f'(x)\}$.

(b) Let $f: \mathbb{R} \to \mathbb{R}$ be given by $f(x) = \max\{0, x\}$. Prove that $f^\circ(x;v) = \max\{0, v\}$. What is $\partial f(0)$?

(c) Find $f^\circ(0;\cdot)$ and $\partial f(0)$ when $f: \mathbb{R}^n \to \mathbb{R}$ is given by $f(x) = \|x\|$.

(d) If f has a local minimum or maximum at x, then $0 \in \partial f(x)$.

We proceed now to derive some of the basic properties of the generalized gradient. A multivalued function F is said to be *upper semicontinuous* at x if for all $\varepsilon > 0$ there exists $\delta > 0$ such that

$$\|x - y\| < \delta \implies F(y) \subset F(x) + \varepsilon B.$$

We denote by $\|\zeta\|_*$ the norm in X^*:

$$\|\zeta\|_* := \sup\{\langle \zeta, v \rangle : v \in X, \|v\| = 1\},$$

and B_* denotes the open unit ball in X^*.

1.5. Proposition. *Let f be Lipschitz of rank K near x. Then:*

(a) *$\partial f(x)$ is a nonempty, convex, weak*-compact subset of X^*, and $\|\zeta\|_* \leq K$ for every $\zeta \in \partial f(x)$.*

(b) *For every v in X we have*

$$f^\circ(x;v) = \max\{\langle \zeta, v \rangle : \zeta \in \partial f(x)\}.$$

(c) *$\zeta \in \partial f(x)$ iff $f^\circ(x;v) \geq \langle \zeta, v \rangle \ \forall v \in X$.*

(d) *If $\{x_i\}$ and $\{\zeta_i\}$ are sequences in X and X^* such that $\zeta_i \in \partial f(x_i)$ for each i, and if x_i converges to x and ζ is a weak* cluster point of the sequence $\{\zeta_i\}$, then we have $\zeta \in \partial f(x)$.*

(e) *If X is finite dimensional, then ∂f is upper semicontinuous at x.*

Proof. We have already noted that $\partial f(x)$ is nonempty and w^*-compact. Each $\zeta \in \partial f(x)$ satisfies $\langle \zeta, v \rangle \leq f^\circ(x; v) \leq K\|v\|$ for all v in X, whence $\|\zeta\|_* \leq K$. The assertions (b), (c) merely reiterate that $f^\circ(x; \cdot)$ is the support function of $\partial f(x)$.

Let us prove the closure property (d). Fix $v \in X$. For each i, we have $f^\circ(x_i; v) \geq \langle \zeta_i, v \rangle$ (in view of (c)). The sequence $\{\langle \zeta_i, v \rangle\}$ is bounded in $\mathbb{R}$, and contains terms that are arbitrarily near $\langle \zeta, v \rangle$. Let us extract a subsequence of $\{\zeta_i\}$ (without relabeling) such that $\langle \zeta_i, v \rangle \to \langle \zeta, v \rangle$. Then passing to the limit in the preceding inequality gives $f^\circ(x; v) \geq \langle \zeta, v \rangle$, since f° is upper semicontinuous in x (Proposition 1.1). Since v is arbitrary, it follows (from (c) again) that $\zeta \in \partial f(x)$.

We turn now to (e). Let $\varepsilon > 0$ be given; then we wish to show that for all y sufficiently near x, we have

$$\partial f(y) \subset \partial f(x) + \varepsilon \overline{B}.$$

If this is not the case, then there is a sequence y_i converging to x and points $\zeta_i \in \partial f(y_i)$ such that $\zeta_i \notin \partial f(x) + \varepsilon \overline{B}$. We can therefore separate ζ_i from the compact convex set in question: for some $v_i \neq 0$ we have

$$\langle \zeta_i, v_i \rangle \geq \max\{\langle \zeta, v_i \rangle : \zeta \in \partial f(x) + \varepsilon \overline{B}\}$$
$$= f^\circ(x; v_i) + \varepsilon \|v_i\|.$$

Because of positive homogeneity, we can take $\|v_i\| = 1$. Note that the sequence $\{\zeta_i\}$ is bounded. Since we are in finite dimensions, we can extract convergent subsequences from $\{\zeta_i\}$ and $\{v_i\}$ (we do not relabel): $\zeta_i \to \zeta$, $v_i \to v$, where $\|v\| = 1$. The inequality above gives in the limit $\langle \zeta, v \rangle \geq f^\circ(x; v) + \varepsilon$, while invoking part (d) yields $\zeta \in \partial f(x)$. But then (c) is contradicted. This completes the proof. □

1.6. Exercise.

(a) Verify the upper semicontinuity of ∂f at 0 for each of the functions in Exercise 1.4(b,c).

(b) Let $\zeta_i \in \partial f(x_i) + \varepsilon_i B_*$, where $x_i \to x$ and $\varepsilon_i \downarrow 0$. Let ζ be a weak*-cluster point of $\{\zeta_i\}$. Prove that $\zeta \in \partial f(x)$.

2 Basic Calculus

We will derive an assortment of formulas that facilitate the calculation of ∂f when f is synthesized from simpler functionals through linear combinations, maximization, composition, and so on. We always assume that the given functions are Lipschitz near the point of interest; as we will see, this property has the useful feature of being preserved under the operations in question.

2.1. Proposition. *For any scalar λ, we have $\partial(\lambda f)(x) = \lambda \partial f(x)$.*

Proof. Note that λf is Lipschitz near x, of rank $|\lambda| K$. When λ is nonnegative, $(\lambda f)^\circ = \lambda f^\circ$, and the result follows immediately. To complete the proof, it suffices to consider now the case $\lambda = -1$. An element ζ of X^* belongs to $\partial(-f)(x)$ iff $(-f)^\circ(x; v) \geq \langle \zeta, v \rangle$ for all v. By Proposition 1.1(c), this is equivalent to: $f^\circ(x; -v) \geq \langle \zeta, v \rangle$ for all v, which is equivalent to $-\zeta$ belonging to $\partial f(x)$ (by Proposition 1.3(c).) □

We now examine the generalized gradient of the sum of two functions f and g, each of which is Lipschitz near x. It is easy to see that $f + g$ is also Lipschitz near x, and we would like to relate $\partial(f+g)(x)$ to $\partial f(x) + \partial g(x)$. We will now do so, and introduce a technique that will be used many times: that of proving an inclusion between closed convex sets by proving an equivalent inequality between support functions.

The support function of $\partial(f+g)(x)$, evaluated at v, is $(f+g)^\circ(x; v)$ (by definition!), while that of $\partial f(x) + \partial g(x)$ is $f^\circ(x; v) + g^\circ(x; v)$ (the support function of a sum of sets is the sum of the support functions). Since the sum of two w^*-compact sets is w^*-compact (addition is w^*-continuous on $X^* \times X^*$), it follows that the general inequality

$$(f+g)^\circ(x; v) \leq f^\circ(x; v) + g^\circ(x; v)$$

noted in Exercise 1.2 is equivalent to the inclusion

$$\partial(f+g)(x) \subset \partial f(x) + \partial g(x),$$

as observed in Proposition 1.3(c).

The extension of this inclusion (a sum rule) to finite linear combinations is immediate.

2.2. Proposition. *Let f_i $(i = 1, 2, \ldots, n)$, be Lipschitz near x, and let λ_i $(i = 1, 2, \ldots, n)$ be scalars. Then $f := \sum_{i=1}^{n} \lambda_i f_i$ is Lipschitz near x, and we have*

$$\partial\left(\sum_{i=1}^{n} \lambda_i f_i\right)(x) \subset \sum_{i=1}^{n} \lambda_i \partial f_i(x).$$

2.3. Exercise. Prove Proposition 2.2, and give an example with $X = \mathbb{R}$ and $n = 2$ for which the inclusion is strict.

2.4. Theorem (Lebourg's Mean Value Theorem). *Let x and y belong to X, and suppose that f is Lipschitz on an open set containing the line segment $[x, y]$. Then there exists a point u in (x, y) such that*

$$f(y) - f(x) \in \langle \partial f(u), y - x \rangle.$$

76   2. Generalized Gradients in Banach Space

Proof. We will need the following special chain rule for the proof. We denote by x_t the point $x + t(y - x)$.

Lemma. *The function* $g\colon [0,1] \to R$ *defined by* $g(t) = f(x_t)$ *is Lipschitz on* $(0,1)$, *and we have*
$$\partial g(t) \subset \langle \partial f(x_t), y - x \rangle.$$

Proof of the Lemma. The fact that g is Lipschitz is plain. The two closed convex sets appearing in the equation are in fact intervals in R, so it suffices to prove that for $v = \pm 1$, we have
$$\max\{\partial g(t)v\} \leq \max\{\langle \partial f(x_t), y - x \rangle v\}.$$

Now the left-hand side is just $g^\circ(t;v)$; that is,
$$\limsup_{\substack{s \to t \\ \lambda \downarrow 0}} \frac{g(s + \lambda v) - g(s)}{\lambda}$$
$$= \limsup_{\substack{s \to t \\ \lambda \downarrow 0}} \frac{f(x + [s + \lambda v](y - x)) - f(x + s(y - x))}{\lambda}$$
$$\leq \limsup_{\substack{y' \to x_t \\ \lambda \downarrow 0}} \frac{f(y' + \lambda v(y - x)) - f(y')}{\lambda}$$
$$= f^\circ(x_t; v(y - x))$$
$$= \max \langle \partial f(x_t), v(y - x) \rangle,$$

which completes the proof of the lemma.

Now to the proof of the theorem. Consider the function θ on $[0,1]$ defined by
$$\theta(t) = f(x_t) + t[f(x) - f(y)].$$
Note that $\theta(0) = \theta(1) = f(x)$, so that there is a point t in $(0,1)$ at which θ attains a local minimum or maximum (by continuity). By Exercise 1.4(d) we have $0 \in \partial \theta(t)$. We may calculate $\partial \theta(t)$ by appealing to Propositions 2.1 and 2.2, and the lemma. We deduce
$$0 \in f(x) - f(y) + \langle \partial f(x_t), y - x \rangle,$$
which is the assertion of the theorem (take $u = x_t$). □

2.5. Theorem (The Chain Rule). *Let* $F\colon X \to \mathbb{R}^n$ *be Lipschitz near* x, *and let* $g\colon \mathbb{R}^n \to \mathbb{R}$ *be Lipschitz near* $F(x)$. *Then the function* $f(x') := g(F(x'))$ *is Lipschitz near* x, *and we have*
$$\partial f(x) \subset \overline{\mathrm{co}}^* \{\partial \langle \gamma, F(\cdot) \rangle(x) \colon \gamma \in \partial g(F(x))\},$$
where $\overline{\mathrm{co}}^*$ *signifies the* w^*-*closed convex hull.*

Proof. The fact that f is Lipschitz is left to the reader to show. Again it is an inclusion between two convex weak*-compact sets that is at issue; the corresponding support function inequality amounts to the statement that for given v, there exists γ in $\partial g(F(x))$, and ζ in the generalized gradient at x of the function $x' \mapsto \langle \gamma, F(x') \rangle$, such that $f^\circ(x; v) \leq \langle \zeta, v \rangle$. We will prove the theorem by producing such a pair γ and ζ.

To begin, we give ourselves sequences $y_i \to x$ and $t_i \downarrow 0$ realizing the lim sup in the definition of $f^\circ(x; v)$; i.e., such that

$$\lim_{i \to \infty} \frac{f(y_i + t_i v) - f(y_i)}{t_i} = f^\circ(x; v).$$

Applying the Mean Value Theorem 2.4 gives, for each i, an element $\gamma_i \in \partial g(z_i)$ such that

$$\frac{f(y_i + t_i v) - f(y_i)}{t_i} = \frac{g(F(y_i + t_i v)) - g(F(y_i))}{t_i}$$

$$= \left\langle \gamma_i, \frac{F(y_i + t_i v) - F(y_i)}{t_i} \right\rangle,$$

where z_i lies on the line segment joining $F(y_i)$ and $F(y_i + t_i v)$. It follows that $z_i \to F(x)$, and that for a suitable subsequence we have $\gamma_i \to \gamma \in \partial g(F(x))$ (we eschew relabeling). This is the required γ; we turn now to exhibiting ζ.

By the Mean Value Theorem again, there exists $\zeta_i \in \partial \langle \gamma, F(\cdot) \rangle(w_i)$ such that

$$\left\langle \gamma, \frac{F(y_i + t_i v) - F(y_i)}{t_i} \right\rangle = \langle \zeta_i, v \rangle,$$

where w_i is on the line segment joining y_i and $y_i + t_i v$. It follows that $w_i \to x$, that the sequence $\{\zeta_i\}$ is bounded in X^*, and that $\{\langle \zeta_i, v \rangle\}$ is bounded in $\mathbb{R}$. We may pass again to a subsequence to arrange for $\langle \zeta_i, v \rangle$ to converge to some limit; having done so, let ζ be a weak*-cluster point of $\{\zeta_i\}$. Then $\langle \zeta_i, v \rangle \to \langle \zeta, v \rangle$ necessarily, and $\zeta \in \partial \langle \gamma, F(\cdot) \rangle(x)$ (Proposition 1.5(d)).

Combining the above, we arrive at

$$\frac{f(y_i + t_i v) - f(y_i)}{t_i} = \left\langle (\gamma_i - \gamma) + \gamma, \frac{F(y_i + t_i v) - F(y_i)}{t_i} \right\rangle$$

$$= \left\langle \gamma_i - \gamma, \frac{F(y_i + t_i v) - F(y_i)}{t_i} \right\rangle + \langle \zeta_i, v \rangle.$$

Now the term $[F(y_i + t_i v) - F(y_i)]/t_i$ is bounded because F is Lipschitz, and we know $\gamma_i \to \gamma$. Therefore passing to the limit yields

$$f^\circ(x; v) = \lim_{i \to \infty} \frac{f(y_i + t_i v) - f(y_i)}{t_i} = \langle \zeta, v \rangle,$$

which confirms that ζ has the required properties. $\square$

78   2. Generalized Gradients in Banach Space

The following exercise illustrates the use of the Chain Rule in a special case; invoke Theorem 2.5 to deduce it.

2.6. Exercise. Let f and g be Lipschitz near x. Then the product function fg is Lipschitz near x and we have

$$\partial(fg)(x) \subset \partial\big(f(x)g(\cdot) + g(x)f(\cdot)\big)(x) \subset f(x)\partial g(x) + g(x)\partial f(x).$$

3   Relation to Derivatives

We remind the reader that some basic definitions and facts about classical differentiability were recalled in Chapter 1, §2. (These carry over to the present Banach space setting when the $\langle \cdot, \cdot \rangle$ is given the duality pairing interpretation.)

3.1. Proposition. *Let f be Lipschitz near x.*

(a) *If f admits a Gâteaux derivative $f'_G(x)$ at x, then $f'_G(x) \in \partial f(x)$.*

(b) *If f is continuously differentiable at x, then $\partial f(x) = \{f'(x)\}$.*

Proof. By definition we have the following relation between $f'_G(x)$ and the one-sided directional derivatives:

$$f'(x; v) = \langle f'_G(x), v \rangle \ \forall v \in \mathbb{R}^n.$$

But clearly, $f'(x; v) \leq f^\circ(x; v)$. That $f'_G(x)$ belongs to $\partial f(x)$ now follows from Proposition 1.5(c).

Now suppose that f is C^1 in a neighborhood of x, and fix $v \in X$. For y near x and $t > 0$ near 0, we have

$$\frac{f(y + tv) - f(y)}{t} = \langle f'(z), v \rangle$$

for some $z \in (y, y+tv)$, by the classical Mean Value Theorem. As $y \to x$ and $t \downarrow 0$, the point z converges to x, and because $f'(\cdot)$ is continuous (as a map between the Banach spaces X and X^*), we derive $f^\circ(x; v) \leq \langle f'(x), v \rangle$. It follows now from Proposition 1.5(c) that $\langle \zeta, v \rangle \leq \langle f'(x), v \rangle$ whenever $\zeta \in \partial f(x)$. Since v is arbitrary, we conclude that $\partial f(x)$ is the singleton $\{f'(x)\}$.   □

Remark. In the end-of-chapter problems we will see that $\partial f(x)$ reduces to a singleton precisely when f is "strictly differentiable" at x, a notion intermediate between Gâteaux and continuous differentiability.

3.2. Theorem. *Let $F: X \to Y$ be continuously differentiable near x, with Y a Banach space, and let $g: Y \to \mathbb{R}$ be Lipschitz near $F(x)$. Then $f := g \circ F$ is Lipschitz near x, and we have*

$$\partial f(x) \subset (F'(x))^* \partial g(F(x)),$$

where $*$ denotes the adjoint. If $F'(x): X \to Y$ is onto, then equality holds.

Proof. The fact that f is Lipschitz near x is straightforward. In terms of support functions, we must prove that given any v, then we have some element ζ of $\partial g(F(x))$ such that

$$f^\circ(x; v) \leq \langle v, F'(x)^* \zeta \rangle = \langle \zeta, F'(x) v \rangle.$$

For any y_i near x and $t_i > 0$ near 0, the difference quotient $[f(y_i + t_i v) - f(y_i)]/t_i$ can be expressed in the form

$$\left\langle \zeta_i, \frac{F(y_i + t_i v) - F(y_i)}{t_i} \right\rangle$$

for some $\zeta_i \in \partial g(z_i)$, where z_i lies in the segment $[F(y_i), F(y_i + t_i v)]$ (we have invoked the Mean Value Theorem 2.4, of course). Much as in the proof of Theorem 2.5, we extract a subsequence from $\{\zeta_i\}$ such that $\langle \zeta_i, v \rangle$ converges, and then let ζ be a cluster point of ζ_i. The required inequality follows from the fact that $[F(y_i + t_i v) - F(y_i)]/t_i$ converges (strongly) to $F'(x) v$.

Now suppose that $F'(x)$ is onto. It follows that F maps every neighborhood of x to a neighborhood of $F(x)$, by a classical theorem due to Graves (we will prove this and other such surjectivity results in Chapter 3). This fact justifies the second equality below:

$$g^\circ(F(x); F'(x)v) = \limsup_{\substack{y \to F(x) \\ t \downarrow 0}} \frac{g(y + tF'(x)v) - g(y)}{t}$$

$$= \limsup_{\substack{x' \to x \\ t \downarrow 0}} \frac{g(F(x') + tF'(x)v) - g(F(x'))}{t}$$

$$= \limsup_{\substack{x' \to x \\ t \downarrow 0}} \frac{g(F(x' + tv)) - g(F(x'))}{t}$$

(since $[F(x' + tv) - F(x') - tF'(x)v]/t$ goes to zero as $x' \to x$ and $t \downarrow 0$, and since g is Lipschitz locally)

$$= f^\circ(x; v).$$

Since v is arbitrary, this implies equality between the two sets figuring in the statement of the theorem, as asserted. □

The fact that equality can hold in the estimate of Theorem 3.2 under a supplementary hypothesis raises as a more general question the possibility of being able to assert equalities in other results too, for example in Proposition 2.2 and Theorem 2.5, or perhaps in Theorem 3.2, but with a supplementary hypothesis on g rather than F. Of course if the data are all smooth then equality holds in all these results, for then they simply reduce to classical differentiation formulas. The interest would lie in identifying a class of functions, not necessarily smooth, giving rise to equalities even when nonsingleton sets are involved. We address this issue in the next section.

4 Convex and Regular Functions

A real-valued function f defined on an open convex subset U of X is termed *convex* provided that for any two points $x, y \in U$ we have

$$f(tx + (1-t)y) \leq tf(x) + (1-t)f(y) \; \forall t \in [0,1].$$

The proof of Proposition 1.7.1 goes through without change in a Banach space to give:

4.1. Proposition. *If f is a convex function on U that is bounded above on a neighborhood of some point in U, then for any x in U, f is Lipschitz near x.*

4.2. Exercise. Let $f: X \to \mathbb{R}$ be convex, and let $\theta: \mathbb{R} \to \mathbb{R}$ be convex and nondecreasing. Prove that $x \mapsto \theta(f(x))$ is convex.

4.3. Proposition. *Let f be convex on U and Lipschitz near $x \in U$. Then the directional derivatives $f'(x;v)$ exist, and we have $f'(x;v) = f^\circ(x;v)$. A vector ζ belongs to $\partial f(x)$ iff*

$$f(y) - f(x) \geq \langle \zeta, y - x \rangle \; \forall y \in U.$$

Proof. It follows directly from the definition of convex function that for small $t > 0$, the function

$$t \mapsto \frac{f(x' + tv) - f(x')}{t}$$

is nondecreasing. This fact, together with the Lipschitz hypothesis, implies the existence and finiteness of the directional derivative for all x' near x, for all v:

$$f'(x'; v) = \inf_{t>0} \frac{f(x' + tv) - f(x')}{t}.$$

Now fix $\delta > 0$, and observe that $f^\circ(x; v)$ can be written as

$$f^\circ(x; v) = \lim_{\varepsilon \downarrow 0} \sup_{\|x'-x\| \leq \varepsilon\delta} \sup_{0 < t < \varepsilon} \frac{f(x' + tv) - f(x')}{t}.$$

The preceding remarks show that an alternative expression for $f^\circ(x; v)$ is

$$\lim_{\varepsilon \downarrow 0} \sup_{\|x'-x\| \leq \varepsilon\delta} \frac{f(x' + \varepsilon v) - f(x')}{\varepsilon}.$$

If K is a Lipschitz constant for f near x, then for all x' in $B(x; \varepsilon\delta B)$, for all ε sufficiently small, we have

$$\left| \frac{f(x' + \varepsilon v) - f(x')}{\varepsilon} - \frac{f(x + \varepsilon v) - f(x)}{\varepsilon} \right| \leq 2\delta K,$$

so that

$$f^\circ(x; v) \leq \lim_{\varepsilon \downarrow 0} \left\{ \frac{f(x + \varepsilon v) - f(x)}{\varepsilon} + 2\delta K \right\} = f'(x; v) + 2\delta K.$$

Since δ is arbitrary we deduce $f^\circ(x; v) \leq f'(x; v)$, and hence equality, since $f^\circ \geq f'$ inherently. Finally, we observe

$$\zeta \in \partial f(x) \iff f^\circ(x; v) \geq \langle \zeta, v \rangle \; \forall v,$$
$$\iff f'(x; v) \geq \langle \zeta, v \rangle \; \forall v,$$
$$\iff \inf_{t > 0} \frac{f(x + tv) - f(x)}{t} \geq \langle \zeta, v \rangle \; \forall v,$$
$$\iff f(y) - f(x) \geq \langle \zeta, y - x \rangle \; \forall y \in U. \qquad \square$$

It turns out that the property of having directional derivatives $f'(x; v)$ that coincide with $f^\circ(x; v)$ is precisely what is required to make our calculus rules "more exact." We give this property a name: the function f is *regular* at x provided that f is Lipschitz near x and admits directional derivatives $f'(x; v)$ at x for all v, with $f'(x; v) = f^\circ(x; v)$. Evidently, functions which are continuously differentiable at x are regular at x, since then $f'(x; v) = \langle f'(x), v \rangle = f^\circ(x; v)$. Also, convex functions which are Lipschitz near x are regular there, by the preceding proposition.

4.4. Exercise. Give an example of a function which is neither C^1 nor convex near x, but which is regular at x.

Let us now illustrate how regularity sharpens certain calculus rules, such as that for the sum of two functions. If f and g are Lipschitz near x, we know (Proposition 2.2) that

$$\partial(f + g)(x) \subset \partial f(x) + \partial g(x).$$

82   2. Generalized Gradients in Banach Space

Suppose now that f and g are regular at x. Then we can argue as follows to get the opposite inclusion: for any v,

$$\max\{\langle \zeta + \xi, v\rangle : \zeta \in \partial f(x), \xi \in \partial g(x)\}$$
$$= f^\circ(x;v) + g^\circ(x;v)$$
$$= f'(x;v) + g'(x;v)$$
$$= (f+g)'(x;v) \leq (f+g)^\circ(x;v)$$
$$= \max\{\langle \zeta, v\rangle : \zeta \in \partial(f+g)(x)\}.$$

This inequality between support functions is equivalent to the inclusion

$$\partial f(x) + \partial g(x) \subset \partial(f+g)(x),$$

so that equality actually holds. A bonus consequence of this argument is the fact that $(f+g)'(x;\cdot)$ and $(f+g)^\circ(x;\cdot)$ coincide, so that $f+g$ inherits regularity from f and g. In fact, it is clear that any (finite) nonnegative linear combination of regular functions is regular.

The following theorem subsumes the case of a finite sum just discussed, and in its regularity conclusions is related to Exercise 4.2. The setting is that of the Chain Rule 2.5, of which this is a refinement.

4.5. Theorem. *Let $F \colon X \to \mathbb{R}^n$ be such that each component function f_i of F is regular at x. Let $g \colon \mathbb{R}^n \to \mathbb{R}$ be regular at $F(x)$, and suppose that each $\gamma \in \partial g(F(x))$ has nonnegative components. Then the function $f(x') := g(F(x'))$ is regular at x, and we have*

$$\partial f(x) = \overline{\mathrm{co}}^*\{\partial\langle \gamma, F(\cdot)\rangle(x) : \gamma \in \partial g(F(x))\}.$$

Proof. We ask the reader to check as a first step that f admits directional derivatives at x:

$$f'(x;v) = g'(F(x); F'(x;v)),$$

where $F'(x;v)$ signifies the vector in $\mathbb{R}^n$ whose ith component is $f_i'(x;v)$. Now consider, for given $v \in X$, the maximum of the inner product of v taken with elements from the right side of the equality asserted by the theorem. That maximum equals

$$\max\{\langle \gamma, F(\cdot)\rangle^\circ(x;v) : \gamma \in \partial g(F(x))\}$$
$$= \max\{\langle \gamma, F(\cdot)\rangle'(x;v) : \gamma \in \partial g(F(x))\}$$

(for $\langle \gamma, F(\cdot) \rangle$ is regular at x, as a nonnegative linear combination of functions regular at x, since γ is nonnegative)

$$\begin{aligned}
&= \max\{\langle \gamma, F'(x;v) \rangle : \gamma \in \partial g(F(x))\} \\
&= g^\circ(F(x); F'(x;v)) \\
&= g'(F(x); F'(x;v)) \quad \text{(since } g \text{ is regular at } F(x)) \\
&= f'(x;v) \quad \text{(as noted above)} \\
&\leq f^\circ(x;v).
\end{aligned}$$

But this last term is the support function of the left side, evaluated at v, implying the opposite inclusion to the one furnished by Theorem 2.5. It follows that the two sets coincide, and that $f^\circ(x;v)$ and $f'(x;v)$ agree. □

4.6. Exercise.

(a) Let $g: \mathbb{R}^n \to \mathbb{R}$ be nondecreasing in its first component, and let $\zeta = (\zeta_1, \zeta_2, \ldots, \zeta_n) \in \partial g(x)$. Prove that $\zeta_1 \geq 0$.

(b) Find conditions on f and g and their values at x under which equality holds in the estimate of Exercise 2.6.

(c) Let f_i ($i = 1, 2, \ldots, n$) be Lipschitz near x, and set $f(x) := \max_{1 \leq i \leq n} f_i(x)$. Prove that f is Lipschitz near x, with

$$\partial f(x) \subset \operatorname{co}\left\{\bigcup_{i \in M(x)} \partial f_i(x)\right\},$$

where $M(x) = \{i \in \{1, 2, \ldots, n\}: f_i(x) = f(x)\}$. Note that we must show that the set appearing on the right is w^*-closed. If each f_i is regular at x, prove that equality holds. (*Hint.* Problem 1.11.14.) Develop the analogous formula when max is replaced by min.

(d) Let $f: \mathbb{R}^2 \to \mathbb{R}$ be given by $f(x,y) := \max\{|x-y|, y-x^2\}$. Find $\partial f(0,0)$.

5 Tangents and Normals

Let S be a nonempty closed subset of X. There is a globally Lipschitz function associated with S that completely characterizes it: its distance function $d_S(\cdot)$ is given by

$$d_S(x) := \inf\{\|x-s\|: s \in S\}.$$

We can apply our Lipschitz calculus to $d_S(\cdot)$ in order to define geometric constructs for S. In this light, a rather natural way to define a direction v *tangent* to S at $x \in S$ is as follows: we require $d_S^\circ(x;v) \leq 0$. (That is,

d_S should not increase in the v direction, as measured by the generalized directional derivative.) We remark that since $d_S^\circ(x;v) \geq 0$ for all v (show), it is equivalent to require $d_S^\circ(x;v) = 0$. We proceed to adopt this definition: the *tangent cone* to S at x, denoted $T_S(x)$, is the set of all those $v \in X$ satisfying $d^\circ(x;v) \leq 0$.

5.1. Exercise. Show that $0 \in T_S(x)$, and that $T_S(x)$ *is a cone.* Prove that $T_S(x)$ is closed and convex.

It is occasionally useful to have the following alternate, direct, characterization of $T_S(x)$ on hand, and reassuring to know that tangency does not depend on the choice of equivalent norms for X (as d_S does):

5.2. Proposition. *An element v of X is tangent to S at x iff, for every sequence x_i in S converging to x and sequence t_i in $(0, \infty)$ decreasing to 0, there exists a sequence v_i in X converging to v such that $x_i + t_i v_i \in S$ for all i.*

Proof. Suppose first that $v \in T_S(x)$, and that sequences $x_i \to x$ (with $x_i \in S$), $t_i \downarrow 0$ are given. We must produce the sequence v_i alluded to in the statement of the theorem. Since $d_S^\circ(x;v) = 0$ by assumption, we have

$$\lim_{i \to \infty} \frac{d_S(x_i + t_i v) - d_S(x_i)}{t_i} = \lim_{i \to \infty} \frac{d_S(x_i + t_i v)}{t_i} = 0.$$

Let s_i be a point in S which satisfies

$$\|x_i + t_i v - s_i\| \leq d_S(x_i + t_i v) + \frac{t_i}{i}$$

and let us set

$$v_i = \frac{s_i - x_i}{t_i}.$$

Then $\|v - v_i\| \to 0$; that is, v_i converges to v. Furthermore, $x_i + t_i v_i = s_i \in S$, as required.

Now for the converse. Let v have the stated property concerning sequences, and choose a sequence y_i converging to x and t_i decreasing to 0 such that

$$\lim_{i \to \infty} \frac{d_S(y_i + t_i v) - d_S(y_i)}{t_i} = d_S^\circ(x;v).$$

Our purpose is to prove this quantity nonpositive, for then $v \in T_S(x)$ by definition. Let s_i in S satisfy

$$\|s_i - y_i\| \leq d_S(y_i) + \frac{t_i}{i}.$$

It follows that s_i converges to x. Thus there is a sequence v_i converging to v such that $s_i + t_i v_i \in S$. But then, since d_S is Lipschitz of rank 1,

$$d_S(y_i + t_i v) \leq d_S(s_i + t_i v_i) + \|y_i - s_i\| + t_i \|v - v_i\|$$
$$\leq d_S(y_i) + t_i \left(\|v - v_i\| + \frac{1}{i} \right).$$

We deduce that the limit above is nonpositive, which completes the proof. □

5.3. Exercise. Let $X = X_1 \times X_2$, where X_1, X_2 are Banach spaces, and let $x = (x_1, x_2) \in S_1 \times S_2$, where S_1, S_2 are subsets of X_1, X_2, respectively. Then

$$T_{S_1 \times S_2}(x) = T_{S_1}(x_1) \times T_{S_2}(x_2).$$

The Normal Cone

In the case of classical manifolds in $\mathbb{R}^n$, the tangent space and the normal space are orthogonal to one another. When convex cones are involved, it is *polarity* that serves to obtain one from the other. We define the *normal cone* to S at x, denoted $N_S(x)$, as follows:

$$N_S(x) := T_S(x)^\circ := \{ \zeta \in X^* : \langle \zeta, v \rangle \leq 0 \; \forall v \in T_S(x) \}.$$

5.4. Proposition.

(a) $N_S(x)$ *is a w^*-closed convex cone.*

(b) $N_S(x) = \mathrm{cl}^* \{ \bigcup_{\lambda \geq 0} \lambda \partial d_S(x) \}.$

(c) $T_S(x)$ *is in turn the polar of $N_S(x)$; that is,*

$$T_S(x) = N_S(x)^\circ = \{ v \in X : \langle \zeta, v \rangle \leq 0 \; \forall \zeta \in N_S(x) \}.$$

Proof. Property (a) is immediate. Let $\zeta \in \partial d_S(x)$, and suppose that $v \in T_S(x)$. Since $d_S^\circ(x; v) \leq 0$ by definition of $T_S(x)$, and since $d_S^\circ(x; \cdot)$ is the support function of $\partial d_S(x)$, we deduce $\langle \zeta, v \rangle \leq 0$. This shows that $\partial d_S(x)$ lies in $N_S(x)$, which implies that the set Σ appearing on the right in (b) is contained in $N_S(x)$. To complete the proof of (b), let ζ be a point in the complement of Σ. By the Separation Theorem, there exists $v \in X$ such that

$$H_\Sigma(v) < \langle \zeta, v \rangle.$$

It follows that $\langle \zeta, v \rangle > 0$ and $H_\Sigma(v) \leq 0$, since Σ is a cone. Therefore $\langle v, \theta \rangle \leq 0 \; \forall \theta \in \partial d_S(x)$, whence $d_S^\circ(x; v) \leq 0$. We conclude that $v \in T_S(x)$. Since $\langle \zeta, v \rangle > 0$, it follows that $\zeta \notin N_S(x)$; (b) is proven.

We turn now to the proof of (c). Let $v \in T_S(x)$. Then $\langle \zeta, v \rangle \leq 0 \; \forall \zeta \in \partial d_S(x)$, which implies $\langle \zeta, v \rangle \leq 0 \; \forall \zeta \in N_S(x)$ in view of (b). Thus $v \in N_S(x)^\circ$. Conversely, let $v \in N_S(x)^\circ$. Then $\langle \zeta, v \rangle \leq 0 \; \forall \zeta \in \partial d_S(x)$, because of (b). But then $d_S^\circ(x; v) \leq 0$ and $v \in T_S(x)$. □

We postpone to later the proof of the fact that T_S and N_S coincide with the classical tangent and normal spaces when S is a smooth manifold. Another special case of interest is the convex one, which we now examine.

5.5. Proposition. *Let S be convex. Then*
$$T_S(x) = \mathrm{cl}\{\lambda(s-x): \lambda \geq 0, s \in S\}$$
and
$$N_S(x) = \{\zeta \in X^*: \langle \zeta, x' - x \rangle \leq 0 \; \forall x' \in S\}.$$

Proof. The convexity of the set S readily implies that of the function $d_S(\cdot)$. It follows then from Proposition 4.3 that $d'_S(x;v)$ exists and coincides with $d^\circ_S(x;v)$. Consequently, $T_S(x)$ consists of those $v \in X$ for which
$$\lim_{t \downarrow 0} \frac{d_S(x+tv)}{t} = 0.$$
This in turn is equivalent to the existence of $s(t) \in S$ such that
$$\frac{\|x + tv - s(t)\|}{t} \to 0 \text{ as } t \downarrow 0.$$
Setting $u(t) := (x + tv - s(t))/t$, this can be expressed in the form
$$v = \left(\frac{1}{t}\right)(s(t) - x) + u(t),$$
where $u(t) \to 0$ as $t \downarrow 0$. This is equivalent to the characterization of $T_S(x)$ given in the statement of the proposition. The expression for $N_S(x)$ then follows immediately from this characterization, together with the fact that $N_S(x)$ is the polar of $T_S(x)$. □

5.6. Exercise. Determine the tangent and normal cones at the origin for each of the following subsets of $\mathbb{R}^2$:

(a) $S_1 := \{(x,y): xy = 0\}$,
(b) $S_2 := \{(x,y): y \geq 2|x|\}$,
(c) $S_3 :=$ the closure of the complement of the set S_2,
(d) $S_4 := \{(x,y): y \leq \sqrt{|x|}\}$,
(e) $S_5 := \{(x,y): y = -\sqrt{|x|}\}$,
(f) $S_6 := \{(x,y): y \leq -\sqrt{|x|}\}$, and
(g) $S_7 := S_2 \cup \{(0,y): y \in \mathbb{R}\}$.

When S is the epigraph of a function, we would expect some relationship to exist between its tangent and normal cones on the one hand, and the generalized gradient of the function on the other. In fact, a complete duality exists, as we now see.

5.7. Theorem. *Let f be Lipschitz near x. Then:*

(a) $T_{\text{epi } f}(x, f(x)) = \text{epi } f^\circ(x; \cdot)$; *and*

(b) $\zeta \in \partial f(x) \iff (\zeta, -1) \in N_{\text{epi } f}(x, f(x))$.

Proof. Suppose first that (v, r) lies in $T_{\text{epi } f}(x, f(x))$. Choose sequences $y_i \to x$, $t_i \downarrow 0$, such that

$$\lim_{i \to \infty} \frac{f(y_i + t_i v) - f(y_i)}{t_i} = f^\circ(x; v).$$

Note that $(y_i, f(y_i))$ is a sequence in epi f converging to $(x, f(x))$. Accordingly, by Proposition 5.2, there exists a sequence (v_i, r_i) converging to (v, r) such that $(y_i, f(y_i)) + t_i(v_i, r_i) \in \text{epi } f$. Thus

$$f(y_i) + t_i r_i \geq f(y_i + t_i v_i).$$

We rewrite this as

$$\frac{f(y_i + t_i v_i) - f(y_i)}{t_i} \leq r_i.$$

Taking limits, we obtain $f^\circ(x; v) \leq r$ as desired.

We now show that for any v, for any $\delta \geq 0$, the point $(v, f^\circ(x; v) + \delta)$ lies in $T_{\text{epi } f}(x, f(x))$; this will complete the proof of (a). Let (x_i, r_i) be any sequence in epi f converging to $(x, f(x))$, and let $t_i \downarrow 0$. We must produce a sequence (v_i, s_i) converging to $(v, f^\circ(x; v) + \delta)$ with the property that $(x_i, r_i) + t_i(v_i, s_i)$ lies in epi f for each i; that is, such that $r_i + t_i s_i \geq f(x_i + t_i v_i)$.

Let us define $v_i = v$ and

$$s_i := \max\left\{ f^\circ(x; v) + \delta, \frac{f(x_i + t_i v) - f(x_i)}{t_i} \right\}.$$

Observe first that $s_i \to f^\circ(x; v) + \delta$, since

$$\limsup_{i \to \infty} \frac{f(x_i + t_i v) - f(x_i)}{t_i} \leq f^\circ(x; v).$$

We have

$$r_i + t_i s_i \geq r_i + [f(x_i + t_i v) - f(x_i)]$$

and $r_i \geq f(x_i)$ (since $(x_i, r_i) \in \text{epi } f$), which together give

$$r_i + t_i s_i \geq f(x_i + t_i v)$$

showing that $(x_i + t_i v, r_i + t_i s_i)$ belongs to epi f, as required.

We turn now to (b). We know that $\zeta \in \partial f(x)$ iff $f^\circ(x;v) \geq \langle \zeta, v \rangle\ \forall v$; that is, precisely when for any v and any $r \geq f^\circ(x;v)$ we have

$$\langle (\zeta, -1), (v, r) \rangle \leq 0.$$

By (a), this last inequality holds for all the (v, r) in question iff it holds for all $(v, r) \in T_{\text{epi } f}(x, f(x))$; that is, precisely when $(\zeta, -1)$ lies in the polar of $T_{\text{epi } f}(x, f(x))$, namely $N_{\text{epi } f}(x, f(x))$. □

Remark. One of the principal advantages of the theory of generalized gradients is the complete duality that it induces between tangency and normality, and between functions and sets (via the epigraph, or via the distance function). Note that in developing the theory, we chose the generalized directional derivative as primitive notion, and used it to define the generalized gradient, the tangent cone (via d_S) and then, by polarity, the normal cone. Alternatively, we could choose tangency as a starting point, using the characterization of Proposition 5.2 as a definition, and proceed from there (how?). We do not know in general how to make normality (or generalized gradients, or subdifferentials) the true starting point of the theory, unless the Banach space X has additional properties, such as in the Hilbert space case considered in Chapter 1. There, normals *were* the starting point. It is now time to connect the proximal theory of that chapter to the present one.

6 Relationship to Proximal Analysis

We now suppose that X is a Hilbert space, so that $\|x\| = \langle x, x \rangle^{1/2}$ for an inner product $\langle \cdot, \cdot \rangle$. In this case, $\partial f(x)$ and $N_S(x)$ are generated by weak limits of their proximal counterparts, and are identified with subsets of X itself:

6.1. Theorem. *Let X be a Hilbert space.*

(a) *If f is Lipschitz near x, then*

$$\partial f(x) = \overline{\text{co}}\Big\{\operatorname*{w-lim}_{i \to \infty} \zeta_i : \zeta_i \in \partial_P f(x_i), x_i \to x\Big\}.$$

(b) *If S is a closed subset of X containing x, then*

$$N_S(x) = \overline{\text{co}}\Big\{\operatorname*{w-lim}_{i \to \infty} \zeta_i : \zeta_i \in N_S^P(x_i), x_i \to x\Big\}.$$

Proof. Assertion (a) is an equality between closed convex sets; we will show that the set on the left is contained in the one on the right by verifying the corresponding inequality between support functions. More precisely then, we will show that given any v, there exists a sequence x_i converging to x, a sequence $\zeta_i \in \partial_P f(x_i)$ converging weakly to a limit ζ, such that $f°(x; v) \leq \langle \zeta, v \rangle$.

Let the sequences $\{y_i\}$, $\{t_i\}$ realize $f°(x; v)$; i.e., be such that $y_i \to x$, $t_i \downarrow 0$, and
$$\lim_{i \to \infty} \frac{f(y_i + t_i v) - f(y_i)}{t_i} = f°(x; v).$$

For each i sufficiently large, f is Lipschitz on a neighborhood of the line segment $[y_i, y_i + t_i v]$, and by the Proximal Mean Value Theorem (Problem 1.11.13) there exists $x_i \in (y_i, y_i + t_i v)$ and $\zeta_i \in \partial_P f(x_i)$ such that
$$f(y_i + t_i v) - f(y_i) < \langle \zeta_i, t_i v \rangle + \frac{1}{i}.$$

Since f is Lipschitz near x, the sequence $\{\zeta_i\}$ is bounded, and we may extract a subsequence (without relabeling) that converges weakly to a limit ζ. It follows that $f°(x; v) \leq \langle \zeta, v \rangle$, as desired.

To prove the opposite inclusion and obtain the result (a), it suffices to prove that any point ζ of the form w-$\lim_{i \to \infty} \zeta_i$, where $\zeta_i \in \partial_P f(x_i)$ and $x_i \to x$, belongs to $\partial f(x)$ (since $\partial f(x)$ is itself closed and convex). To see this, observe first that $\partial_P f(x_i) \subset \partial f(x_i)$ (why?); then it suffices to invoke Proposition 1.5(d). Thus (a) is proved.

We turn to (b). To show that $N_S(x)$ is contained in the set on the right side of the formula asserted there, it suffices to show that $\partial d_S(x)$ is contained in the right side, in view of Proposition 5.4. In turn, in light of (a), this would follow from proving that any ζ of the form w-$\lim \zeta_i$, where $\zeta_i \in \partial_P d_S(x_i)$ and $x_i \to x$, belongs to the right side. But this is evident from Theorem 1.6.4.

To complete the proof of the theorem, it suffices now to prove that any point ζ of the form w-$\lim \zeta_i$, where $\zeta_i \in N_S^P(x_i)$ and $x_i \to x$, belongs to $N_S(x)$. We may suppose $\|\zeta\| = 1$ without loss of generality, and (by passing to a subsequence) that $\|\zeta_i\| \to \lambda > 0$.

Since $\zeta_i \in N_S^P(x_i)$, there exists $\sigma_i \geq 0$ such that
$$\langle \zeta_i, x' - x_i \rangle \leq \sigma_i \|x' - x_i\|^2 \ \forall x' \in S.$$

Thus the function $x' \mapsto \langle -\zeta_i, x' \rangle + \sigma_i \|x' - x_i\|^2$ attains a minimum over S at $x' = x_i$. Note that for x' in a neighborhood of x_i, the function in question has Lipschitz rank $\|\zeta_i\| + 1/i$. We invoke Proposition 1.6.3 to deduce that locally, x_i minimizes the function
$$x' \mapsto \langle -\zeta_i, x' \rangle + \sigma_i \|x' - x_i\|^2 + \left(\|\zeta_i\| + \frac{1}{i} \right) d_S(x').$$

We now deduce

$$\zeta_i \in \left(\|\zeta_i\| + \frac{1}{i}\right)\partial_P d_S(x_i) \subset \left(\|\zeta_i\| + \frac{1}{i}\right)\partial d_S(x_i),$$

or $\zeta_i/(\|\zeta_i\|+1/i) \in \partial d_S(x_i)$. But $\zeta_i/(\|\zeta_i\|+1/i)$ converges weakly to ζ/λ, so by Proposition 1.5(d) we obtain $\zeta \in \lambda \partial d_S(x) \subset N_S(x)$ (by Proposition 5.4). □

6.2. Remark. In terms of the limiting subdifferential $\partial_L f$ and the limiting normal cones N_S^L defined in §1.10, the theorem asserts that in a Hilbert space we have $N_S(x) = \overline{\text{co}}\, N_S^L(x)$, and, when f is Lipschitz near x, $\partial f(x) = \overline{\text{co}}\, \partial_L f(x)$. We will establish in Chapter 3 that when $X = \mathbb{R}^n$, a function f which is regular at x satisfies $\partial_L f(x) = \partial f(x)$. In view of the closed convex hulls that appear in Theorem 6.1, we sometimes adopt the alternative notations $\partial_C f$, N_S^C, and T_S^C for the generalized gradient, normal cone, and tangent cone, particularly when other constructs such as ∂_P and ∂_L are present simultaneously. We remark as well that $\partial_C f$ and T_S^C are often referred to as Clarke's generalized gradient and tangent cone, respectively.

Theorem 6.1 gives us access in the Hilbert space setting to the results of Chapter 1. It follows, for example (from Exercise 1.10.3(a) and the theorem) that $N_S(x) \neq \{0\}$ if x lies on the boundary of S and X is finite dimensional. Another illustration is the following:

6.3. Exercise. Let $S \in \mathbb{R}^n$ have the form

$$S := \{x \in \mathbb{R}^n : f_i(x) = 0, i = 1, 2, \ldots, k\},$$

where each $f_i \colon \mathbb{R}^n \to \mathbb{R}$ is C^2. Let $x \in S$, and suppose that the set $\{f_i'(x) \colon i = 1, 2, \ldots, k\}$ is linearly independent. Prove that

$$N_S^C(x) = \text{span}\{f_i'(x) \colon i = 1, 2 \ldots, k\} = N_S^P(x),$$
$$T_S^C(x) = \{v \in \mathbb{R}^n : \langle f_i'(x), v \rangle = 0, i = 1, 2, \ldots, k\}.$$

(*Hint.* Proposition 1.1.9.)

6.4. Exercise. Confirm the formula of Theorem 6.1(b) in each case of Exercise 5.6. (Note that the formula offers a simpler approach to finding N_S^C in some cases.)

7 The Bouligand Tangent Cone and Regular Sets

Let S be a closed subset of the Banach space X, $x \in S$. The Bouligand (or *contingent*) tangent cone to S at x, denoted $T_S^B(x)$, is defined as follows:

$$T_S^B(x) := \left\{\lim_{i \to \infty} \frac{x_i - x}{t_i} : x_i \xrightarrow{S} x, t_i \downarrow 0\right\}.$$

(Recall that $x_i \xrightarrow{S} x$ means that $x_i \in S$ $\forall i = 1, 2, \ldots$, and that $\lim_{i \to \infty} x_i = x$.) This very natural concept of tangency can be characterized by means of the distance function, as can the tangent cone $T_S(x)$ of §5.

7.1. Exercise.

(a) Prove that $v \in T_S^B(x)$ iff

$$\liminf_{t \downarrow 0} \frac{d_S(x + tv)}{t} = 0.$$

(b) Deduce that $T_S^C(x) \subset T_S^B(x)$.

(c) Calculate $T_S^B(0)$ for each of the sets S of Exercise 5.6. Note that in contrast to $T_S^C(x)$, $T_S^B(x)$ may not be convex.

(d) When X is a Hilbert space, show that

$$T_S^B(x) \subset \left(N_S^P(x)\right)^\circ.$$

It is natural to inquire into the possible coincidence of $T_S^B(x)$ and $T_S^C(x)$. This turns out to be an interesting question meriting a definition. We will say that a set S is *regular* at $x \in S$ provided that $T_S^B(x) = T_S^C(x)$.

7.2. Exercise. Show that a convex set S is regular at each of its points. (*Hint.* $d_S(\cdot)$ is convex.)

We have used the term "regular" both for functions (§4) and now for sets; here is why the use of the same term is appropriate:

7.3. Proposition. *Let f be Lipschitz near x. Then f is regular at x iff epi f is regular at $(x, f(x))$.*

Proof. Let us suppose first that f is regular at x. We wish to show that $T_{\text{epi} f}^B(x, f(x))$ and $T_{\text{epi} f}^C(x, f(x))$ agree; equivalently, that an element (v, r) of $T_{\text{epi} f}^B(x, f(x))$ must lie in $T_{\text{epi} f}^C(x, f(x)) = \text{epi } f^\circ(x; \cdot)$ (by Theorem 5.7), which is to say that we have $f^\circ(x; v) \leq r$. Since $f'(x; v) = f^\circ(x; v)$, it suffices to produce a sequence $t_i \downarrow 0$ such that

$$\liminf_{i \to \infty} \frac{f(x + t_i v) - f(x)}{t_i} \leq r.$$

Now by Exercise 7.1(a), we have

$$\liminf_{t \downarrow 0} \frac{d_{\text{epi} f}((x, f(x)) + t(v, r))}{t} \leq 0.$$

That this condition implies the existence of the stated sequence t_i is easy to establish.

92   2. Generalized Gradients in Banach Space

For the converse, suppose now that epi f is regular at $(x, f(x))$. To establish that f is regular at x, we will show that for any $v \in X$, we have

$$\liminf_{t \downarrow 0} \frac{f(x+tv) - f(x)}{t} \geq f^\circ(x; v)$$

(why does this imply that f is regular at x?). In turn, this follows if we show that for any sequence $t_i \downarrow 0$, we have

$$\lambda := \lim_{i \to \infty} \frac{f(x + t_i v) - f(x)}{t_i} \geq f^\circ(x; v)$$

whenever this limit λ exists. But we have

$$(v, \lambda) = \lim_{i \to \infty} \frac{(x + t_i v, f(x + t_i v)) - (x, f(x))}{t_i},$$

which shows that $(v, \lambda) \in T^B_{\mathrm{epi}\, f}(x, f(x)) = T^C_{\mathrm{epi}\, f}(x, f(x))$ (since epi f is regular at $(x, f(x))$) $= \mathrm{epi}\, f^\circ(x; \cdot)$ (by Theorem 5.7). It follows that $f^\circ(x; v) \leq \lambda$, as required. $\square$

The principal role of set regularity is to provide "more exact" estimates in formulas for tangent and normal cones, just as functional regularity led to sharper results for generalized gradients. We illustrate this in the following result, which will be extended to the case of mixed equalities and inequalities in Chapter 3.

7.4. Proposition. *Let S be a subset of $\mathbb{R}^n$ given as follows:*

$$S := \{x \in \mathbb{R}^n : f_j(x) = 0, j = 1, 2, \ldots, k\},$$

where each $f_j \colon \mathbb{R}^n \to \mathbb{R}$ is locally Lipschitz and admits one-sided directional derivatives $f'_j(x; v)$ for each v. Then

$$T^B_S(x) \subset \{v \in \mathbb{R}^n : f'_j(x; v) = 0, j = 1, 2, \ldots, k\}.$$

If in addition each f_j is C^1 near x and the vectors $\{f'_j(x)\}_{j=1}^k$ are linearly independent, then

$$\{v \colon \langle f'_j(x), v \rangle = 0, j = 1, 2, \ldots, k\} \subset T^C_S(x);$$

in this latter case, equality holds in both estimates, and the set S is regular at x.

Proof. Any $v \in T^B_S(x)$ is of the form $\lim_{i \to \infty}(x_i - x)/(t_i)$, where $x_i \xrightarrow{S} x$ and $t_i \downarrow 0$. Then $f_j(x_i) = 0$ $(j = 1, 2, \ldots, k)$, and we deduce

$$f'_j(x; v) = \lim_{i \to \infty} \frac{f_j(x + t_i v) - f_j(x)}{t_i} = \lim_{i \to \infty} \frac{f_j(x + t_i v)}{t_i}$$

$$= \lim_{i \to \infty} \frac{f_j(x_i)}{t_i} = 0,$$

since $|f_j(x + t_i v) - f_j(x_i)|/t_i \leq K\|x + t_i v - x_i\|/t_i \to 0$, where K is a Lipschitz constant for f_j. This gives the stated estimate for $T_S^B(x)$.

Now let us posit the additional hypotheses on the f_j. Then Proposition 1.9(a) applies, for all x' sufficiently near x, and yields

$$N_S^P(x') \subset \text{span}\{f_j'(x')\}_{j=1}^k.$$

We can now appeal to Theorem 6.1(b) to obtain

$$N_S^C(x) \subset \text{span}\{f_j'(x)\}_{j=1}^k.$$

Polarity transforms this estimate for $N_S^C(x)$ to precisely the one for $T_S^C(x)$ that is claimed. Since $T_S^C(x) \subset T_S^B(x)$ is always true, equality holds in both estimates of the proposition, and so S is regular at x. □

8 The Gradient Formula in Finite Dimensions

The celebrated theorem of Rademacher asserts that if a function $f: \mathbb{R}^n \to \mathbb{R}$ is Lipschitz on an open set U, then it is differentiable almost everywhere (a.e.) on U (in the sense of Lebesgue measure); we prove this theorem in Chapter 3. It turns out that the derivative of f can be used to generate its generalized gradient, as depicted in the following formula, one of the most useful computational tools in nonsmooth analysis. It shows that in $\mathbb{R}^n$, $\partial f(x)$ can be generated by the values of $\nabla f(x')$ at nearby points x' at which $f'(x')$ exists, and furthermore that points x' belonging to any prescribed set of measure zero can be ignored in the construction without changing the result. This latter aspect of $\partial f(x)$ is referred to as being "blind to sets of measure zero."

8.1. Theorem (Generalized Gradient Formula). *Let $x \in \mathbb{R}^n$, and let $f: \mathbb{R}^n \to \mathbb{R}$ be Lipschitz near x. Let Ω be any subset of zero measure in $\mathbb{R}^n$, and let Ω_f be the set of points in $\mathbb{R}^n$ at which f fails to be differentiable. Then*

$$\partial f(x) := \text{co}\{\lim \nabla f(x_i): x_i \to x, x_i \notin \Omega, x_i \notin \Omega_f\}.$$

Proof. The meaning of the formula is the following: consider any sequence $\{x_i\}$ converging to x while avoiding both Ω and points at which f is not differentiable, and such that the sequence $\{\nabla f(x_i)\}$ converges; then the convex hull of all such limits is $\partial f(x)$. As usual in a Hilbert space setting, we identify $\partial f(x)$ with a subset of $\mathbb{R}^n$, the space itself.

Let us note to begin with that there are "plenty" of sequences x_i which converge to x and avoid $\Omega \cup \Omega_f$, since the latter has measure 0 near x. Further, because ∂f is locally bounded near x (Proposition 1.5) and $\nabla f(x_i)$

belongs to $\partial f(x_i)$ for each i (Proposition 3.1) the sequence $\{\nabla f(x_i)\}$ admits a convergent subsequence by the Bolzano–Weierstrass Theorem. The limit of any such sequence must belong to $\partial f(x)$ by the closure property of ∂f proved in Proposition 1.5. It follows that the set

$$\{\lim \nabla f(x_i) \colon x_i \to x, x_i \notin \Omega \cup \Omega_f\}$$

is contained in $\partial f(x)$ and is nonempty and bounded, and in fact compact, since it is rather obviously closed. Since $\partial f(x)$ is convex, we deduce that the left-hand side of the formula asserted by the theorem contains the right. Now, the convex hull of a compact set in R^n is compact, so to complete the proof we need only show that the support function of the left-hand side (i.e., $f^\circ(x; \cdot)$) never exceeds that of the right. This is what the following lemma does:

Lemma. *For any $v \neq 0$ in R^n, for any $\varepsilon > 0$, we have*

$$f^\circ(x; v) - \varepsilon \leq \limsup\{\nabla f(y) \cdot v \colon y \to x, y \notin \Omega \cup \Omega_f\}.$$

To prove this, let the right-hand side be α. Then by definition, there is a $\delta > 0$ such that the conditions

$$y \in x + \delta B, \quad y \notin \Omega \cup \Omega_f,$$

imply $\nabla f(y) \cdot v \leq \alpha + \varepsilon$. We also choose δ small enough so that f is Lipschitz on $B(x; \delta)$ and $\Omega \cup \Omega_f$ has measure 0 in $x + \delta B$. Now consider the line segments $L_y = \{y + tv \colon 0 < t < \delta/(2|v|)\}$. Since $\Omega \cup \Omega_f$ has measure 0 in $x + \delta B$, it follows from Fubini's Theorem that for almost every y in $x + (\delta/2)B$, the line segment L_y meets $\Omega \cup \Omega_f$ in a set of 0 one-dimensional measure. Let y be any point in $x + (\delta/2)B$ having this property, and let t lie in $(0, \delta/(2|v|))$. Then

$$f(y + tv) - f(y) = \int_0^t \nabla f(y + sv) \cdot v \, ds,$$

since f' exists a.e. on L_y. Since we have $\|y + sv - x\| < \delta$ for $0 < s < t$, it follows that $\nabla f(y + sv) \cdot v \leq \alpha + \varepsilon$, whence

$$f(y + tv) - f(y) \leq t(\alpha + \varepsilon).$$

Since this is true for all y within $\delta/2$ of x except those in a set of measure 0, and for all t in $(0, \delta/(2|v|))$, and since f is continuous, it is in fact true for all such y and t. We deduce

$$f^\circ(x; v) \leq \alpha + \varepsilon,$$

which completes the proof. □

8.2. Corollary.

$$f°(x; v) = \limsup_{y \to x} \{\nabla f(y) \cdot v : y \notin \Omega \cup \Omega_f\}.$$

8.3. Exercise. Use the theorem to calculate $\partial f(0,0)$ where f on R^2 is given by

$$f(x,y) = \max\{\min[x, -y], y - x\},$$

as follows. Define

$$C_1 = \{(x,y): y \leq 2x \text{ and } y \leq -x\},$$
$$C_2 = \{(x,y): y \leq x/2 \text{ and } y \geq -x\},$$
$$C_3 = \{(x,y): y \geq 2x \text{ or } y \geq x/2\}.$$

Then $C_1 \cup C_2 \cup C_3 = R^2$, and we have

$$f(x,y) = \begin{cases} x & \text{for } (x,y) \in C_1, \\ -y & \text{for } (x,y) \in C_2, \\ y - x & \text{for } (x,y) \in C_3. \end{cases}$$

Note that the boundaries of these three sets form a set Ω of measure 0, and that if (x,y) does not lie in Ω, then f is differentiable and $\nabla f(x,y)$ is one of the points $(1,0)$, $(0,-1)$, or $(-1,1)$. Then $\partial f(0,0)$ is the triangle obtained as the convex hull of these three points.

The derivative of the distance function $d_S(\cdot)$ of a closed set S in $\mathbb{R}^n$ admits an interpretation along the lines of its proximal subgradients (Theorem 1.6.1).

8.4. Proposition. *Let $d'_S(x)$ exist and be different from 0. Then $x \notin S$, $\text{proj}_S(x)$ is a singleton $\{s\}$, and $\nabla d_S(x) = (x-s)/\|x-s\|$.*

Proof. If x lies in S, then $d'_S(x) = 0$ since d_S has a minimum at x; thus $x \notin S$. Let $s \in \text{proj}_S(x)$. We will show that $\nabla d_S(x) = (x-s)/\|x-s\|$, thereby establishing also that $\text{proj}_S(x)$ contains no other points. For $t \in (0,1)$, the closest point in S to $x + t(s-x)$ is s (see Proposition 1.1.3), whence

$$d_S(x + t(s-x)) = (1-t)\|x - s\|.$$

Subtracting $d_S(x) = \|x - s\|$ from both sides, dividing by t, and letting $t \downarrow 0$ produces

$$d'_S(x; s-x) = \langle \nabla d_S(x), s-x \rangle = -\|s - x\|.$$

Now $\|\nabla d_S(x)\| \leq 1$, since d_S is Lipschitz of rank 1, so the last equation implies $\nabla d_S(x) = (x-s)/\|x-s\|$. □

8.5. Exercise.

(a) Let $x \in \mathrm{bdry}(S)$. Use Theorem 8.1, Exercise 5.4, and Proposition 8.4 to deduce the following formula for $N_S(x)$ when S is a closed subset of $\mathbb{R}^n$ and Ω is any set of measure 0:

$$N_S(x) = \overline{\mathrm{co}}\Big\{\lim_{i \to \infty} \lambda_i(x_i - s_i) : \lambda_i \geq 0, x_i \notin \Omega \cup S,$$
$$x_i \to x, \mathrm{proj}_S(x_i) = \{s_i\}\Big\}.$$

It follows that $N_S(x)$ contains nonzero elements.

(b) Use this characterization to calculate the normal cones of Exercise 5.6.

9 Problems on Chapter 2

9.1. A function $f\colon X \to \mathbb{R}$ is said to be *strictly (Hadamard) differentiable* at x if there exists an element $\zeta \in X^*$ such that for each v in X we have

$$\lim_{\substack{x' \to x \\ t \downarrow 0}} \frac{f(x' + tv) - f(x')}{t} = \langle \zeta, v \rangle,$$

and provided the convergence is uniform for v in compact sets.

(a) f is strictly differentiable at x iff f is Lipschitz near x and $\partial f(x)$ is a singleton.

(b) If f is regular at x and $f'(x)$ exists, then f is strictly differentiable at x.

(c) If f is regular at x and $f'(x)$ exists, and if g is Lipschitz near x, then

$$\partial(f + g)(x) = \{f'(x)\} + \partial g(x).$$

(d) If X is finite dimensional and U is an open subset of X, then $f \in C^1(U)$ iff f is locally Lipschitz on U and $\partial f(x)$ is a singleton for each $x \in U$.

9.2. Let $\varphi \colon [0,1] \to \mathbb{R}$ belong to $L^\infty[0,1]$ and define a (Lipschitz) function $f\colon [0,1] \to \mathbb{R}$ via $f(x) := \int_0^x \varphi(t)\, dt$. Prove that $\partial f(x)$ is the interval $[\varphi^-(x), \varphi^+(x)]$, where $\varphi^-(x)$ and $\varphi^+(x)$ are the essential infimum and essential supremum, respectively, of φ at x; thus, for instance,

$$\varphi^+(x) := \inf\{M : \exists\, \varepsilon > 0 \ni \varphi(x') \leq M \text{ a.e. for } x' \in [x - \varepsilon, x + \varepsilon]\}.$$

9.3. For f as in Problem 9.2, show that there is a point $z \in (0,1)$ such that for any $\delta > 0$, there exist x and y within δ of z such that

$$\varphi(x) - \delta \leq \int_0^1 \varphi(t)\, dt \leq \varphi(y) + \delta.$$

(When φ is continuous this holds for $\delta = 0$: the Intermediate Value Theorem.)

9.4. Let $X = X_1 \times X_2$, where X_1 and X_2 are Banach spaces, and let $f(x_1, x_2)$ on X be Lipschitz near (x_1, x_2). We denote by $\partial_1 f(x_1, x_2)$ the (partial) generalized gradient of $f(\cdot, x_2)$ at x_1, and similarly for $\partial_2 f$.

(a) Let $\pi_1 \partial f(x_1, x_2)$ be the set

$$\{\zeta_1 \in X_1^* : \text{ for some } \zeta_2 \in X_2^*, (\zeta_1, \zeta_2) \in \partial f(x_1, x_2)\}.$$

Prove that $\partial_1 f(x_1, x_2) \subset \pi_1 \partial f(x_1, x_2)$, with equality when f is regular at (x_1, x_2). (*Hint.* Consider Theorem 3.2 and $F(x) = (x, x_2)$.)

(b) Calculate $\partial_1 f(0,0)$, $\partial_2 f(0,0)$ for the function f of Exercise 8.3. Note that

$$\partial_1 f(0,0) \times \partial_2 f(0,0) \not\subset \partial f(0,0) \not\subset \partial_1 f(0,0) \times \partial_2 f(0,0).$$

9.5. Let X be a Banach space, and let $f: X \to \mathbb{R}$ be Lipschitz near x. Suppose that x minimizes f over a closed set S. Then $0 \in \partial f(x) + N_S(x)$. (*Hint.* Use exact penalization as in Proposition 1.6.3.)

9.6. Let $f: X \to \mathbb{R}$ be Lipschitz near x, where X is a Hilbert space, and suppose $0 \notin \partial f(x)$. Let ζ be the element of least norm in $\partial f(x)$. Prove that $-\zeta$ is a "direction of descent" in the following sense: for all $t > 0$ sufficiently small we have

$$f(x - t\zeta) \leq f(x) - t\frac{\|\zeta\|^2}{2}.$$

9.7. Show by an example in one dimension that unlike ∂f, $\partial_L f$ is not "blind to sets of measure 0"; that is, we cannot neglect an arbitrary set of measure 0 in the formula

$$\partial_L f(x) = \left\{\lim_{i \to \infty} \zeta_i : \zeta_i \in \partial_P f(x_i), x_i \xrightarrow{f} x\right\},$$

even when f is Lipschitz.

9.8. For $f: \mathbb{R}^n \to \mathbb{R}$ locally Lipschitz, let

$$Df(x; v) := \liminf_{t \downarrow 0} \frac{f(x + tv) - f(x)}{t}$$

denote the *Dini subderivate*. Prove that

$$f^\circ(x; v) = \limsup_{y \to x} Df(y; v).$$

9.9. (Analysis of a Line-Fitting Problem.) We wish to minimize over $\mathbb{R}^2$ the function

$$f(\alpha, \beta) := |\alpha N + \beta| + \sum_{i=0}^{N-1} |\alpha i + \beta - i|,$$

where N is a given positive integer. (We interpret this as finding the best line to fit the given data points $(1, 1), (2, 2), \ldots, (N-1, N-1)$, and $(N, 0)$.)

(a) Let $g(\alpha, \beta) := |\alpha c + \beta - k|$ for given c and k. Prove that

$$\partial g(\alpha, \beta) = \begin{cases} \{(c, 1)\} & \text{if } \alpha c + \beta - k > 0, \\ \{(-c, -1)\} & \text{if } \alpha c + \beta - k < 0, \\ \{\lambda(c, 1) : |\lambda| \leq 1\} & \text{if } \alpha c + \beta - k = 0. \end{cases}$$

(b) Use (a) to interpret the necessary condition $0 \in \partial f(\alpha, \beta)$ in the form

$$0 = \lambda_N(N, 1) + \sum_{i=0}^{N-1} \lambda_i(i, 1)$$

for certain constants λ_i ($i = 0, 1, \ldots, N$) with values depending on (α, β).

(c) For $N \geq 3$, invoke the convexity of f to deduce that $\alpha = 1$, $\beta = 0$ minimizes f.

(d) If $N > 4$, no other (α, β) satisfy the necessary condition of (b). (Thus the anomalous point $(N, 0)$ is ignored by the nonsmooth criterion, which is not the case in the usual least squares fit.)

9.10. Let S in $\mathbb{R}^n$ be of the form $\{x : f(x) \leq 0\}$, where $f : \mathbb{R}^n \to \mathbb{R}$ is locally Lipschitz. Suppose that $x \in S$, and $0 \notin \partial f(x)$. Prove that $\operatorname{int} T_S(x) \neq \emptyset$, and that

$$T_S(x) \supset \{v \in \mathbb{R}^n : f^\circ(x; v) \leq 0\}.$$

Show that equality holds if f is regular at x, in which case S is regular at x. (*Remark.* Sets which admit tangent cones having nonempty interior have useful properties; they will be studied in Chapter 3.)

9.11. We will prove a formula for the tangent and normal cones of the intersection of two closed subsets S_1 and S_2 of $\mathbb{R}^n$. (We remind the reader that the limiting normal cone N_S^L was introduced in §1.10, and the alternative notation N_S^C and T_S^C for N_S and T_S in Remark 6.2.)

(a) Prove that $T_C^B(x) \subset T_S^B(x)$ if $C \subset S$. (Show that this monotonicity property fails for T^C.) Deduce that
$$T_{S_1 \cap S_2}^B(x) \subset T_{S_1}^B(x) \cap T_{S_2}^B(x).$$

(b) Note that $T_S^B(x)^\circ \subset N_S^C(x)$ generally, and deduce
$$N_{S_1 \cap S_2}^C(x) \supset T_{S_1}^B(x)^\circ + T_{S_2}^B(x)^\circ.$$

(c) Suppose henceforth that $N_{S_1}^L(x) \cap \left(-N_{S_2}^L(x)\right) = \{0\}$ (a condition called *transversality*). Prove that $N_{S_1 \cap S_2}^L(x) \subset N_{S_1}^L(x) + N_{S_2}^L(x)$. (*Hint.* See the argument that proved Proposition 1.10.1.) Give a counterexample to this conclusion when transversality fails.

(d) **Theorem.** *If transversality holds at x, then*
$$N_{S_1 \cap S_2}^C(x) \subset N_{S_1}^C(x) + N_{S_2}^C(x), \quad T_{S_1 \cap S_2}^C(x) \supset T_{S_1}^C(x) \cap T_{S_2}^C(x),$$

and equality holds if both S_1 and S_2 are regular at x, in which case $S_1 \cap S_2$ is also regular at x.

9.12. Let $f \colon \mathbb{R}^n \to \mathbb{R}$ be Lipschitz near x.

(a) $f'(x)$ exists iff $T_{\mathrm{gr}\, f}^B(x, f(x))$ is a hyperplane.

(b) f is strictly differentiable at x iff $T_{\mathrm{gr}\, f}^C(x, f(x))$ is a hyperplane.

(c) f is strictly differentiable at x iff $T_{\mathrm{epi}\, f}^C(x, f(x))$ is a half-space. (*Hint.* Theorem 5.7.)

(d) The fact that $T_{\mathrm{epi}\, f}^B(x, f(x))$ is a half-space does not imply that $f'(x)$ exists.

9.13. (Danskin's Theorem.) Let a continuous function $g \colon \mathbb{R}^n \times M \to \mathbb{R}$ be given, where M is a compact metric space. We suppose that for a neighborhood Ω of a given point $x \in \mathbb{R}^n$, the derivative $g_x(x', u)$ exists and is continuous (jointly) for $(x', u) \in \Omega \times M$. We set
$$f(x') := \max_{u \in M} g(x', u).$$
Then f is Lipschitz and regular near x, and we have
$$\partial_C f(x) = \mathrm{co}\{g_x(x, u) \colon u \in M(x)\},$$
where $M(x) := \{u \in M \colon g(x, u) = f(x)\}$. Deduce Danskin's formula
$$f'(x; v) = \max\{\langle g_x(x, u), v \rangle \colon u \in M(x)\}.$$

100   2. Generalized Gradients in Banach Space

9.14. Let X be a Banach space, and set $f(x) := \|x\|$. If $\zeta \in \partial_C f(x)$ for some $x \neq 0$, then $\|\zeta\|_* = 1$.

9.15. Let $f \colon \mathbb{R}^m \times \mathbb{R}^n \to \mathbb{R}$ be locally Lipschitz, and such that for each $x \in \mathbb{R}^m$, the function $y \mapsto f(x, y)$ is convex. Then if $(0, \zeta) \in \partial_C f(x, y)$, it follows that $\zeta \in \partial_C f(x, \cdot)(y)$. (Problem 9.4 shows that this is false in general if the partial convexity hypothesis is deleted.)

9.16. We define a *continuous* function $f \colon [-1, 1] \to \mathbb{R}$ as follows:

(i) $f(x) = 0$ when $x = \pm 2^{-n}$, $n = 0, 1, 2, \ldots$.

(ii) $f(x) = x$ when $x = \pm 3/2^{2n}$, $n = 0, 1, 2, 3, \ldots$.

(iii) $f(x) = -x$ when $x = \pm 3/2^{2n+1}$, $n = 1, 2, 3, \ldots$.

(iv) Between any two successive values of x cited above, f is affine.

(a) Prove that f is Lipschitz, and that the directional derivatives $f'(0; 1)$ and $f'(0; -1)$ fail to exist.

(b) Show that the subderivate $Df(0; 1)$ (see Problem 9.8) equals -1.

(c) Show that $\partial_P f(0) = \emptyset$ and $\partial_L f(0) = \partial_C f(0) = [-3, 3]$.

9.17. Let A, C, D be convex w^*-closed bounded subsets of X^* such that $A + C \subset D + C$. Prove that $A \subset D$.

9.18. Let $F \colon X \to \mathbb{R}^n$ be locally Lipschitz, let $A \subset \mathbb{R}^n$ be compact. We set
$$\theta(x) := \max_{\alpha \in A} \langle \alpha, F(x) \rangle.$$

(a) Show that θ is locally Lipschitz.

(b) If the maximum defining $\theta(x)$ is attained at a *single* $\alpha \in A$, prove that
$$\partial \theta(x) \subset \partial \langle \alpha, F(\cdot) \rangle(x).$$

9.19. Let f, g_i, h_j be locally Lipschitz functions on a Hilbert space X, $i \in I := \{1, 2, \ldots, m\}$, $j \in J := \{1, 2, \ldots, n\}$, and let C be a closed subset of X. We assume that $\bar{x}$ is a local solution of the following optimization problem:
$$\text{minimize}\{f(x) \colon g_i(x) \leq 0 \ (i \in I), h_j(x) = 0 \ (j \in J), \ x \in C\}.$$

We will prove the Fritz John necessary conditions, namely the following *multiplier rule*: There exist $\lambda_0 \in \mathbb{R}$, $\gamma \in \mathbb{R}^m$, $\lambda \in \mathbb{R}^n$ with
$$\lambda_0 = 0 \text{ or } 1, \quad (\lambda_0, \gamma, \lambda) \neq 0, \quad \gamma \geq 0, \quad \langle \gamma, g(\bar{x}) \rangle = 0,$$
such that
$$0 \in \partial_C \{\lambda_0 f + \langle \gamma, g \rangle + \langle \lambda, h \rangle\}(\bar{x}) + N_C^L(\bar{x}).$$

(a) We may assume that C is bounded, that f is globally Lipschitz and bounded below, and that $\bar{x}$ is a global solution. We set

$$A := \{(\lambda_0, \gamma, \lambda) =: \alpha \in \mathbb{R} \times \mathbb{R}^m \times \mathbb{R}^n : \lambda_0 \geq 0, \gamma \geq 0, \|(\lambda_0, \gamma, \lambda)\}\| = 1\}.$$

For $\varepsilon > 0$ fixed, we define

$$F_\varepsilon(x) := \big(f(x) - f(\bar{x}) + \varepsilon, g(x), h(x)\big), \, \theta_\varepsilon(x) := \max_{\alpha \in A} \langle \alpha, F_\varepsilon(x) \rangle.$$

Then $\theta_\varepsilon > 0$ on C, and $\theta_\varepsilon(\bar{x}) = \varepsilon$.

(b) There exists $x_\varepsilon \in C \cap B(\bar{x}; \sqrt{\varepsilon})$ and a constant K such that

$$0 \in \partial_L \theta_\varepsilon(x_\varepsilon) + K \, \partial_L d_C(x_\varepsilon) + 4\sqrt{\varepsilon}B.$$

(*Hint.* Invoke the Minimization Principle Theorem 1.4.2.)

(c) There exists $\alpha_\varepsilon \in A$ such that

$$0 \in \partial_C \langle \alpha_\varepsilon, F_\varepsilon(\cdot) \rangle (x_\varepsilon) + K \, \partial_L d_C(x_\varepsilon) + 4\sqrt{\varepsilon}B.$$

(*Hint.* See the previous problem.)

(d) Pass to the limit as $\varepsilon \downarrow 0$.

9.20. Let X and Y be Banach spaces and $A \colon X \to Y$ a continuous linear operator. Let Ω and Z be closed convex subsets of X and Y, respectively, and consider for each y in Y the problem:

$$\text{minimize } f(x) \colon x \in \Omega, \quad Ax \in Z + y,$$

where $f \colon X \to \mathbb{R}$ is given. Let $V(y)$ denote the value of this problem for a given y.

Now suppose that there exists a bounded subset K of X such that f is Lipschitz on K, and such that for all y sufficiently near 0, we have

$$x \in \Omega, \quad Ax \in Z + y \implies x \in K.$$

(a) If the condition $0 \in \text{int}\{A\Omega - Z\}$ is satisfied, then $V(\cdot)$ is Lipschitz near 0.

(b) If in addition $\bar{x}$ is a solution of the problem obtained when $y = 0$, then there exist $\zeta \in \partial f(\bar{x})$ and $\theta \in \partial V(0)$ such that

$$-\zeta + A^*\theta \in N_\Omega(\bar{x}) \quad \text{and} \quad \theta \in -N_Z(A\bar{x}).$$

(*Hint.* For any x in Ω and z in Z, we have

$$V(Ax - z) \leq f(x).)$$

3
Special Topics

Evidemment, évidemment
On danse encore
Sur les accords
Qu'on aimait tant....
Mais pas comme avant.

—Michel Berger et France Gall, *Evidemment*

In this chapter we study a number of different issues, each of interest in its own right. All the results obtained here build upon, and in some cases complement, those of the preceding chapters, and several of them address problems discussed in the Introduction. This is the case, for example, of the first section on constrained optimization. Some of the results of §§5 and 6 will be called upon in Chapter 4.

1 Constrained Optimization and Value Functions

Consider the problem of minimizing $f(x)$ over those points x satisfying $h(x) = 0$, where $f\colon \mathbb{R}^n \to \mathbb{R}$ and $h\colon \mathbb{R}^n \to \mathbb{R}^m$ are given (let us say smooth) functions. As we did in Chapter 0, let us associate with this single problem in constrained optimization a family of related problems $P(\alpha)$ parametrized by $\alpha \in \mathbb{R}^m$:

Problem $P(\alpha)$:   minimize $f(x)$ subject to $h(x) + \alpha = 0$.

The *value function* $V(\cdot)$ associated with this scheme is the function whose value at α is the minimum (or infimum) in the problem $P(\alpha)$:

$$V(\alpha) := \inf\{f(x) : h(x) + \alpha = 0\}.$$

In general, V will take values in $[-\infty, \infty]$, the value $+\infty$ corresponding to cases in which the *feasible set* for $P(\alpha)$, i.e., the set

$$\Phi(\alpha) := \{x \in \mathbb{R}^n : h(x) + \alpha = 0\},$$

is empty.

As we pointed out in Chapter 0, the very definition of V implies that for any $x \in \mathbb{R}^n$ we have $f(x) \geq V(-h(x))$. If $\Sigma(\alpha)$ denotes the (possibly empty) set of solutions to $P(\alpha)$ (i.e., those $x \in \Phi(\alpha)$ such that $V(\alpha) = f(x)$), then equality holds in the preceding for any $x_0 \in \Sigma(0)$. Thus, if $V'(0)$ exists, we derive the conclusion

$$f'(x_0) + \nabla V(0)^* h'(x_0) = 0,$$

that is to say, the Lagrange Multiplier Rule. (Our convention is to identify $h'(x_0)$ with the $m \times n$ Jacobian matrix; the gradient of V, ∇V, is an $m \times 1$ column vector, as an element of $\mathbb{R}^m$; $*$ denotes transpose.) As we have said, V may not be differentiable at 0, so that this line of reasoning, while attractive, is problematic. A key to a more realistic approach is to observe that the argument above can be salvaged if we possess a vector ζ belonging to the proximal subdifferential of V at 0, $\partial_P V(0)$. Let us see why.

The proximal subgradient inequality (Theorem 1.2.5) asserts that for some $\sigma \geq 0$, for all α sufficiently near 0, we have

$$V(\alpha) - V(0) + \sigma\|\alpha\|^2 \geq \langle \zeta, \alpha \rangle.$$

Our given point $x_0 \in \Sigma(0)$ satisfies $f(x_0) = V(0)$. If x is sufficiently near x_0, then $h(x)$ is close to $h(x_0) = 0$, and so the choice $\alpha = -h(x)$ is legitimate in the preceding inequality. Finally, observe again that $f(x) \geq V(-h(x))$. Substituting into the proximal subgradient inequality and rearranging leads to

$$f(x) + \langle \zeta, h(x) \rangle + \sigma\|h(x)\|^2 \geq f(x_0),$$

an inequality which holds for all x near x_0. This amounts to saying that the function

$$x \mapsto f(x) + \langle \zeta, h(x) \rangle + \sigma\|h(x)\|^2$$

admits a local minimum at $x = x_0$, whence

$$f'(x_0) + \zeta^* h'(x_0) = 0.$$

We recover the Multiplier Rule in terms similar to the above, but, with $V'(0)$ replaced by any element ζ of $\partial_P V(0)$.

An important objection persists, however, for there is no guarantee that $\partial_P V(0)$ is nonempty. The theorem below deals with this contingency by invoking the Proximal Density Theorem to find points α_i near 0 at which $\partial_P V(\alpha_i) \neq \emptyset$, and passing to the limit as $\alpha_i \to 0$. This approach requires only that $V(\cdot)$ be lower semicontinuous, which (unlike differentiability!) is easy to guarantee a priori. For this purpose we introduce the following *growth hypothesis*:

1.1. Growth Hypothesis. *For every r, $s \in \mathbb{R}$, the following set is bounded*:
$$\{x \in \mathbb{R}^n : f(x) \leq r, \|h(x)\| \leq s\}.$$

1.2. Exercise. Under Growth Hypothesis 1.1, prove the following facts:

(a) $V(\alpha) < \infty$ precisely when $\Phi(\alpha) \neq \emptyset$, and in that case $\Sigma(\alpha)$ is nonempty.

(b) $V \colon \mathbb{R}^m \to (-\infty, \infty]$ is lower semicontinuous, and so $V \in \mathcal{F}(\mathbb{R}^m)$ if V is somewhere finite.

We will see presently how the following theorem is the generator of necessary conditions and solvability results.

1.3. Theorem. *Let f and h be C^1, let* Growth Hypothesis 1.1 *hold, and suppose $V(0) < +\infty$. Then there exists a sequence $\{\alpha_i\}$ converging to 0, with $V(\alpha_i) \to V(0)$, and points $\zeta_i \in \partial_P V(\alpha_i)$, $x_i \in \Sigma(\alpha_i)$, such that*
$$f'(x_i) + \zeta_i^* h'(x_i) = 0, \quad i = 1, 2, \ldots.$$

Proof. We use Theorem 1.3.1 to deduce the existence of a sequence $\{\alpha_i\}$ such that $\alpha_i \to 0$, $V(\alpha_i) \to V(0)$, and such that $\partial_P V(\alpha_i) \neq \emptyset$. Pick $\zeta_i \in \partial_P V(\alpha_i)$ and $x_i \in \Sigma(\alpha_i)$ (this set is nonempty by Exercise 1.2). Then, as in the argument presented above, the proximal subgradient inequality for V at α_i translates to the condition
$$f(x) + \langle \zeta_i, h(x) \rangle + \sigma_i \|h(x) - h(x_i)\|^2 \geq f(x_i) + \langle \zeta_i, \alpha_i \rangle$$
for all x near x_i. Setting the derivative equal to 0 gives the final conclusion of the theorem. □

The question now becomes that of passing to the limit in the context of the sequences provided by the theorem. How we proceed will be conditioned by the specific goal we have in mind. Suppose to begin with that we require information on the *sensitivity* of the problem $P(0)$ with respect to perturbations of the equality constraint. This might arise, for example, from a numerical algorithm which, applied to $P(0)$, can be shown to yield a solution not to $P(0)$ itself, but to a perturbed problem $P(\alpha)$. The question then is: How different is $V(\alpha)$ from $V(0)$? Differential information about

V at 0 becomes relevant. To state a result along these lines, we introduce the *multiplier set* $M(x)$ corresponding to x:

$$M(x) := \{\zeta \in \mathbb{R}^m : f'(x) + \zeta^* h'(x) = 0\}.$$

The hypotheses of Theorem 1.3 remain in force.

1.4. Corollary. *Suppose that for every $x \in \Sigma(0)$, the Jacobian $h'(x)$ is of maximal rank. Then $V(\cdot)$ is Lipschitz near 0, and we have*

$$\emptyset \neq \partial_L V(0) \subset \bigcup_{x \in \Sigma(0)} M(x).$$

Proof. We consider the sequences provided by the theorem, and deduce from the conditions $f(x_i) \to V(0)$, $h(x_i) = \alpha_i \to 0$ that, in light of Growth Hypothesis 1.1, the sequence $\{x_i\}$ is bounded. We extract a subsequence converging to a limit x_0, without relabeling. It follows from the continuity of f and h that $f(x_0) = V(0)$, $h(x_0) = 0$, whence $x_0 \in \Sigma(0)$. Now extract a further subsequence to arrange either $\|\zeta_i\| \to \infty$ or else $\zeta_i \to \zeta_0 \in \mathbb{R}^m$. Let us immediately dispose of the first possibility: Divide across by $\|\zeta_i\|$ in the equation

$$f'(x_i) + \zeta_i^* h'(x_i) = 0,$$

take a further subsequence to arrange to have $\zeta_i/\|\zeta_i\|$ converge to a (necessarily nonzero) limit $\lambda \in \mathbb{R}^m$, and arrive at $\lambda^* h'(x_0) = 0$, contradicting the maximal rank hypothesis. Thus $\{\zeta_i\}$ is necessarily bounded; the arbitrariness of this sequence of proximal subgradients (having $V(\alpha_i) \to V(0)$) implies that for some $\varepsilon > 0$, for some K, for all $\alpha \in B(0,\varepsilon)$ for which $|V(\alpha) - V(0)| < \varepsilon$ and for all $\zeta \in \partial_P V(\alpha)$, we have $\|\zeta\| \leq K$. This is a proximal criterion guaranteeing that V is Lipschitz on a neighborhood of 0 (Problem 1.11.11), as stated. To conclude the proof, observe that any $\zeta_0 \in \partial_L V(0)$ is the limit of a sequence ζ_i as above, and that we have, along the sequence,

$$f'(x_i) + \zeta_i^* h'(x_i) = 0.$$

It suffices to pass to the limit to see that $\zeta_0 \in M(x_0)$. □

Remark. We should not expect an *exact* formula for $\partial_L V$ in terms of multipliers, since there is in general a real distinction between actual minimization (as measured by V) and stationarity conditions (provided by multipliers). In the "fully convex" case of the problem the distinction vanishes, and the formula becomes exact; see the end-of-chapter problems.

Given a *particular* solution x_0 to $P(0)$, it is not clear at first glance that Corollary 1.4 will imply the Multiplier Rule, namely that $M(x_0) \neq \emptyset$. Yet it does, as we now see.

1.5. Exercise.

(a) Let x_0 solve $P(0)$, where $h'(x_0)$ has maximal rank. Prove that $M(x_0) \neq \emptyset$. (*Hint.* Modify the problem by adding $\|x - x_0\|^2$ to f, so that x_0 becomes the *unique* solution to $P(0)$; How is $M(x_0)$ modified?)

(b) Show that the rank hypothesis cannot be deleted by considering $n = m = 1$, $f(x) = x$, $h(x) = x^2$.

Another issue that comes to mind is that of relaxing the smoothness hypothesis on f and h. An analysis of the argument used to prove Corollary 1.4 shows that an extension to nonsmooth data would hinge upon being able to pass to the limit in the relation

$$0 \in \partial_P\{f(\cdot) + \langle \zeta_i, h(\cdot)\rangle\}(x_i).$$

Here is the required fact.

1.6. Exercise. Let $\theta \colon \mathbb{R}^n \to \mathbb{R}^k$ be locally Lipschitz, and let $\{\lambda_i\}$ in $\mathbb{R}^k$ and $\{x_i\}$ in $\mathbb{R}^n$ be sequences converging to λ and x, respectively. Suppose that $0 \in \partial_L\{\langle \lambda_i, \theta\rangle(\cdot)\}(x_i)$ for each i. Prove that $0 \in \partial_L\{\langle \lambda, \theta\rangle(\cdot)\}(x)$. (*Hint.* Proposition 1.10.1.)

We now redefine the multiplier set $M(x)$ for nonsmooth data as the set of those $\zeta \in \mathbb{R}^m$ for which $0 \in \partial_L\{f(\cdot) + \langle \zeta, h(\cdot)\rangle\}(x)$. We say that x is *normal* provided that $0 \in \partial_L\langle \zeta, h(\cdot)\rangle(x)$ implies $\zeta = 0$ (note that this is equivalent to the rank of $h'(x)$ being maximal if h is C^1). Otherwise we call x *abnormal*. Then, upon noting that Exercise 1.2 requires only continuity of f and h, and calling upon Exercise 1.6 at a certain point, our sequential argument yields the following analogue of Corollary 1.4, whose detailed proof is left as an exercise.

1.7. Theorem. *Let f and h be locally Lipschitz, and let* Growth Hypothesis 1.1 *hold. Suppose that $V(0) < \infty$, and that every $x \in \Sigma(0)$ is normal. Then $V(\cdot)$ is Lipschitz near 0, and*

$$\emptyset \neq \partial_L V(0) \subset \bigcup_{x \in \Sigma(0)} M(x).$$

This can be used just as in the smooth case (i.e., in Exercise 1.5(a)) to deduce the multiplier rule for a given solution x_0 in the case of locally Lipschitz data. The result is often stated in an equivalent form without the normality hypothesis, but with a conclusion that holds in one of two alternative forms, abnormal or normal, as in the following:

1.8. Exercise. Let x_0 solve $P(0)$, where f and h are locally Lipschitz and Growth Hypothesis 1.1 holds. Then there exists $(\lambda_0, \zeta) \in \mathbb{R} \times \mathbb{R}^m$ with $(\lambda_0, \zeta) \neq (0,0)$ and $\lambda_0 = 0$ or 1, such that

$$0 \in \partial_L\{\lambda_0 f(\cdot) + \langle \zeta, h(\cdot)\rangle\}(x_0).$$

A result such as Theorem 1.7 incorporates *solvability* information: Since $V(\cdot)$ is Lipschitz near 0, therefore $V(\alpha)$ is finite for α near 0, and so the feasible set $\Phi(\alpha)$ is nonempty for α near 0. That is to say, for α near 0 the equation $h(x) = -\alpha$ admits at least one solution. The next theorem improves on this somewhat. Notice that optimization plays no role in its formulation; however, the proof is in essence an appeal to Theorem 1.7.

1.9. Theorem. *Let f and h be locally Lipschitz. Suppose that x_0 is a normal point satisfying $h(x_0) = 0$. Then for some K, for all α sufficiently near 0, the equation $h(x) + \alpha = 0$ admits a solution x_α satisfying $\|x_\alpha - x_0\| \leq K\|\alpha\|$.*

Proof. Consider the problem $P(\alpha)$ of minimizing $\|x - x_0\|$ subject to $h(x) + \alpha = 0$. Clearly x_0 is the unique solution of $P(0)$, and Growth Hypothesis 1.1 is satisfied. Apply Theorem 1.7, and let K be a Lipschitz constant for $V(\cdot)$ in a neighborhood of 0, say $B(0; \varepsilon)$. Then, if $\|\alpha\| < \varepsilon$ and if x_α is a solution to $P(\alpha)$, we have $h(x_\alpha) + \alpha = 0$ and

$$\|x_\alpha - x_0\| = V(\alpha) \leq V(0) + K\|\alpha\| = K\|\alpha\|$$

as required. □

Some remarks on this result are in order. First, note than when h is smooth, it reduces to a classical corollary of the Implicit Function Theorem: If the Jacobian matrix $h'(x_0)$ is of maximal rank, then $h(x) + \alpha = 0$ is locally solvable. We will address the issue of constructing implicit and inverse functions in §3, in a more general context, as well as how to quantify precisely the parameters of such a result. Next, observe that even when h is smooth, the proof of the theorem involves the nonsmooth cost function $f(x) = \|x - x_0\|$. A smooth choice such as $\|x - x_0\|^2$ would give only a weaker (Hölder) bound $\|x_\alpha - x_0\| \leq K\|\alpha\|^{1/2}$. Finally, we remark that §3 will give a criterion in terms of the *generalized Jacobian* of h by which the normality assumption can be verified more directly.

Inequality Constraints

Besides equality constraints, many optimization problems feature inequality constraints. Let us consider now the problem of minimizing $f(x)$ subject to $g(x) \leq 0$, where $g: \mathbb{R}^n \to \mathbb{R}$ is given and is locally Lipschitz, along with f. Much as before, we define $P(\beta)$ to be the perturbed problem having constraint $g(x) + \beta \leq 0$, and associate to $P(\beta)$ the feasible set $\Phi(\beta)$, the solution set $\Sigma(\beta)$, and the value function $V(\beta)$. The appropriate Growth Hypothesis is that the set $\{x: f(x) \leq r, g(x) \leq s\}$ be bounded for each r and $s \in \mathbb{R}$.

The first real difference with the equality case arises when we analyze the proximal subgradient inequality for V. Let $\gamma \in \partial_P V(0)$, for example, and

let $x_0 \in \Sigma(0)$. Then, as before,
$$V(\beta) - V(0) + \sigma|\beta|^2 \geq \langle \gamma, \beta \rangle = \gamma\beta$$
for all β near 0. Note that $V(\beta) \leq V(0)$ for $\beta \leq 0$ (the function V is nondecreasing by its very nature), so that $\langle \gamma, \beta \rangle \leq \sigma|\beta|^2$ for all small $\beta \leq 0$. This implies that $\gamma \geq 0$. Suppose now that $g(x_0)$ is strictly less than 0. It follows that $V(\beta) = V(0)$ for any $\beta \in [g(x_0), 0]$ (why?), and the proximal subgradient inequality yields $\langle \gamma, \beta \rangle \leq \sigma|\beta|^2$ for all β near 0, so that $\gamma = 0$.

To summarize thus far, we have proved that at least one of γ or $g(x_0)$ must be zero, a conclusion which can also be stated in the form $\gamma g(x_0) = 0$.

If $g(x_0) < 0$, then x_0 is a local minimum for f, and $0 \in \partial_L f(x_0)$. Otherwise, we proceed to substitute $\beta = -g(x)$ into the proximal subgradient inequality, for x near x_0. We deduce
$$f(x) + \sigma|g(x)|^2 + \gamma g(x) \geq f(x_0),$$
with equality when $x = x_0$. This implies $0 \in \partial_L\{f(\cdot) + \gamma g(\cdot)\}(x_0)$, a condition that actually holds in either of the two cases, since in the first case we have $\gamma = 0$.

These observations lead us to define the new multiplier set $M(x)$ for the present context as those scalars $\gamma \geq 0$ for which $\gamma g(x) = 0$ and $0 \in \partial_L\{f(\cdot) + \gamma g(\cdot)\}(x)$. A point x satisfying $g(x) \leq 0$ is now called *normal* if
$$\gamma \geq 0, \quad \gamma g(x) = 0, \quad 0 \in \partial_L\{\gamma g(\cdot)\}(x) \implies \gamma = 0.$$

With these modified definitions, the following result is visually identical to Theorem 1.7.

1.10. Exercise. With f, g as above, suppose that $V(0) < \infty$, and that every $x \in \Sigma(0)$ is normal. Then $V(\cdot)$ is Lipschitz near 0, and
$$\emptyset \neq \partial_L V(0) \subset \bigcup_{x \in \Sigma(0)} M(x).$$

Here are the counterparts of the necessary conditions (Exercise 1.8) and the solvability consequences (Theorem 1.9) for the inequality case being considered.

1.11. Exercise. Let x_0 solve $P(0)$, where f and g are locally Lipschitz and the (modified) Growth Hypothesis holds. Then there exist (λ_0, γ) not both 0, with $\lambda_0 = 0$ or 1 and $\gamma \geq 0$, such that
$$\gamma g(x_0) = 0 \quad \text{and} \quad 0 \in \partial_L\{\lambda_0 f(\cdot) + \gamma g(\cdot)\}(x_0).$$

1.12. Exercise. Let f and g be locally Lipschitz. Suppose that x_0 is a normal point satisfying $g(x_0) \leq 0$. Then for some K, for all β sufficiently near 0, the inequality $g(x) + \beta \leq 0$ admits a solution x_β satisfying $\|x_\beta - x_0\| \leq K \max\{0, \beta\}$.

1.13. Exercise. Show that for $n = 1$, $x_0 = 0$, the function $h(x) := -|x|$ is *not* normal in the sense of the word that prevails in Theorem 1.9, but that the same function $g(x) := -|x|$ *is* normal as the term is employed in Exercise 1.12. Note that the conclusion of Exercise 1.12 holds while that of Theorem 1.9 fails.

The *general case* of mixed equality and inequality constraints is that of minimizing $f(x)$ over the points $x \in \mathbb{R}^n$ satisfying $h(x) = 0$, $g(x) \leq 0$, where $h: \mathbb{R}^n \to \mathbb{R}^m$ and $g: \mathbb{R}^n \to \mathbb{R}^p$ are given functions and where a vector inequality such as $g(x) \leq 0$ is understood component-wise. We will assume that all these functions are locally Lipschitz, and that the set

$$\{x \in \mathbb{R}^n : f(x) \leq r, g(x) \leq s, \|h(x)\| \leq t\}$$

is bounded for each r, t in $\mathbb{R}$, and s in $\mathbb{R}^p$. Let $P(\alpha, \beta)$ denote the problem of minimizing $f(x)$ subject to $h(x) + \alpha = 0$, $g(x) + \beta \leq 0$, and let $V(\alpha, \beta)$ be the corresponding value function.

The *Lagrangian* of the problem is the function $\mathcal{L}: \mathbb{R}^n \times \mathbb{R}^p \times \mathbb{R}^m \to \mathbb{R}$ defined by

$$\mathcal{L}(x, \gamma, \zeta) := f(x) + \langle \gamma, g(x) \rangle + \langle \zeta, h(x) \rangle.$$

For any point x feasible for $P(0,0)$, the set $M(x)$ of *multipliers* for x consists of those $(\gamma, \zeta) \in \mathbb{R}^p \times \mathbb{R}^m$ satisfying

$$\gamma \geq 0, \quad \langle \gamma, g(x) \rangle = 0, \quad 0 \in \partial_L \mathcal{L}(\cdot, \gamma, \zeta)(x).$$

The point x is now called *normal* if

$$\gamma \geq 0, \quad \langle \gamma, g(x) \rangle = 0,$$
$$0 \in \partial_L \{\langle \gamma, g(\cdot) \rangle + \langle \zeta, h(\cdot) \rangle\}(x) \implies \gamma = 0, \ \zeta = 0,$$

and otherwise is called *abnormal*. We suppose that $V(0,0) < \infty$. In light of the preceding, the genesis of the following theorem should be clear. Its detailed proof is a substantial exercise.

1.14. Theorem.

(a) *If every $x \in \Sigma(0,0)$ is normal, then V is Lipschitz near $(0,0)$ and we have*

$$\emptyset \neq \partial_L V(0,0) \subset \bigcup_{x \in \Sigma(0,0)} M(x).$$

(b) *If x is any solution to $P(0,0)$, then either x is abnormal or else $M(x) \neq \emptyset$.*

1.15. Corollary. *Let x_0 be a point satisfying $h(x_0) = 0$, $g(x_0) \leq 0$, and suppose that x_0 is normal. Then for some $K > 0$, for all α and β near 0, there exists a point $x_{\alpha,\beta}$ such that*

$$h(x_{\alpha,\beta}) = 0, \quad g(x_{\alpha,\beta}) \leq 0, \quad \|x_{\alpha,\beta} - x_0\| \leq K\left(\|\alpha\| + \sum_{i=1}^{p} \max\{0, \beta_i\}\right).$$

We remark that the value function approach illustrated above has proven to be a robust and effective line of attack for many other more complex optimization problems than the finite-dimensional ones considered here, including a variety of problems in control theory.

2 The Mean Value Inequality

In this section we will prove a "multidirectional" Mean Value Theorem, a result that is novel even in the context of smooth multivariable calculus, and one that is a cornerstone of nonsmooth analysis. It is instructive to begin by recalling the statement of the classical Mean Value Theorem, which plays a fundamental role in analysis by relating the values of a function to its derivative at an intermediate point. For the purpose of comparisons to come, it is also worth noting the optimization-based proof of this familiar result. Recall that given two points x and $y \in X$, the open line segment connecting them is denoted (x, y):

$$(x, y) := \{tx + (1-t)y : 0 < t < 1\}.$$

If $t = 0$ and $t = 1$ are allowed, we obtain the closed line segment $[x, y]$. Throughout this section, X is a Hilbert space.

2.1. Proposition. *Suppose $x, y \in X$ and that $f \in \mathcal{F}$ is Gâteaux differentiable on a neighborhood of $[x, y]$. Then there exists $z \in (x, y)$ such that*

$$f(y) - f(x) = \langle f'_G(z), y - x \rangle. \tag{1}$$

Proof. Define $g \colon [0, 1] \to \mathbb{R}$ by

$$g(t) := f(ty + (1-t)x) - tf(y) - (1-t)f(x).$$

Then g is continuous on $[0, 1]$, differentiable on $(0, 1)$, and satisfies $g(0) = g(1) = 0$. It follows that there exists a point $\bar{t} \in (0, 1)$ that is either a maximum or a minimum of g on $[0, 1]$, and thus satisfies $g'(\bar{t}) = 0$. We calculate $g'(\bar{t})$ by referring directly to the definition of the Gâteaux derivative at $z := \bar{t}y + (1 - \bar{t})x = x + \bar{t}(y - x)$, and conclude that

$$0 = g'(\bar{t}) = \langle f'_G(z), (y - x) \rangle - (f(y) - f(x)).$$

Hence (1) holds. □

A typical application of the Mean Value Theorem is to deduce monotonicity. For example: suppose that $\langle f'_G(z), y - x \rangle \leq 0$ for all $z \in (x, y)$. Then it follows immediately that $f(y) \leq f(x)$. In this, and in fact in most applications, it suffices to invoke the Mean Value Theorem in its *inequality* form; i.e., the form in which (1) is replaced by

$$f(y) - f(x) \leq \langle f'_G(z), y - x \rangle.$$

When f is nondifferentiable, the inequality form provides the appropriate model for generalization. To illustrate, consider the function $f(x) := -|x|$ on $\mathbb{R}$. There exists $z \in (-1, 1)$ and $\zeta \in \partial_P f(z)$ such that

$$0 = f(1) - f(-1) \leq \langle \zeta, 1 - (-1) \rangle = 2\zeta,$$

but equality (i.e., $\zeta = 0$) never holds. In a nonsmooth setting, it turns out as well that some tolerance must be present in the resulting inequality. For example, if $f(x_1, x_2) := -|x_2|$, then $f(-1, 0) = f(1, 0)$, but the difference (0) between these two values of f cannot be estimated by $\partial_P f(z)$ for any z on the line segment between $(-1, 0)$ and $(1, 0)$, since $\partial_P f(z)$ is empty for all such z. We will need to allow z to stray a bit from the line segment.

Besides nondifferentiability, the Mean Value Inequality that we will present possesses a novel, perhaps surprising feature: *multidirectionality*; i.e., an estimate that is maintained uniformly over many directions. This new aspect greatly enhances the range of applications of the result, even in a smooth setting. Let us describe it in a particular two-dimensional case, in which we seek to compare the value of a Gâteaux differentiable continuous function f at $(0, 0)$ to the set of its values on the line segment

$$Y := \{(1, t) : 0 \leq t \leq 1\}.$$

For each $t \in [0, 1]$, the usual Mean Value Theorem applied to the two points $(0, 0)$ and $y_t := (1, t)$ asserts the existence of a point z_t lying in the open segment connecting $(0, 0)$ and y_t such that

$$f(y_t) - f(0) \leq \langle f'_G(z_t), y_t \rangle.$$

We deduce

$$\min_{y \in Y} (f(y) - f(0)) \leq \langle f'_G(z_t), y_t \rangle. \tag{2}$$

The new Mean Value Inequality asserts that for some z in the triangle which is the convex hull of $\{0\} \cup Y$, we have

$$\min_{y \in Y} (f(y) - f(0)) \leq \langle f'_G(z), y_t \rangle \ \forall t \in [0, 1]. \tag{3}$$

While these conclusions appear quite similar, there is an important difference: (3) holds for the *same* given z, *uniformly* for all y_t, in contrast to (2), where z_t depended on y_t.

To further underline the difference, consider now the case in which Y is the closed unit ball $\overline{B}$ in X. For a continuous Gâteaux differentiable function f, the new Mean Value Inequality asserts that for some $z \in \overline{B}$, we have

$$\inf_{\overline{B}} f - f(0) \leq \langle f'_G(z), y \rangle \ \forall y \in \overline{B}.$$

Assume now that $\|f'_G\| \geq 1$ on $\overline{B}$. Then, taking $y = -f'_G(z)/\|f'_G(z)\|$ in the inequality leads to

$$\inf_{\overline{B}} f - f(0) \leq -1$$

(this is where we exploit the uniformity of the conclusion). Applied to $-f$, the same reasoning yields

$$\sup_{\overline{B}} f - f(0) \geq 1,$$

whence

$$\sup_{\overline{B}} f - \inf_{\overline{B}} f \geq 2.$$

The reader is invited to ponder how the usual Mean Value Theorem is inadequate to yield this conclusion. Further, let us note that the failure of f' to exist even at a single point can make a difference and will need to be taken account of: Witness $f(x) := \|x\|$, for which we have

$$\sup_{\overline{B}} f - \inf_{\overline{B}} f = 1,$$

in contrast to the above, even though $\|f'(x)\| = 1$ for all $x \neq 0$.

A Smooth Finite-Dimensional Version

The general theorem that we will present features nonsmoothness, infinite dimensions, and multidirectionality. The structure of its proof can be conveyed much more easily in the absence of the first two factors. Accordingly, we will begin by proving a version of the theorem for a differentiable function on $\mathbb{R}^n$. A preliminary result we will require is the following.

2.2. Proposition. *Suppose $Y \subset X$ is closed and convex, that $f \in \mathcal{F}$ attains its minimum over Y at $\bar{y}$, and that f has directional derivatives in all directions at $\bar{y}$. Then*

$$f'(\bar{y}; y - \bar{y}) \geq 0 \ \forall y \in Y. \tag{4}$$

In particular, if f is Gâteaux differentiable at $\bar{y}$, then

$$\langle f'_G(\bar{y}); y - \bar{y} \rangle \geq 0 \ \forall y \in Y.$$

Proof. Let $y \in Y$, and define $g\colon [0,1] \to \mathbb{R}$ by $g(t) = f\bigl(\bar{y} + t(y-\bar{y})\bigr)$. Note that $\bar{y} + t(y-\bar{y}) = ty + (1-t)\bar{y}$ belongs to Y for each $t \in [0,1]$, since Y is convex. Since $\bar{y}$ minimizes f over Y, we have $g(0) \leq g(t)$ for all t. Thus by letting $t \downarrow 0$ and using the assumption that the directional derivative $f'(\bar{y}, y-\bar{y})$ exists, we conclude

$$0 \leq \lim_{t \downarrow 0} \frac{g(t) - g(0)}{t} = f'(\bar{y}; y - \bar{y}).$$

Therefore (4) holds, since $y \in Y$ is arbitrary. □

For $Y \subset X$ and $x \in X$, we denote by $[x, Y]$ the convex hull of $\{x\} \cup Y$:

$$[x, Y] := \{tx + (1-t)y \colon t \in [0,1], y \in Y\}. \tag{5}$$

In particular, if $Y = \{y\}$, then $[x, Y]$ is simply the closed line segment connecting x and y. Note that below we admit the possibility that $x \in Y$.

2.3. Theorem. *Suppose $Y \subset \mathbb{R}^n$ is closed, bounded, and convex. Let $x \in \mathbb{R}^n$, and let $f \in \mathcal{F}$ be Gâteaux differentiable on a neighborhood of $[x, Y]$. Then there exists $z \in [x, Y]$ such that*

$$\min_{y' \in Y} f(y') - f(x) \leq \langle f'(z), y - x \rangle \quad \forall y \in Y. \tag{6}$$

Proof. Set

$$r := \min_{y' \in Y} f(y') - f(x), \tag{7}$$

and let U be an open neighborhood of $[x, Y]$ on which f is Gâteaux differentiable. Define $g\colon [0,1] \times U \to \mathbb{R}$ by

$$g(t, y) = f\bigl(x + t(y-x)\bigr) - rt. \tag{8}$$

Since g is lower semicontinuous and $[x, Y]$ is compact, we have $r \neq -\infty$ and there exists $(\bar{t}, \bar{y}) \in [0,1] \times Y$ at which g attains its minimum over $[0,1] \times Y$. We will show that (6) holds with z given by

$$z := x + \bar{t}(\bar{y} - x) \in [x, Y], \tag{9}$$

unless $\bar{t} = 1$, in which case we set $z = x$. The multidirectional inequality (6) will be deduced from necessary conditions for the attainment of a minimum. Let us first assume that $\bar{t} \in (0, 1)$. Then the function

$$t \mapsto g(t, \bar{y}),$$

attains its minimum over $t \in [0, 1]$ at $t = \bar{t}$, and thus has a vanishing derivative there. We calculate this derivative using (8) and (9) to conclude

$$0 = \frac{\partial}{\partial t}\bigg|_{t = \bar{t}} g(t, \bar{y}) = \langle f'(z), \bar{y} - x \rangle - r. \tag{10}$$

On the other hand, the function
$$y \mapsto g(\bar{t}, y),$$
attains its minimum over $y \in Y$ at $\bar{y}$. The necessary conditions in Proposition 2.2 state that
$$\left\langle \left.\frac{\partial}{\partial y}\right|_{y=\bar{y}} g(\bar{t}, y), y - \bar{y} \right\rangle \geq 0 \ \forall y \in Y. \tag{11}$$

It is easily calculated using (8) and (9) again that
$$\left.\frac{\partial}{\partial y}\right|_{y=\bar{y}} g(\bar{t}, y) = \bar{t} f'(z).$$

Substituting this value into (11) and dividing by $\bar{t} > 0$ gives us that
$$\langle f'(z), y - \bar{y} \rangle \geq 0 \ \forall y \in Y. \tag{12}$$

We now combine (10) and (12) to conclude that
$$r \leq \langle f'(z), \bar{y} - x \rangle + \langle f'(z), y - \bar{y} \rangle = \langle f'(z), y - x \rangle,$$
which holds for all $y \in Y$. This statement is precisely (6) and finishes the proof of the theorem in this instance.

The case $\bar{t} = 0$ is handled more simply than above. For each $(t, y) \in (0, 1] \times Y$, we note from (8) and the definition of a minimum that
$$f(x) = g(0, \bar{y}) \leq g(t, y) = f(x + t(y - x)) - rt. \tag{13}$$

After dividing (13) by t and rearranging terms, we have for all $y \in Y$ that
$$r \leq \lim_{t \downarrow 0} \frac{1}{t} \big(f(x + t(y - x)) - f(x) \big) = \langle f'(z), y - x \rangle,$$
since $z = x$, and this is the same as (6).

Finally we consider the case in which $\bar{t} = 1$. We claim $g(1, \bar{y}) = f(x)$. Indeed, $g(1, \bar{y}) \leq g(0, \bar{y}) = f(x)$, while on the other hand,
$$g(1, \bar{y}) = f(\bar{y}) - r = f(\bar{y}) - \min_{y \in Y} f(y) + f(x) \geq f(x).$$

It now follows that $f(x)$ is the minimum value of g on $[0, 1] \times Y$. But $f(x) = g(0, y)$ for any $y \in Y$, and thus $(0, y)$ is also a minimizer of g. Thus if $\bar{t} = 1$, then we equally could have chosen the minimizer with $\bar{t} = 0$, and we finish by deferring to that case, treated above. □

2.4. Exercise.

(a) Prove that the point z in Theorem 2.3 also satisfies
$$f(z) \leq \min_{w \in [x,Y]} f(w) + |r|,$$
where r is given by (7).

(b) Consider the special case of Theorem 2.3 in which $n = 2$, $x = (0,0)$, and
$$Y = \{(1,t) \colon -1 \leq t \leq 1\}.$$
Suppose that $f(0,0) = 0$ and that $f(1,t) = 1 \ \forall t \in [-1,1]$. Prove that some point (u,v) in the triangle $[0,Y]$ satisfies
$$f_x(u,v) \geq 1 + |f_y(u,v)|.$$

(c) With x and Y as in part (b), and with $f(x,y) := x - y + xy$, find all the points z that satisfy the conclusion of Theorem 2.3.

The General Case of the Mean Value Inequality

We now wish to extend the theorem to the case in which Y is a closed, bounded, convex subset of the Hilbert space X, and f is merely lower semicontinuous.

The first technical difficulty in proving the theorem in this greater generality is the same as that previously encountered in the proof of the Density Theorem. Namely, a lower semicontinuous function that is bounded below may not attain its minimum over a closed bounded set. We may note that the proof of Theorem 2.3 above is valid in infinite dimensions without modification if in addition Y is assumed to be compact, or f is assumed to be weakly lower semicontinuous. This is because these hypotheses imply the existence of the minimizer $(\bar{t}, \bar{y})$. But the additional assumptions are not actually required to prove the theorem in infinite dimensions, as we will see. The idea is to follow the above proof but to obtain the minimizer via a minimization principle (Theorem 1.4.1). However, there is a price to pay for this device to work effectively: The statement of the result itself needs a slight modification, a certain slackening of the inequality in the conclusion. The result is not true without this modification, and it is essentially required in the proof to avoid the possibility that $\bar{t} = 1$. (We may check that if $\bar{t} = 1$, no relevant information is obtained from the necessary conditions.) In the above proof, the case $\bar{t} = 1$ was handled by deferring to the case $\bar{t} = 0$, but after adding the perturbation term, this is no longer possible.

The second technical difficulty arises in replacing derivatives by proximal subgradients. When f is nondifferentiable, Proposition 2.2 is no longer available to us, and in the minimization process we have recourse instead to a smooth penalty term to replace the constraint that z lie in $[x, Y]$.

2 The Mean Value Inequality

This will now allow the minimizer z to fail to lie in $[x, Y]$, although it can be brought arbitrarily near to that set (an example will show that this tolerance is necessary). Further, the quantity r defined by (7) must be replaced by one that is more stable under such approximations, namely

$$\hat{r} := \lim_{\delta \downarrow 0} \inf_{w \in Y + \delta B} \{f(w) - f(x)\}. \tag{14}$$

2.5. Exercise. If X is finite dimensional, show that the $\hat{r}$ defined by (14) reduces to the r defined by (7). (Examples can be adduced to show that this may fail in infinite dimensions.)

2.6. Theorem (Mean Value Inequality). *Let Y be a closed, convex, bounded subset of X, and let $x \in \operatorname{dom} f$, where $f \in \mathcal{F}$. Suppose that $\hat{r}$ defined above is not $-\infty$, and let any $\bar{r} < \hat{r}$ and $\varepsilon > 0$ be given. Then there exist $z \in [x, Y] + \varepsilon B$ and $\zeta \in \partial_P f(z)$ such that*

$$\bar{r} < \langle \zeta, y - x \rangle \quad \forall y \in Y.$$

Further, we can choose z to satisfy

$$f(z) < \inf_{[x,Y]} f + |\bar{r}| + \varepsilon.$$

We remark that the case $\hat{r} = \infty$ is not excluded; it can only arise if $Y \cap \operatorname{dom} f = \emptyset$.

Proof of the Mean Value Inequality

Proof. Without loss of generality, we assume that $x = 0$ and $f(x) = 0$. Suppose $\bar{r} < \hat{r}$ and $\varepsilon > 0$. We will first fix some constants that will be featured in the estimates to come. The hypotheses of the theorem justify the existence of such choices. We write $\|Y\|$ for the quantity $\sup\{\|y\| : y \in Y\}$.

Let $r \in \mathbb{R}$ and positive constants δ, M, k be chosen so that

$$\bar{r} < r < \hat{r}, \quad |r| < |\bar{r}| + \varepsilon, \tag{15}$$

$$\delta \leq \varepsilon \text{ and } y \in Y + \delta B \implies f(y) \geq r + \delta, \tag{16}$$

$$z' \in [0, Y] + \delta \overline{B} \implies f(z') \geq -M, \tag{17}$$

$$k > \frac{1}{\delta^2}\{M + |r| + 1 + 2\|Y\| + \delta\}. \tag{18}$$

Now we define $g \colon [0, 1] \times X \times X \longrightarrow (-\infty, \infty]$ by

$$g(t, y, z) := f(z) + k\|ty - z\|^2 - tr.$$

To simplify the notation somewhat, let us set

$$S := Y \times \{[0, Y] + \delta \overline{B}\},$$

118   3. Special Topics

and write $s = (y, z)$ for elements in S. We can easily verify that

$$t \mapsto \inf_{s \in S} g(t, s), \tag{19}$$

is continuous on $[0, 1]$, and thus attains its infimum at some point $\bar{t} \in [0, 1]$. The choice of a large k value forces $\bar{t}$ away from 1, which is the assertion of the first claim.

Claim 1. $\bar{t} < 1$.

Proof. Let $s = (y, z) \in S$, and note that if $\|y - z\| \leq \delta$, then $f(z) \geq r + \delta$ by (16). A lower bound for $g(1, s)$ can be obtained as follows:

$$g(1, s) = f(z) + k\|y - z\|^2 - r$$
$$\geq \begin{cases} f(z) + k\|y - z\|^2 - r & \text{if } \|y - z\| \leq \delta, \\ f(z) + k\delta^2 - r & \text{if } \|y - z\| > \delta, \end{cases}$$
$$\geq \min\{r + \delta - r, -M + (M + |r| + 1) - r\} > 0.$$

On the other hand, we have an upper bound of the function in (19) by setting $t = 0$ and noting that

$$\inf_{s \in S} g(0, s) \leq g(0, y, 0) = f(0) + k\|0\|^2 = 0. \tag{20}$$

Thus the infimum of $t \mapsto \inf_{s \in S} g(t, s)$ taken over $t \in [0, 1]$ cannot occur at $t = 1$, and we have $\bar{t} < 1$.

We now address the difficulty that the minimum of $s \mapsto g(\bar{t}, s)$ taken over $s \in S$ may not be attained at any point in S. We resort to a minimization principle to obtain a minimum of $g(\bar{t}, \cdot)$ perturbed by a linear function.

Note that $g(\bar{t}, \cdot) \in \mathcal{F}$ and is bounded below on the closed bounded set S. By Theorem 1.4.1, for each small $\eta > 0$, there exists $\xi := (\xi_y, \xi_z) \in X \times X$ and $\bar{s} := (\bar{y}, \bar{z}) \in S$ so that $\|\xi\| < \eta$ and the function

$$s \mapsto g(\bar{t}, s) + \langle \xi, s \rangle, \tag{21}$$

attains its minimum over $s \in S$ at $\bar{s}$. Of course ξ and $\bar{s}$ depend on η, as does the choice of ζ appearing in the conclusions of the theorem, the latter being defined as

$$\zeta := 2k(\bar{t}\bar{y} - \bar{z}) - \xi_z.$$

We will see that the conclusions of the theorem hold for these choices of ζ and $\bar{z}$, provided that η is chosen sufficiently small.

Claim 2. *If η is small enough, we have $\bar{z} \in [0, Y] + \delta B$.*

2 The Mean Value Inequality 119

Proof. Suppose on the contrary that $d_{[0,Y]}(\bar{z}) \geq \delta$. Observe first that

$$|\langle \xi, s \rangle| \leq \eta(2\|Y\| + \delta) \ \forall s \in S. \tag{22}$$

Then since $\bar{t}\bar{y} \in [0, Y]$, we have by (22) and (17) that

$$g(\bar{t}, \bar{s}) + \langle \xi, \bar{s} \rangle = f(\bar{z}) + k\|\bar{t}\bar{y} - \bar{z}\|^2 - r\bar{t} + \langle \xi, \bar{s} \rangle$$
$$\geq -M + k\delta^2 - |r| - \eta(2\|Y\| + \delta),$$

which by the choice of k in (18) can be made greater than 1 if η is sufficiently small. On the other hand, recall that $\bar{s}$ achieves the infimum in the function (21). Therefore if $\eta \leq (2\|Y\| + \delta)^{-1}/2$, we have by (20) and (22)

$$g(\bar{t}, \bar{s}) + \langle \xi, \bar{s} \rangle \leq g(0, \bar{y}, 0) + \langle \xi, \bar{s} \rangle$$
$$\leq \eta(2\|\bar{y}\| + \delta)$$
$$\leq 1/2.$$

Thus we conclude $\bar{z} \in [0, Y] + \delta B$ as claimed.

Claim 3. $\zeta \in \partial_P f(\bar{z})$.

Proof. Since $z \mapsto f(z) + k\|\bar{t}\bar{y} - z\|^2 + \langle \xi_z, z \rangle$ has a minimum over $[0, Y] + \delta \bar{B}$ at $z = \bar{z}$, and $\bar{z}$ is a point in the interior of this set (Claim 2), we have that

$$\zeta = -\left(k\|\bar{t}\bar{y} - (\cdot)\|^2 + \langle \xi_z, (\cdot) \rangle\right)'(\bar{z}) \in \partial_P f(\bar{z}).$$

We next turn to proving the upper bound on $f(\bar{z})$.

Claim 4. *The estimate $f(\bar{z}) \leq \inf_{z \in [0,Y]} f(z) + |r| + \varepsilon$ holds, provided η is small enough.*

Proof. Using (22), we calculate

$$f(\bar{z}) - \bar{t}r \leq f(\bar{z}) + k\|\bar{t}\bar{y} - \bar{z}\|^2 - \bar{t}r + \langle \xi, (\bar{y}, \bar{z}) \rangle + \eta(2\|Y\| + \delta)$$
$$= \inf_{s \in S}\{g(\bar{t}, s) + \langle \xi, s \rangle\} + \eta(2\|Y\| + \delta)$$
$$\leq \inf_{(y,z) \in S}\{f(z) + k\|ty - z\|^2 - tr + \langle \xi, (y, z) \rangle\} + \eta(2\|Y\| + \delta)$$
$$\leq \inf_{z \in [0,Y]} f(z) + \max\{0, -r\} + 2\eta(2\|Y\| + \delta).$$

The final inequality is justified since the last infimum is only over $z \in [0, Y]$, and included in the next to last infimum is the case when $ty = z \in [0, Y]$, making it therefore an infimum over a larger set. Hence

$$f(\bar{z}) \leq \inf_{z \in [0,Y]} f(z) + \max\{\bar{t}, 1 - \bar{t}\}|r| + 2\eta(2\|Y\| + \delta)$$
$$\leq \inf_{z \in [0,Y]} f(z) + |\bar{r}| + \varepsilon,$$

provided η is sufficiently small, where the second relation in (15) has been used.

120   3. Special Topics

Claim 5. *Suppose $\bar{t} = 0$, and $0 < \eta < \left[(r - \bar{r})/6(2\|Y\| + \delta)\right]^2$. Then $\bar{r} < \langle \zeta, y \rangle$ for all $y \in Y$.*

Proof. In the present case with $\bar{t} = 0$, the definition of ζ reduces to $\zeta = -2k\bar{z} - \xi_z$. Let us note (using (22)) that

$$\inf_{s \in S} g(0, s) \geq \inf_{s \in S}\{g(0, s) + \langle \xi, s \rangle\} - \eta(2\|Y\| + \delta)$$
$$= f(\bar{z}) + k\|\bar{z}\|^2 + \langle \xi, (\bar{y}, \bar{z}) \rangle - \eta(2\|Y\| + \delta) \quad (23)$$
$$\geq f(\bar{z}) + k\|\bar{z}\|^2 - 2\eta(2\|Y\| + \delta).$$

Now we let $t := \min\{\sqrt{\eta}, (r - \bar{r})/3k(\|Y\|^2 + \delta)\} > 0$. We have for any $y \in Y$ that

$$\inf_{s \in S} g(t, s) \leq f(\bar{z}) + k\|ty - \bar{z}\|^2 - tr. \quad (24)$$

Since $\bar{t} = 0$ by definition provides a minimum of (19), it follows from (23) and (24) that for all $y \in Y$

$$0 \leq \inf_{s \in S} g(t, s) - \inf_{s \in S} g(0, s)$$
$$\leq k\|ty - \bar{z}\|^2 - tr - k\|\bar{z}\|^2 + 2\eta(2\|Y\| + \delta)$$
$$= t[\langle -2k\bar{z}, y \rangle - r] + t^2 k\|y\|^2 + 2\eta(2\|Y\| + \delta)$$
$$= t[\langle \zeta, y \rangle - r] + t^2 k\|y\|^2 + t\langle \xi_z, y \rangle + 2\eta(2\|Y\| + \delta)$$
$$\leq t[\langle \zeta, y \rangle - r] + t^2 k\|Y\|^2 + t\eta\|Y\| + 2\eta(2\|Y\| + \delta).$$

Dividing through by t and using our choices of t and η, we obtain

$$\langle \zeta, y \rangle \geq r - tk\|Y\|^2 - \eta\|Y\| - \frac{2\eta}{t}(2\|Y\| + \delta)$$
$$> \bar{r},$$

which concludes the proof of this claim.

To handle the situation where $\bar{t} > 0$, we first prove an estimate regarding $\bar{y}$.

Claim 6. *If η is sufficiently small, then $\langle \zeta, \bar{y} \rangle > (r + \bar{r})/2$.*

Proof. In a manner similar to the beginning of the proof of Claim 5, we have

$$\inf_{s \in S} g(\bar{t}, s) \geq f(\bar{z}) + k\|\bar{t}\bar{y} - \bar{z}\|^2 - 2\eta(2\|Y\| + \delta), \quad (25)$$

and

$$\inf_{s \in S} g(1, s) \leq f(\bar{z}) + k\|(1 - \bar{t})\bar{y} + (\bar{t}\bar{y} - \bar{z})\|^2 - r. \quad (26)$$

Recall that $\bar{t}$ provides an infimum of the function in (21). By subtracting (25) from (26), we obtain

$$0 \leq \inf_{s \in S} g(1,s) - \inf_{s \in S} g(\bar{t},s)$$
$$\leq k\|(1-\bar{t})\bar{y} + (\bar{t}\bar{y} - \bar{z})\|^2 - r - k\|\bar{t}\bar{y} - \bar{z}\|^2 + 2\eta(2\|Y\| + \delta)$$
$$= (1-\bar{t})[\langle 2k(\bar{t}\bar{y} - \bar{z}), \bar{y}\rangle - r] + 2\eta(2\|Y\| + \delta).$$

This in turn implies

$$\langle 2k(\bar{t}\bar{y} - \bar{z}), \bar{y}\rangle \geq r - \frac{2\eta}{1-\bar{t}}(2\|Y\| + \delta).$$

Consequently we have

$$\langle \zeta, \bar{y}\rangle = \langle 2k(\bar{t}\bar{y} - \bar{z}) - \xi_z, \bar{y}\rangle$$
$$\geq \langle 2k(\bar{t}\bar{y} - \bar{z}), \bar{y}\rangle - \eta\|Y\|$$
$$\geq r - \eta\left[\frac{2}{1-\bar{t}}(2\|Y\| + \delta) + \|Y\|\right].$$

Hence $\langle \zeta, \bar{y}\rangle > (r + \bar{r})/2$ if η is chosen small enough.

The proof of the theorem will ensue from the next claim.

Claim 7. *If in addition to the requirements on η in all of the earlier claims, we also have $\eta < \bar{t}(r - \bar{r})/(8\|Y\|)$, then*

$$\bar{r} < \langle \zeta, y\rangle \ \forall y \in Y.$$

Proof. We have that the function

$$y \mapsto k\|\bar{t}y - \bar{z}\|^2 + \langle \xi_y, y\rangle,$$

achieves its infimum over $y \in Y$ at $y = \bar{y}$. Since Y is convex, we have by Proposition 2.2 that for all $y \in Y$,

$$\left\langle \frac{\zeta_y}{\bar{t}}, y - \bar{y}\right\rangle \geq -2\frac{\eta}{\bar{t}}\|Y\|.$$

Therefore

$$\langle \zeta, y - \bar{y}\rangle = \langle 2k(\bar{t}\bar{y} - \bar{z}) - \xi_z, y - \bar{y}\rangle$$
$$\geq -2\frac{\eta}{\bar{t}}\|Y\| - 2\eta\|Y\|$$
$$\geq -4\frac{\eta}{\bar{t}}\|Y\| \geq \frac{\bar{r} - r}{2}.$$

Finally, from this and Claim 6 we conclude that for any $y \in Y$ that

$$\begin{aligned}\langle \zeta, y \rangle &\geq \langle \zeta, \bar{y} \rangle - \frac{r - \bar{r}}{2} \\ &> \frac{r + \bar{r}}{2} - \frac{r - \bar{r}}{2} \\ &= \bar{r},\end{aligned}$$

which completes the proof of the theorem. □

2.7. Exercise.

(a) Show that if $x \in Y$, then the point z in Theorem 2.6 satisfies $f(z) < f(x) + \varepsilon$.

(b) Deduce from Theorem 2.6 the (unidirectional) Proximal Mean Value Theorem: *If $f \in \mathcal{F}$ and $x, y \in \mathrm{dom}\, f$ are given, then for any $\varepsilon > 0$ there exists $z \in [x, y] + \varepsilon B$ such that, for some $\zeta \in \partial_P f(z)$ we have*

$$f(y) - f(x) < \langle \zeta, y - x \rangle + \varepsilon.$$

(c) If now the point y in part (b) is not in $\mathrm{dom}\, f$, deduce that there are points z arbitrarily near $[x, y]$ admitting proximal subgradients $\zeta \in \partial_P f(z)$ for which $\langle \zeta, y - x \rangle$ (and hence $\|\zeta\|$) is arbitrarily large.

(d) Let f be Lipschitz on a neighborhood of $[x, y]$. Prove the existence of $z \in [x, y]$ and $\zeta \in \partial_L f(z)$ such that

$$f(y) - f(x) \leq \langle \zeta, y - x \rangle.$$

The Decrease Principle

As the first of several applications of the Mean Value Inequality we derive a result which plays an important role in the next section. To motivate it, consider a differentiable function f whose derivative at x_0 is nonvanishing: $f'(x_0) \neq 0$. Since x_0 cannot therefore be a local minimum of f, we conclude that for any open ball $x_0 + \rho B$ around x_0, we must have

$$\inf_{x \in x_0 + \rho B} f(x) < f(x_0).$$

This fact can be extended locally and quantified, as follows:

2.8. Theorem. *Let $f \in \mathcal{F}$. Suppose that for some $\delta > 0$ and $\rho > 0$ we have the following condition:*

$$z \in x_0 + \rho B, \quad \zeta \in \partial_P f(z) \implies \|\zeta\| \geq \delta.$$

Then

$$\inf_{x \in x_0 + \rho B} f(x) \leq f(x_0) - \rho \delta.$$

Proof. We may suppose that $x_0 \in \text{dom } f$. We will show that for any $s \in (0, \rho)$, for all $\varepsilon > 0$ arbitrarily small, we have

$$\inf_{x_0 + \rho B} f \leq f(x_0) - s\delta + \varepsilon, \tag{27}$$

a fact which implies the estimate of the theorem. Clearly there is no loss of generality in assuming that the left side of (27) is finite, since otherwise the conclusion of the theorem is immediate. Given $s \in (0, \rho)$, set $Y := x_0 + s\overline{B}$, and observe that the corresponding quantity $\hat{r}$ defined via (14) satisfies (for $x := x_0$)

$$\infty > \hat{r} \geq \inf_{x_0 + \rho B} f - f(x_0).$$

Choose $\varepsilon \in (0, \rho - s)$, and set $\bar{r} := \hat{r} - \varepsilon$. Applying Theorem 2.6 gives

$$\inf_{x_0 + \rho B} f - f(x_0) \leq \inf_{x_0 + s\overline{B}} f - f(x_0)$$

$$\leq \hat{r} = \bar{r} + \varepsilon$$

$$< \langle \zeta, su \rangle + \varepsilon \ \forall u \in \overline{B},$$

where $\zeta \in \partial_P f(z)$ and $z \in Y + \varepsilon B = x_0 + s\overline{B} + \varepsilon B \subset x_0 + \rho B$. By hypothesis, we have $\|\zeta\| \geq \delta > 0$. Choosing $u = -\zeta/\|\zeta\|$ in the preceding inequality therefore gives precisely (27). □

Theorem 2.8 implies a hybrid form of Theorems 1.4.1 and 1.4.2, the two minimization principles proved in Chapter 1.

2.9. Corollary. *Let $f \in \mathcal{F}$ be bounded below, and let $x \in X$, $\varepsilon > 0$ be such that*

$$f(x) < \inf_X f + \varepsilon.$$

Then for any $\lambda > 0$ there exists $z \in x + \lambda B$ with $f(z) < \inf_X f + \varepsilon$ and $\zeta \in \partial_P f(z)$ such that $\|\zeta\| < \varepsilon/\lambda$.

2.10. Exercise. Prove the corollary. Deduce also that for some $\sigma \geq 0$, the function

$$x' \mapsto f(x') - \langle \zeta, x' \rangle + \sigma \|x' - z\|^2$$

attains a local minimum at $x' = z$. Compare this conclusion to those of Theorems 1.4.1 and 1.4.2. Note also that if f is differentiable then we get $\|f'(z)\| \leq \sqrt{\varepsilon}$ by taking $\lambda = \sqrt{\varepsilon}$, which improves upon Exercise 1.4.3.

Strong Versus Weak Monotonicity

Let C be a nonempty compact convex subset of X. A function f is said to be *strongly decreasing* relative to C if

$$\forall t > 0, \ \forall y \in x + tC, \ \text{we have } f(y) \leq f(x).$$

We say that f is *weakly decreasing* relative to C if

$$\forall t > 0 \quad \exists y \in x + tC \text{ such that } f(y) \leq f(x).$$

The lower and upper *support functions* of C, $h_C(\cdot)$, and $H_C(\cdot)$, respectively, are defined as follows:

$$h_C(\zeta) := \min\{\langle \zeta, c \rangle : c \in C\}, \quad H_C(\zeta) := \max\{\langle \zeta, c \rangle : c \in C\}.$$

The following characterizations of weak and strong decrease foreshadow certain corresponding properties of control trajectories that figure prominently in Chapter 4.

2.11. Theorem. *Let $f \in \mathcal{F}$. Then f is strongly decreasing relative to C iff*

$$H_C(\zeta) \leq 0 \ \forall \zeta \in \partial_P f(x), \ \forall x \in X;$$

f is weakly decreasing relative to C iff

$$h_C(\zeta) \leq 0 \ \forall \zeta \in \partial_P f(x), \ \forall x \in X.$$

Proof. The condition involving H_C can be expressed more compactly in the form

$$H_C\big(\partial_P f(x)\big) \leq 0 \ \forall x.$$

Suppose that this holds, and let $y \in x + tC$ for $t > 0$. We wish to prove that $f(y) \leq f(x)$; there is no loss of generality in assuming $x \in \text{dom } f$. If $y \in \text{dom } f$ too, then by the Proximal Mean Value Theorem (Exercise 2.7(b)), for any $\varepsilon > 0$ we can estimate $f(y) - f(x)$ from above by a term of the form $\langle \zeta, y - x \rangle + \varepsilon$, $\zeta \in \partial_P f(z)$. But $y - x \in tC$, so that $\langle \zeta, y - x \rangle \leq H_C(\zeta) \leq 0$ by assumption. It follows that $f(y) \leq f(x)$. There remains the possibility that $f(y) = +\infty$. But then $\langle \zeta, y - x \rangle$ can be made arbitrarily large by Exercise 2.7(c), which contradicts $H_C(\zeta) \leq 0$.

For the converse, let f be strongly decreasing and let $\zeta \in \partial_P f(x)$. For any $c \in C$ and $t > 0$ we then have $f(x + tc) \leq f(x)$. Combined with the proximal subgradient inequality for ζ, this implies $\langle \zeta, c \rangle \leq 0$. Thus $H_C\big(\partial_P f(x)\big) \leq 0$.

We now suppose that $h_C\big(\partial_P f(x)\big) \leq 0 \ \forall x$. Let $t > 0$ be given, and apply Theorem 2.6 with $Y := x + tC$ (compact) and $\varepsilon > 0$. We obtain

$$\min_{x+tC} f - f(x) \leq \langle \zeta, tc \rangle + \varepsilon \ \forall c \in C.$$

Now take the minimum over $c \in C$ on the right side of this inequality. The result is

$$\min_{x+tC} f - f(x) \leq t h_C(\zeta) + \varepsilon.$$

Since $h_C(\zeta) \leq 0$ by assumption, and since ε is arbitrary, we deduce the existence of $y \in x + tC$ such that $f(y) \leq f(x)$; i.e., f is weakly decreasing.

Now suppose that f is weakly decreasing, and let $\zeta \in \partial_P f(x)$. For each $t > 0$ there exists $c \in C$ for which $f(x + tc) \leq f(x)$. Let $t_i \downarrow 0$ and extract a subsequence of the corresponding points c_i (without relabeling) so that they converge to $c \in C$. For all i sufficiently large, the proximal subgradient inequality gives

$$0 \geq f(x + t_i c_i) - f(x) \geq \langle \zeta, t_i c_i \rangle - \sigma \|t_i c_i\|^2.$$

Dividing by t_i and passing to the limit yields $\langle \zeta, c \rangle \leq 0$, whence $h_C(\zeta) \leq 0$. $\square$

2.12. Exercise.

(a) Show that the characterization of weak decrease continues to hold for a set C which is closed, bounded, and convex (not necessarily compact) if the function f is assumed to be weakly lower semicontinuous.

(b) Let $X = \mathbb{R}^2$, and let $C = \{(t, 1-t) \colon 0 \leq t \leq 1\}$. Under what condition on $f'(x)$ is a C^1 function $f \colon \mathbb{R}^2 \to \mathbb{R}$ weakly decreasing relative to C?

3 Solving Equations

Consider an equation

$$f(x, \alpha) = 0, \tag{1}$$

where f is a given function of two variables. It is a familiar consideration in mathematics to seek to solve this equation for x, while viewing α as a parameter. Typically this is done in a neighborhood of a given point (x_0, α_0) for which (1) is satisfied, and the important issues are these: For a given α sufficiently near to α_0, does there continue to be at least one value of x for which (1) holds? How does this set of solutions vary with α? When can we be sure there is a suitably "nice" function $x(\alpha)$ which "tracks" solutions as α varies; i.e., such that $f(x(\alpha), \alpha) = 0$, and $x(\alpha_0) = x_0$?

We will develop a simple and powerful way to deal with such issues via nonsmooth analysis, and in so doing recover some classical results as well as establish new ones. For a given *nonnegative* function f, we are interested, for given α, in the points $x \in X$ (a Hilbert space) for which (1) holds, and we wish to limit attention to just those x lying in a prescribed subset Ω of X. We define, then, the feasible set $\Phi(\alpha)$ as follows:

$$\Phi(\alpha) := \{x \in \Omega \colon f(x, \alpha) = 0\}. \tag{2}$$

The following result is the key to our approach; in its statement, $\partial_P f(x, \alpha)$ refers to the proximal subdifferential of the function $x \mapsto f(x, \alpha)$.

3. Special Topics

3.1. Theorem (Solvability Theorem). *Let A be a parameter set, and suppose that for each $\alpha \in A$, the function $x \mapsto f(x, \alpha)$ is nonnegative and belongs to $\mathcal{F}(X)$. Suppose that for some $\delta > 0$ and subset V of X, we have the following nonstationarity condition:*

$$\alpha \in A, \quad x \in V, \quad f(x, \alpha) > 0, \quad \zeta \in \partial_P f(x, \alpha) \implies \|\zeta\| \geq \delta.$$

Then for all $(x, \alpha) \in X \times A$ we have

$$\min\{d(x; \operatorname{comp} V), d(x; \operatorname{comp} \Omega), d(x; \Phi(\alpha))\} \leq \frac{f(x, \alpha)}{\delta}. \qquad (3)$$

Proof. If the conclusion is false, then for some $(x, \alpha) \in X \times A$, $f(x, \alpha)$ is finite, and there exists $\rho > 0$ such that

$$\min\{d(x; \operatorname{comp} V), d(x; \operatorname{comp} \Omega), d(x; \Phi(\alpha))\} > \rho > \frac{f(x, \alpha)}{\delta}.$$

It follows that

$$x + \rho B \subset V, \quad x + \rho B \subset \Omega, \quad d(x; \Phi(\alpha)) > \rho.$$

Thus we have $f(x', \alpha) > 0$ for all $x' \in x + \rho B$. By hypothesis, then, the proximal subgradients of $f(\cdot, \alpha)$ on the open set $x + \rho B$ have norm bounded below by δ. We invoke the Decrease Principle (Theorem 2.8) to conclude

$$0 \leq \inf_{x' \in x + \rho B} f(x', \alpha) \leq f(x, \alpha) - \rho\delta < 0.$$

This contradiction proves the theorem. $\square$

We remark that the context in which the theorem is most often used is that in which a given point (x_0, α_0) has been identified at which $f = 0$, and in which V and Ω are neighborhoods of x_0. Then (3) asserts that

$$d(x; \Phi(\alpha)) \leq \frac{f(x, \alpha)}{\delta} \qquad (4)$$

whenever $f(x, \alpha)$ is sufficiently small and x is sufficiently near x_0. In cases where f is continuous, we then derive (4) for all (x, α) sufficiently near (x_0, α_0). Of course, part of the interest of (4) is that it implies the nonemptiness of $\Phi(\alpha)$ (since $d(x, \Phi(\alpha)) = +\infty$ when $\Phi(\alpha)$ is empty).

The following application of the Solvability Theorem illustrates these remarks, together with the reduction of vector-valued equations to a single scalar one. We consider the equation

$$F(x, \alpha) = 0, \qquad (5)$$

where F is a mapping from $X \times M$ to Y, with M a metric space and Y another Hilbert space. We assume that some open set U of X exists such that F is continuous on $U \times M$, and such that the partial derivative $F'_x(x, \alpha)$ exists for all $(x, \alpha) \in U \times M$, and is continuous there (jointly in (x, α)).

3.2. Theorem (Graves–Lyusternik). *Let $(x_0, \alpha_0) \in U \times M$ be a point satisfying $F(x_0, \alpha_0) = 0$, and suppose that $F'_x(x_0, \alpha_0)$ is onto:*

$$F'_x(x_0, \alpha_0) X = Y.$$

Let Ω be any neighborhood of x_0. Then for some $\delta > 0$, for all (x, α) sufficiently near (x_0, α_0), we have

$$d\bigl(x; \Phi(\alpha)\bigr) \leq \frac{\|F(x, \alpha)\|}{\delta},$$

where $\Phi(\alpha) := \{x' \in \Omega \colon F(x', \alpha) = 0\}$.

Proof. By the Open Mapping Theorem of Banach, there exists $\delta > 0$ such that $F'_x(x_0, \alpha_0) \overline{B}_X \supset 2\delta \overline{B}_Y$. Since $(x, \alpha) \mapsto F'_x(x, \alpha)$ is continuous, there exist neighborhoods V and A of x_0 and α_0 such that the operator norm of $F'_x(x_0, \alpha_0) - F'_x(x, \alpha)$ is no greater than δ when $(x, \alpha) \in V \times A$. For such (x, α) we have

$$2\delta \overline{B}_Y \subset F'_x(x_0, \alpha_0) \overline{B}_X \subset F'_x(x, \alpha) \overline{B}_X + \bigl[F'_x(x_0, \alpha_0) - F'_x(x, \alpha)\bigr] \overline{B}_X$$
$$\subset F'_x(x, \alpha) \overline{B}_X + \delta \overline{B}_Y.$$

This implies that

$$F'_x(x, \alpha) \overline{B}_X \supset \delta \overline{B}_Y \ \forall (x, \alpha) \in V \times A.$$

It follows that for any unit vector y in Y, for any $(x, \alpha) \in V \times A$, we have $\|F'_x(x, \alpha)^* y\| \geq \delta$. Indeed, there exists a vector $v \in \overline{B}_X$ such that $F'_x(x, \alpha) v = \delta y$. Then

$$\|F'_x(x, \alpha)^* y\| \geq \langle F'_x(x, \alpha)^* y, v \rangle = \langle y, F'_x(x, \alpha) v \rangle = \langle y, \delta y \rangle = \delta,$$

as claimed.

We are now ready to apply the Solvability Theorem 3.1, with $f(x, \alpha) := \|F(x, \alpha)\|$. We need to verify the nonstationarity condition of the theorem. If $f(x, \alpha) > 0$, then the only possible proximal subgradient of $f(\cdot, \alpha)$ at x is its derivative there, namely $F'_x(x, \alpha)^* F(x, \alpha) / \|F(x, \alpha)\|$ (by the classical Chain Rule). We have shown above that such a vector has norm at least δ, as required.

Once the Solvability Theorem has been applied, there remains only to observe that (3) reduces to the required estimate for all (x, α) sufficiently near (x_0, α_0), since then we will have

$$\min\{d(x; \text{comp } V), d(x; \text{comp } \Omega)\} > \frac{f(x, \alpha)}{\delta}. \qquad \square$$

We remark that the conclusion of the theorem subsumes as special cases stability results with respect to either a parameter change or to a change in the variable x. If α alone changes (somewhat), then x_0 is no longer a solution to $F(x,\alpha) = 0$, but is not far from one; if only x changes, then there is a solution for the original equation which approximates x to the same extent to which x is infeasible (i.e., $\|F(x,\alpha_0)\|$). (We had noted the first of these two properties in a different context in Theorem 1.9.) The term $\|F(x,\alpha)\|$ which appears in the estimate given by Theorem 3.2 is nondifferentiable at (x_0, α_0), and this is necessarily the case: No term of the form $\theta(\|F(x,\alpha)\|)$ with $\theta(0) = 0$ and θ differentiable can provide the estimate, as simple examples show.

A Mixed Equality/Constraint System

The following result will be used in the next theorem, which will illustrate the use of *extended-valued* functions f in connection with solvability, as well as the use of a *constraint qualification*. This refers to a nondegeneracy hypothesis on the data incorporating the constraints of the problem, similar to the normality hypothesis introduced in §3.1.

3.3. Lemma. *Let F and (x_0, α_0) satisfy all the hypotheses of Theorem 3.2, where we take X and Y finite dimensional. Let S be a closed subset of X containing x_0, and assume in addition that the following constraint qualification holds:*

$$0 \in F'_x(x_0, \alpha_0)^* y + N_S^L(x_0) \implies y = 0. \tag{6}$$

Then there exist $\delta > 0$ and neighborhoods V and A of x_0 and α_0 with the following property: for all $(x, \alpha) \in V \times A$, for any unit vector y, any vector ζ belonging to

$$F'_x(x, \alpha)^* y + N_S^L(x)$$

satisfies $\|\zeta\| \geq \delta$.

Proof. If the assertion is false, then there exist sequences $\{x_i\}$ converging to x_0, $\{\alpha_i\}$ converging to α_0, $\{y_i\}$ unit vectors such that, for some sequence $\{\zeta_i\}$ converging to 0, we have

$$\zeta_i \in F'_x(x_i, \alpha_i)^* y_i + N_S^L(x_i).$$

We can assume that the sequence y_i converges to a limit y_0, necessarily a unit vector too. Since the multifunction $x \mapsto N_S^L(x)$ has closed graph (X being finite dimensional), we obtain in the limit $0 \in F'_x(x_0, \alpha_0) y + N_S^L(x_0)$, contradicting the hypothesis (6). □

3.4. Theorem. *Let (x_0, α_0) be a solution of the equality/constraint system*

$$F(x, \alpha) = 0, \quad x \in S,$$

3 Solving Equations   129

under the hypotheses of Lemma 3.3, and in particular in the presence of the constraint qualification (6). Then for some $\delta > 0$, for all (x, α) sufficiently near (x_0, α_0) having $x \in S$, we have

$$d(x; \Phi_S(\alpha)) \leq \frac{\|F(x,\alpha)\|}{\delta},$$

where $\Phi_S(\alpha) := \{x \in S : F(x, \alpha) = 0\}$. In particular, $\Phi_S(\alpha)$ is nonempty for α sufficiently near α_0.

3.5. Exercise.

(a) Prove Theorem 3.4 with the help of Lemma 3.3, setting

$$f(x, \alpha) := \|F(x, \alpha)\| + I_S(x)$$

in the Solvability Theorem and recalling Proposition 1.10.1.

(b) When $x_0 \in \mathrm{int}\, S$ in the context of Lemma 3.3, show that (6) holds iff the matrix $F'_x(x_0, \alpha_0)$ has maximal rank.

(c) A given function $F \colon \mathbb{R}^3 \times M \to \mathbb{R}^2$ satisfies the hypotheses of Theorem 3.2, with

$$F'_x(x_0, \alpha_0) = \begin{bmatrix} 1 & 0 & 0 \\ 1 & 1 & -1 \end{bmatrix}.$$

Prove that for every α sufficiently near α_0 there exists $x = (x_1, x_2, x_3)$ in $\mathbb{R}^3$ with $x_2 \geq 0$, $x_3 \geq 0$ such that $F(x, \alpha) = 0$. In fact, show that we can also require that $x_2 x_3 = 0$.

Implicit and Inverse Functions

Under additional structural assumptions, the equation $F(x, \alpha) = 0$ can be shown to uniquely define x as a function of α (locally); the implicit function inherits regularity from F. We make the same assumptions as in Theorem 3.2, but now with $X = Y$.

3.6. Theorem.

(a) *Let $F'_x(x_0, \alpha_0)$ be onto and one-to-one. Then there exist neighborhoods Ω of x_0 and W of α_0 and a unique continuous function $\hat{x}(\cdot) \colon W \to \Omega$ with $\hat{x}(\alpha_0) = x_0$ such that $F(\hat{x}(\alpha), \alpha) = 0 \ \forall \alpha \in W$.*

(b) *If in addition F is Lipschitz in a neighborhood of (x_0, α_0), then $\hat{x}$ is Lipschitz.*

(c) *If in addition M is a Hilbert space and $F(x, \alpha)$ is C^1 (in both variables) near (x_0, α_0), then $\hat{x}$ is C^1.*

Proof. It is a well-known consequence of the Closed Graph Theorem that $F'_x(x_0, \alpha_0)^{-1}$ is a continuous linear operator, and this can be used to show that $F'_x(x_0, \alpha_0)^*$ is onto (with inverse $(F'_x(x_0, \alpha_0)^{-1})^*$). Thus for some $\eta > 0$, we have $F'_x(x_0, \alpha_0)^* \overline{B} \supset 2\eta \overline{B}$. For some convex neighborhood $\Omega \times N \subset U \times M$ of (x_0, α_0), we have

$$\|F'_x(x, \alpha) - F'_x(x_0, \alpha_0)\| < \eta \ \forall (x, \alpha) \in \Omega \times N.$$

Now let $x, x' \in \Omega$ and $\alpha \in N$, $x' \neq x$. Then, for any vector θ in $\overline{B}$ we have, by the Mean Value Theorem, for some $z \in [x, x']$:

$$\begin{aligned}\langle \theta, F(x', \alpha) - F(x, \alpha)\rangle &= \langle \theta, F'_x(z, \alpha)(x' - x)\rangle \\ &\geq \langle \theta, F'_x(x_0, \alpha_0)(x' - x)\rangle \\ &\quad + \langle \theta, (F'_x(z, \alpha) - F'_x(x_0, \alpha_0))(x' - x)\rangle \\ &\geq \langle F'_x(x_0, \alpha_0)^*\theta, x' - x\rangle - \eta\|x' - x\| \\ &= \eta\|x' - x\|,\end{aligned}$$

where θ has been chosen in $\overline{B}$ to satisfy

$$F'_x(x_0, \alpha_0)^*\theta = 2\eta \frac{(x' - x)}{\|x' - x\|}.$$

It follows that

$$\|F(x', \alpha) - F(x, \alpha)\| \geq \eta\|x' - x\| \ \forall (x, \alpha) \in \Omega \times N. \tag{7}$$

We now invoke Theorem 3.2 to deduce the existence of a neighborhood $W \subset N$ of α_0 such that

$$\Phi(\alpha) := \{x \in \Omega : F(x, \alpha) = 0\} \neq \emptyset \ \forall \alpha \in W.$$

It follows from (7) that $\Phi(\alpha)$ is a singleton $\{\hat{x}(\alpha)\}$ for each $\alpha \in W$, and that $\hat{x}$ is the unique function from W to Ω such that $F(\hat{x}(\alpha), \alpha) = 0 \ \forall \alpha \in W$.

We now show that $\hat{x}(\cdot)$ is continuous. From (7) we have

$$\eta\|\hat{x}(\alpha') - \hat{x}(\alpha)\| \leq \|F(\hat{x}(\alpha'), \alpha') - F(\hat{x}(\alpha), \alpha')\| = \|F(\hat{x}(\alpha), \alpha')\|,$$

which tends to 0 as α' tends to α. This establishes continuity. In addition, if F is Lipschitz near (x_0, α_0), then for α and α' sufficiently near α_0 the last term above satisfies

$$\|F(\hat{x}(\alpha), \alpha')\| = \|F(\hat{x}(\alpha), \alpha') - F(\hat{x}(\alpha), \alpha)\| \leq K d_M(\alpha, \alpha'),$$

for some suitable Lipschitz constant K, where d_M is the metric on M. We deduce that $\hat{x}(\cdot)$ is Lipschitz near α_0, proving (b).

Finally, let M be a Hilbert space and let F be C^1 in (x,α). The Taylor expansion gives

$$0 = F(\hat{x}(\alpha),\alpha) = F(x_0,\alpha_0) + F'_x(x_0,\alpha_0)(\hat{x}(\alpha) - x_0) + F'_\alpha(x_0,\alpha_0)(\alpha - \alpha_0) + o(\hat{x}(\alpha) - x_0, \alpha - \alpha_0),$$

where

$$o(x - x_0, \alpha - \alpha_0)/\|(x - x_0, \alpha - \alpha_0)\| \to 0 \text{ as } (x,\alpha) \to (x_0,\alpha_0).$$

Since $\hat{x}(\cdot)$ is Lipschitz near α_0, $o(\hat{x}(\alpha)-x_0,\alpha-\alpha_0)$ can be written $o(\alpha-\alpha_0)$. Then we can solve the Taylor expansion explicitly for $\hat{x}(\alpha)$ and deduce from the resulting expression that $\hat{x}(\cdot)$ is differentiable at α_0, with derivative $-F'_x(x_0,\alpha_0)^{-1}F'_\alpha(x_0,\alpha_0)$. But this argument holds not just at α_0 but at any α in a neighborhood, whence $\hat{x}(\cdot)$ is C^1 near α_0. □

A special case of the theorem is that of inverse functions:

3.7. Corollary. *Let F satisfy the hypotheses of the theorem and have the form $F(x,\alpha) = G(x) - \alpha$ (so that $X = Y = M$). Then there exist neighborhoods W of α_0 and Λ of x_0 and a C^1 function $\hat{x}$ on W such that*

$$G(\hat{x}(\alpha)) = \alpha \ \forall \alpha \in W, \quad \hat{x}(G(x)) = x \ \forall x \in \Lambda.$$

Proof. The function $\hat{x}(\cdot)$ of the theorem is C^1 and satisfies $G(\hat{x}(\alpha)) = \alpha$ for $\alpha \in W$, the neighborhood of α_0 provided by the theorem. Now pick $\Lambda \subset \Omega$ so that $G(x) \in W \ \forall x \in \Lambda$. Then, as shown in the theorem, $\Phi(G(x))$ is a singleton when $x \in \Lambda$; that is, there is a unique $x' \in \Omega$ such that $G(x') = G(x)$. But then this x', namely $\hat{x}(G(x))$, must equal x. □

Systems of Mixed Equalities and Inequalities

We consider now the solutions of a parametrized system of the form

$$h(x) = \alpha, \quad g(x) \leq \beta, \tag{8}$$

where $x \in X$ (a Hilbert space), and where the vector $(\alpha,\beta) \in \mathbb{R}^m \times \mathbb{R}^p$ is the parameter. Let x_0 satisfy the system for $(\alpha,\beta) = (0,0)$, and let us suppose that the functions $g: X \to \mathbb{R}^p$ and $h: X \to \mathbb{R}^m$ are Lipschitz in a neighborhood Ω of x_0. We set

$$\Phi(\alpha,\beta) := \{x \in \Omega: \text{ system (8) is satisfied}\}.$$

3.8. Theorem. *Suppose that the following constraint qualification is satisfied at x_0:*

$$\gamma \geq 0, \quad \langle \gamma, g(x_0) \rangle = 0, \quad 0 \in \partial_L\{\langle \gamma, g(\cdot)\rangle + \langle \lambda, h(\cdot)\rangle\}(x_0)$$
$$\implies \gamma = 0 \quad \text{and} \quad \lambda = 0.$$

132   3. Special Topics

Then for some $\delta > 0$, for all x sufficiently near x_0, and for all (α, β) sufficiently near $(0,0)$, we have

$$d(x; \Phi(\alpha, \beta)) \leq \frac{1}{\delta} \max_{i,j}\{(g_i(x) - \beta_i)_+, |h_j(x) - \alpha_j|\},$$

where $(g_i(x) - \beta_i)_+$ signifies $\max(g_i(x) - \beta_i, 0)$. In particular, $\Phi(\alpha, \beta)$ is nonempty for (α, β) near $(0,0)$.

Proof. We set

$$f(x, \alpha, \beta) := \max_{i,j}\{(g_i(x) - \beta_i)_+, |h_j(x) - \alpha_j|\} \geq 0.$$

Then x is a solution of (8) iff $f(x, \alpha, \beta) = 0$. Note that f is Lipschitz in a neighborhood of $(x_0, 0, 0)$, as observed in Problem 1.11.17.

Lemma. *There exist neighborhoods V of x_0 and A of $(0,0)$ and $\delta > 0$ such that*

$$x \in V, \quad (\alpha, \beta) \in A, \quad f(x, \alpha, \beta) > 0, \quad \zeta \in \partial_P f(x, \alpha, \beta) \implies \|\zeta\| \geq \delta.$$

Proof. If this fails to be the case, then there exist sequences $\{x_k\}$, $\{(\alpha_k, \beta_k)\}$, $\{\zeta_k\}$ such that

$$x_k \to x_0, \quad (\alpha_k, \beta_k) \to (0,0), \quad f(x_k, \alpha_k, \beta_k) > 0,$$
$$\zeta_k \in \partial_P f(x_k, \alpha_k, \beta_k), \quad \|\zeta_k\| \to 0.$$

The proximal subgradient ζ_k can be characterized via Problem 1.11.17, for k large enough so that x_k enters some prescribed neighborhood on which f is Lipschitz in x. Note that in the formula provided there, only the terms $(g_i(x) - \beta_i)_+$ and $|h_j(x) - \alpha_j|$ which are strictly positive can contribute. Locally, these coincide with $g_i(x) - \beta_i$ and $\pm(h_j(x) - \alpha_j)$. The upshot is that for all k large enough, ζ_k belongs to the set

$$\partial_L\{\langle \gamma^k, g(\cdot)\rangle + \langle \lambda^k, h(\cdot)\rangle\}(x_k),$$

where

$$\gamma_i^k \geq 0, \quad \langle \gamma^k, g(x_k) - \beta_k\rangle = 0, \quad \sum_i \gamma_i^k + \sum_j |\lambda_j^k| = 1.$$

If we now extract subsequences as required to make each sequence $\{\gamma^k\}$, $\{\lambda^k\}$ converge, passage to the limit (see Exercise 1.6) produces a nonzero (γ, λ) with precisely the properties ruled out by the constraint qualification. This contradiction proves the lemma.

Armed with the lemma, the application of the Solvability Theorem 3.1 is immediate and gives the required result. □

When the inequalities are absent in (8), it is natural to seek nonsmooth analogues of the implicit function assertions of Theorem 3.6. We will examine for illustrative purposes the issue of an inverse function induced by an equation $G(x) = \alpha$, which was treated in the case of smooth data in Corollary 3.7. The solvability aspect is at hand, as a special case of the preceding theorem:

3.9. Exercise. Let $G\colon X \to \mathbb{R}^n$ be Lipschitz on a neighborhood of the point x_0, where $G(x_0) = \alpha_0$. Suppose that the following constraint qualification holds:

$$0 \in \partial_L \langle \zeta, G(\cdot) \rangle (x_0) \implies \zeta = 0. \tag{9}$$

Then, for any neighborhood Ω of x_0, there exists $\delta > 0$ such that for all (x, α) sufficiently near (x_0, α_0) we have

$$d\bigl(x; \Phi(\alpha)\bigr) \leq \frac{\|G(x) - \alpha\|}{\delta},$$

where $\Phi(\alpha) := \bigl\{ x \in \Omega \colon G(x) = \alpha \bigr\}$.

Show how this reduces to a special case of Theorem 3.2 when G is C^1.

Generalized Jacobians and Inverse Functions

What is required now to deduce the existence of an inverse is a condition akin to (7) which implies that G is locally one-to-one. A convenient way to formulate such a condition, which will also turn out to be a more readily verifiable criterion implying the constraint qualification (9), is in terms of the *generalized Jacobian* ∂G, a tool which is applicable when the underlying space is finite dimensional.

Let $X = \mathbb{R}^m$. We define $\partial G(x)$ as follows:

$$\partial G(x) := \mathrm{co}\Bigl\{ \lim_{i \to \infty} G'(x_i) \colon G'(x_i) \text{ exists}, \ x_i \to x \Bigr\},$$

a formula that brings to mind the Generalized Gradient Formula of Theorem 2.8.1. (We remark that, in keeping with the earlier formula, an arbitrary set of measure 0 can be avoided in calculating $\partial G(x)$, without the results below being affected.) We identify $\partial G(x)$ with a convex set of $n \times m$ matrices, and we say that $\partial G(x)$ is *of maximal rank* provided that every matrix in $\partial G(x)$ is of maximal rank. When $n = m$, it is alternate terminology to say that $\partial G(x)$ is *nonsingular*.

3.10. Exercise.

(a) Let $G\colon \mathbb{R}^m \to \mathbb{R}^n$ be Lipschitz near x. Prove that $\partial G(x)$ is compact, convex, and nonempty, where the space of $n \times m$ matrices is identified with $\mathbb{R}^{nm}$. Prove that ∂G is upper semicontinuous: For all $\varepsilon > 0$ there exists $r > 0$ such that

$$x' \in x + rB \implies \partial G(x') \subset \partial G(x) + \varepsilon B_{n \times m},$$

where $B_{n \times m}$ denotes the open unit ball in the space of $n \times m$ matrices.

(b) Show that for any $\zeta \in \mathbb{R}^n$, we have $\partial \langle \zeta, G(\cdot) \rangle (x) = \zeta^* \partial G(x)$.

(c) Calculate $\partial G(0)$ where $n = m = 2$ and

$$G(x, y) = \big(|x| + y, 2x + |y|\big),$$

and show that it is nonsingular.

(d) If G has component functions $g_1, g_2, \ldots, g_n$, show that

$$\partial G(x) \subset \partial g_1(x) \times \partial g_2(x) \times \cdots \times \partial g_n(x),$$

and give an example in which strict containment occurs.

3.11. Proposition. *Let $G \colon \mathbb{R}^n \to \mathbb{R}^n$ be Lipschitz near x_0, and suppose that $\partial G(x_0)$ is nonsingular. Then the constraint qualification (9) holds, and for some $\eta > 0$, for some neighborhood Ω of x_0, we have*

$$\|G(x) - G(y)\| \geq \eta \|x - y\| \quad \forall x, y \in \Omega.$$

Proof. Let 0 belong to $\partial_L \langle \zeta, G(\cdot) \rangle (x_0)$. Then

$$0 \in \partial \langle \zeta, G(\cdot) \rangle (x_0) = \zeta^* \partial G(x_0)$$

(by Exercise 3.10). Since $\partial G(x_0)$ is nonsingular, we deduce $\zeta = 0$, thereby verifying the constraint qualification. The next step in the proof requires the following:

Lemma. *There are positive numbers r and η with the following property. Given any unit vector $v \in \mathbb{R}^n$, there is a unit vector w in $\mathbb{R}^n$ such that, whenever x lies in $x_0 + rB$ and the matrix M belongs to $\partial G(x)$, then we have*

$$\langle w, Mv \rangle \geq \eta.$$

Proof. To see this, let S be the unit sphere in $\mathbb{R}^n$ and note that the subset $\partial G(x_0) S$ of $\mathbb{R}^n$ is compact and does not contain 0. Hence,

$$d\big(0; \partial G(x_0) S\big)/2 =: \eta > 0.$$

For some $\varepsilon > 0$ small enough, we will have

$$d\big(0; \big(\partial G(x_0) + \varepsilon B_{n \times n}\big) S\big) > \eta.$$

Now for some $r > 0$, for all $x \in x_0 + rB$, we have $\partial G(x) \subset \partial G(x_0) + \varepsilon B_{n \times n}$, in light of Exercise 3.10, where r is small enough so that G is Lipschitz on

$x_0 + r\overline{B}$. Now let any unit vector v be given. It follows from the above that the convex set $(\partial G(x_0) + \varepsilon B_{n\times n})v$ is distance at least η from 0. By the Separation Theorem, there exists a unit vector w such that $\langle \zeta, w \rangle \geq \eta$ for all $\zeta \in (\partial G(x_0) + \varepsilon B_{n\times n})v$, and all the more for any ζ of the form Mv, $M \in \partial G(x)$, $x \in x_0 + rB$. This establishes the lemma.

Now let x and y belong to $\Omega := x_0 + rB$. We will complete the proof of the proposition by showing that $\|G(x) - G(y)\| \geq \eta \|x - y\|$. We may suppose $x \neq y$. Set
$$v = \frac{y - x}{\|y - x\|}, \quad \lambda = \|y - x\|,$$
so that $y = x + \lambda v$.

Let π be the plane perpendicular to v and passing through x. The set P of points x' in $x_0 + rB$ where $G'(x')$ fails to exist is of measure 0, and hence by Fubini's Theorem, for almost every x' in π, the ray
$$x' + tv, \quad t \geq 0,$$
meets P in a set of 0 one-dimensional measure. Choose an x' with the above property and sufficiently close to x so that $x' + tv$ lies in $x_0 + rB$ for every t in $[0, \lambda]$. Then the function
$$t \to G(x' + tv)$$
is Lipschitzian for t in $[0, \lambda]$ and has almost everywhere on this interval derivative $G'(x' + tv)v$. Thus
$$G(x' + \lambda v) - G(x') = \int_0^\lambda G'(x' + tv)v\, dt.$$

Let w correspond to v in the lemma above. We deduce
$$w \cdot [G(x' + \lambda v) - G(x')] = \int_0^\lambda w \cdot [G'(x' + tv)v]\, dt \geq \int_0^\lambda \eta\, dt = \lambda \eta.$$

Recalling the definition of λ, we arrive at:
$$\|G(x' + \lambda v) - G(x')\| \geq \eta \|x - y\|.$$

This conclusion has been obtained for almost all x' in a neighborhood of x. Since G is continuous, the proposition is proved. □

We now possess all the necessary components to prove the *Lipschitz Inverse Function Theorem*.

3.12. Theorem. *If $G: \mathbb{R}^n \to \mathbb{R}^n$ is Lipschitz near x_0 and $\partial G(x_0)$ is nonsingular, then there exist neighborhoods W of $G(x_0)$ and Λ of x_0 and a Lipschitz function $\hat{x}$ on W such that*
$$G(\hat{x}(\alpha)) = \alpha \ \forall \alpha \in W, \quad \hat{x}(G(x)) = x \ \forall x \in \Lambda.$$

3.13. Exercise.

(a) Prove Theorem 3.12.

(b) Show that the theorem fails (even for $n = 1$) if the hypothesis that $\partial G(x_0)$ be nonsingular is replaced by the condition that $G'(x)$ be nonsingular whenever it exists.

(c) Confirm the conclusion of the theorem for the example of Exercise 3.10(c).

4 Derivate Calculus and Rademacher's Theorem

The first constructs of nonsmooth analysis were put forward by Dini in the nineteenth century. His well-known derivates were defined for a real-valued function of a single variable, but the basic idea can be extended beyond that setting and developed so as to include corresponding notions of subdifferential, tangent, and normal. We will take a brief look at this theory now, and relate it to the results of the two preceding chapters. The setting will be that of a function $f \in \mathcal{F}(\mathbb{R}^n)$; for $x \in \operatorname{dom} f$, we define the *subderivate* of f at x in the direction v, denoted $Df(x; v)$, as follows:

$$Df(x;v) := \liminf_{\substack{w \to v \\ t \downarrow 0}} \frac{f(x+tw)-f(x)}{t}.$$

4.1. Exercise.

(a) If f is Lipschitz of rank K near x, then $Df(x;v)$ agrees with

$$\liminf_{t \downarrow 0} \frac{f(x+tv)-f(x)}{t},$$

and we have $Df(x;0) = 0$. The function $v \mapsto Df(x;v)$ is Lipschitz of rank K on $\mathbb{R}^n$, and $Df(x;v) \leq f^\circ(x;v)$. Equality holds for all v iff f is regular at x.

(b) In general, $Df(x;0)$ is either 0 or $-\infty$, and we have $Df(x;\lambda v) = \lambda Df(x;v)$ for any $\lambda > 0$. Give an example of a continuous $f \colon \mathbb{R} \to \mathbb{R}$ having $Df(0;0) = -\infty$.

(c) If $g \in \mathcal{F}(\mathbb{R}^n)$ has $x \in \operatorname{dom} g$, then

$$D(f+g)(x;v) \geq Df(x;v) + Dg(x;v)$$

(interpreting $\infty - \infty$ as $-\infty$).

(d) If $\zeta \in \partial_P f(x)$, then $Df(x;v) \geq \langle \zeta, v \rangle \ \forall v$.

(e) The function $v \mapsto Df(x;v)$ is lower semicontinuous.

The following theorem is the key to relating derivate calculus to proximal analysis.

4.2. Theorem (Subbotin). *Let $f \in \mathcal{F}(\mathbb{R}^n)$, $x \in \text{dom } f$, and let E be a nonempty compact convex subset of $\mathbb{R}^n$. Suppose that for some scalar ρ we have*
$$Df(x;e) > \rho \ \forall e \in E.$$
Then, for any $\varepsilon > 0$, there exist $z \in x + \varepsilon B$ and $\zeta \in \partial_P f(z)$ such that
$$|f(z) - f(x)| < \varepsilon, \quad \langle \zeta, e \rangle > \rho \ \forall e \in E.$$

Proof. We claim that for all $t > 0$ sufficiently small, we have
$$f(x + te + t^2 u) - f(x) > \rho t + t^2 \ \forall e \in E, \ \forall u \in \overline{B}. \tag{1}$$

Were this not so, there would be sequences $t_i \downarrow 0$, $e_i \in E$, and $u_i \in \overline{B}$ such that
$$\frac{f(x + t_i e_i + t_i^2 u_i) - f(x)}{t_i} \leq \rho + t_i.$$

Invoking the compactness of E, there is a subsequence (we eschew relabeling) of e_i converging to a point $e_0 \in E$. Then $e_i + t_i u_i$ converges to e_0 as well, and we deduce $Df(x; e_0) \leq \rho$, contradicting the hypothesis and establishing the claim.

Now pick $t > 0$ so that (1) holds, so that $tE + t^2 B \subset \varepsilon B$, so that $t < \varepsilon/(2\rho)$, and finally also small enough so that $f(x') > f(x) - \varepsilon$ for all $x' \in x + tE + t^2 \overline{B}$ (this is possible because f is lower semicontinuous). We will proceed to apply Theorem 2.6, the Mean Value Inequality, for $Y := x + tE$. We remark that in light of (1), the $\hat{r}$ of the theorem satisfies $\hat{r} \geq \rho t + t^2$, so $\bar{r} = \rho t$ is a suitable choice in applying the theorem. We pick for the ε of the theorem any positive $\varepsilon' < \min[\varepsilon/2, t^2]$. The point z provided by the theorem lies in $[x, Y] + \varepsilon' B$, which gives
$$z \in x + tE + t^2 B \subset x + \varepsilon B,$$
whence $f(z) > f(x) - \varepsilon$ by choice of t. Also, Theorem 2.6 asserts
$$f(z) < f(x) + \bar{r} + \varepsilon' < f(x) + \rho t + \frac{\varepsilon}{2} < f(x) + \varepsilon.$$

Thus $|f(z) - f(x)| < \varepsilon$ as desired. Finally we have, for the designated ζ in $\partial_P f(z)$,
$$\langle \zeta, te \rangle > \bar{r} := \rho t \ \forall e \in E,$$
which is the other assertion of the theorem. $\square$

4.3. Exercise. Show that the theorem is false if the convexity of E is deleted from the hypotheses, by considering $X = \mathbb{R}^2$, $f(u) := \|u\|$, $x = 0$, $E =$ unit circle.

138   3. Special Topics

The D-Subdifferential

The classical formula $f'(x;v) = \langle \nabla f(x), v \rangle$ encapsulates the duality between derivate (directional derivative) constructions and differential ones. Motivated by this, as well as by the duality between $f°(x;\cdot)$ and $\partial f(x)$ observed in Chapter 2, we proceed to define a subdifferential that corresponds to the subderivate $Df(x;\cdot)$. We say that ζ is a *directional subgradient* or *D-subgradient* of f at x provided that $x \in \text{dom } f$ and

$$Df(x;v) \geq \langle \zeta, v \rangle \ \forall v \in \mathbb{R}^n.$$

We denote the set of all such ζ by $\partial_D f(x)$ and refer to it as the D-*subdifferential*. It follows from Exercise 4.1(d) that $\partial_D f(x)$ contains $\partial_P f(x)$, and so must be nonempty on a dense subset of dom f by the Proximal Density Theorem 1.3.1.

4.4. Exercise.

(a) Show that $\partial_D f(x)$ is closed and convex, and reduces to $\{f'(x)\}$ if f is (Fréchet) differentiable at x.

(b) Prove that $\partial_D f(x)$ is bounded if f is Lipschitz near x.

(c) We have

$$\partial_D(f_1 + f_2)(x) \supseteq \partial_D f_1(x) + \partial_D f_2(x).$$

Equality holds if one of the functions is differentiable at x.

(d) For $f(x) = -\|x\|$, show that $\partial_D f(0) = \emptyset$.

(e) Recall that for $f(x) = -|x|^{3/2}$ we have $\partial_P f(x) = \emptyset$; show that $\partial_D f(0) = \{0\}$.

(f) There is no true duality between $Df(x;\cdot)$ and $\partial_D f(x)$: It is not true in general that $Df(x;v) = \sup\{\langle \zeta, v \rangle : \zeta \in \partial_D f(x)\}$.

Approximation by P-Subgradients

We now prove that although $\partial_D f(x)$ can be strictly larger than $\partial_P f(x)$, the essential difference between the two in a local sense (as opposed to pointwise) is slight.

4.5. Proposition. *Let $\zeta \in \partial_D f(x)$. Then for any $\varepsilon > 0$ there exist $z \in x + \varepsilon B$ and $\xi \in \partial_P f(z)$ such that $|f(x) - f(z)| < \varepsilon$ and $\|\zeta - \xi\| < \varepsilon$.*

Proof. It follows from the definition of $\partial_D f(x)$ that we have

$$D\varphi(x;v) \geq 0 \ \forall v \in \mathbb{R}^n,$$

where

$$\varphi(y) := f(y) - \langle \zeta, y \rangle.$$

Then for arbitrary fixed $\delta > 0$,
$$D\varphi(x; v) > -\delta \ \forall v \in \overline{B}.$$

We now apply Subbotin's Theorem 4.2 in order to obtain the existence of z and $\xi' \in \partial_P \varphi(z)$ such that
$$z \in x + \delta B, \quad |\varphi(z) - \varphi(x)| < \delta,$$

and
$$\langle \xi', v \rangle > -\delta \ \forall v \in \overline{B}. \tag{2}$$

It follows that
$$|f(z) - f(x)| \leq |\varphi(z) - \varphi(x)| + \|\zeta\| \|x - z\| < (1 + \|\zeta\|)\delta, \tag{3}$$

and from (2), that $\|\xi'\| < \delta$. But
$$\partial_P \varphi(z) = \partial_P f(z) - \zeta,$$

which implies the existence of $\xi \in \partial_P f(z)$ such that $\xi' = \xi - \zeta$. Hence, bearing (3) in mind, if we choose
$$\delta < \frac{\varepsilon}{1 + \|\zeta\|},$$

then z and ξ fulfill the requirements of the proposition. □

4.6. Exercise. Let $f'(x)$ exist. Then for any $\varepsilon > 0$ there exist $z \in x + \varepsilon B$ and $\zeta \in \partial_P f(z)$ such that $|f(x) - f(z)| < \varepsilon$ and $\|f'(x) - \zeta\| < \varepsilon$.

Directional Calculus

We have seen in the preceding proposition that the graph of the multifunction $\partial_P f$ is dense in that of $\partial_D f$. This fact can be used to derive rather easily the basic calculus of $\partial_D f$ by appealing to the known results for $\partial_P f$. Here are a few examples, the first of which goes back to a result of Dini that can be viewed as the first theorem of nonsmooth analysis.

4.7. Exercise.

(a) Let $f \in \mathcal{F}(\mathbb{R})$. Then the following are equivalent:
  (i) f is decreasing;
  (ii) $Df(x; 1) \leq 0 \ \forall x$; and
  (iii) $\zeta \leq 0 \ \forall \zeta \in \partial_D f(x) \ \forall x$.

(Compare with Theorem 0.1.1.)

(b) Let $f_1, f_2 \in \mathcal{F}(\mathbb{R}^n)$, and let $\zeta \in \partial_D(f_1+f_2)(x)$. Then for any $\varepsilon > 0$ there exist points x_1 and x_2 in $x+\varepsilon B$ with $|f_i(x)-f_i(x_i)| < \varepsilon$ ($i=1,2$) such that

$$\zeta \in \partial_D f_1(x_1) + \partial_D f_2(x_2) + \varepsilon B.$$

(c) A function $f \in \mathcal{F}(\mathbb{R}^n)$ is Lipschitz of rank K near $x \in \mathrm{dom}\, f$ iff there is a neighborhood U of x such that

$$Df(x';v) \leq K\|v\| \;\forall v \in \mathbb{R}^n, \;\forall x' \in U.$$

The fact that D-subgradients can be approximated by proximal ones implies that D-subgradients generate the same limiting subdifferential $\partial_L f(x)$ as that studied in §1.10, and also has a bearing on the generalized gradient $\partial f(x)$ and regularity (§2.4), as we now see. We remind the reader that $\partial_C f$ is an alternative notation for the generalized gradient (see Remark 2.6.2).

4.8. Proposition. *Let $f \in \mathcal{F}(\mathbb{R}^n)$ and $x \in \mathrm{dom}\, f$.*

(a) $\partial_L f(x) = \{\lim_{i\to\infty} \zeta_i : \zeta_i \in \partial_D f(x_i), x_i \xrightarrow{f} x\}$.

(b) *If f is Lipschitz near x, then*

$$\partial_D f(x) \subset \partial_L f(x) \subset \partial_C f(x),$$

with equality throughout iff f is regular at x (in particular, when f is convex).

Proof. Part (a) is immediate from Proposition 4.5, and implies the first inclusion in (b); the second has been noted earlier in connection with Theorem 2.6.1. Suppose now that f is regular at x, and let $\zeta \in \partial_C f(x)$. Then, for any $v \in \mathbb{R}^n$,

$$\begin{aligned}\langle \zeta, v\rangle &\leq f^\circ(x;v) &&\text{(by definition of } \partial_C f(x)) \\ &= f'(x;v) &&\text{(by definition of regularity)} \\ &= Df(x;v) &&\text{(in light of Exercise 4.1(a)).}\end{aligned}$$

Thus $\zeta \in \partial_D f(x)$, and we have $\partial_D f(x) = \partial_L f(x) = \partial_C f(x)$.

For the converse, observe that the equality of the three sets in question implies that $Df(x;v) = f^\circ(x;v)$ for all v, since then

$$\begin{aligned}Df(x;v) &\geq \max\{\langle\zeta,v\rangle : \zeta \in \partial_D f(x)\} \\ &= \max\{\langle\zeta,v\rangle : \zeta \in \partial_C f(x)\} = f^\circ(x;v)\end{aligned}$$

and since the opposite inequality always holds. That this implies regularity at x was noted in Exercise 4.1. $\square$

4 Derivate Calculus and Rademacher's Theorem 141

Tangents and Normals

A natural way to define a tangent direction v to a set S at $x \in S$ is to require $Dd_S(x; v) = 0$ (or equivalently, ≤ 0), where d_S is the distance function associated with S. The set of such v constitutes the *D-tangent cone* $T_S^D(x)$. Taking a now familiar path, we define the *D-normal cone* by polarity: $N_S^D(x) := T_S^D(x)^\circ$. The following exercise includes the geometric counterpart of Proposition 4.8.

4.9. Exercise. Let S be a nonempty closed subset of $\mathbb{R}^n$.

(a) $T_S^D(x)$ coincides with the Bouligand tangent cone $T_S^B(x)$ of §2.7.

(b) We do not have full duality: In general, $T_S^D(x)$ is not the polar of $N_S^D(x)$.

(c) $\zeta \in N_S^D(x)$ iff
$$\limsup_{x' \xrightarrow{S} x} \frac{\langle \zeta, x' - x \rangle}{\|x' - x\|} \leq 0. \tag{4}$$

(d) Any unit vector ζ satisfying (4) belongs to $\partial_D d_S(x)$.

(e) $N_S^D(x)$ is the cone generated by $\partial_D d_S(x)$ (compare with Theorem 1.6.4, Proposition 2.5.4, and Problem 1.11.27.)

(f) $N_S^D(x) \subset N_S^L(x) \subset N_S^C(x)$, with equality throughout iff S is regular at x. (*Hint.* To prove the first inclusion, use part (e), approximate $\partial_D d_S$ by $\partial_P d_S$, and recall Theorems 2.6.1 and 1.6.4.)

Other Characterizations of $\partial_D f$

An alternate characterization of $\partial_D f$ that bypasses Df is provided by the following:

4.10. Proposition. $\zeta \in \partial_D f(x)$ iff
$$\liminf_{\substack{u \to 0 \\ u \neq 0}} \frac{f(x+u) - f(x) - \langle \zeta, u \rangle}{\|u\|} \geq 0. \tag{5}$$

Proof. Let (5) hold, and let $\{v_i\}$ be a sequence converging to v and $\{t_i\}$ a sequence decreasing to 0. Then (5) implies
$$\liminf_{i \to \infty} \frac{f(x + t_i v_i) - f(x)}{t_i} \geq \langle \zeta, v \rangle,$$

whence $Df(x; v) \geq \langle \zeta, v \rangle \ \forall v$. Thus $\zeta \in \partial_D f(x)$.

For the converse, let $\zeta \in \partial_D f(x)$ and suppose that (5) fails. Then there is a sequence $\{u_i\}$ and $\varepsilon > 0$ such that $u_i \to 0$ and
$$f(x + u_i) - f(x) \leq \langle \zeta, u_i \rangle - \varepsilon \|u_i\|.$$

Note that $u_i \neq 0$. We can extract a subsequence (without relabeling) so that the vectors $v_i := u_i/\|u_i\|$ converge to a limit v. Setting $t_i := \|u_i\|$, the last inequality becomes

$$\frac{f(x + t_i v_i) - f(x)}{t_i} \leq \langle \zeta, v_i \rangle - \varepsilon,$$

which implies

$$Df(x; v) \leq \langle \zeta, v \rangle - \varepsilon,$$

contradicting $\zeta \in \partial_D f(x)$ and completing the proof. □

We remark that the set of subgradients defined via (5) is sometimes referred to as the *Fréchet subdifferential* $\partial_F f(x)$, an object which can differ from $\partial_D f(x)$ in infinite dimensions.

4.11. Exercise.

(a) Show that (5) holds iff there exists a nonnegative function $o(r)$ such that $o(r)/r \to 0$ as $r \downarrow 0$ and

$$f(x + u) - f(x) + o(\|u\|) \geq \langle \zeta, u \rangle,$$

for all u near 0. Show that $o(\cdot)$ can always be taken to be increasing. By considering

$$\bar{o}(t) := t \max\{o(r)/r : 0 < r \leq t\},$$

show that o can also be assumed to have the property that $o(t)/t$ is increasing.

(b) $\zeta \in \partial_D f(x)$ iff for every $\varepsilon > 0$ there exists a neighborhood N_ε of x such that x minimizes the function

$$g_\varepsilon(y) := f(y) - f(x) - \langle \zeta, y - x \rangle + \varepsilon \|y - x\|$$

over N_ε.

(c) Show that the function o of part (a) can also be assumed to be continuous. (*Hint.* Show that the following function has all the required properties:

$$\hat{o}(t) := 2t \int_t^{2t} \frac{o(r)}{r^2} dr.)$$

The following is another, less intrinsic but interesting characterization of $\partial_D f$. It is the definition of subdifferential most often put forward in the literature on viscosity solutions of differential equations, and $\partial_D f$ is sometimes referred to as the "viscosity subdifferential."

4.12. Proposition. *Let $f \in \mathcal{F}$. Then $\zeta \in \partial_D f(x)$ iff there exists a continuous function $g \colon \mathbb{R}^n \to \mathbb{R}$ which is differentiable at x with $g'(x) = \zeta$, and such that $f - g$ has a local minimum at x.*

4 Derivate Calculus and Rademacher's Theorem

Proof. First suppose that there exists g with the properties mentioned. Then $0 \in \partial_D(f - g)(x)$, and Exercise 4.4(c) implies $g'(x) \in \partial_D f(x)$.

Now suppose that $\zeta \in \partial_D f(x)$. Consider the (ultimate) function $o(\cdot)$ in Exercise 4.11(a). Since $o(\cdot)$ is increasing, it is continuous except at countably many points. Since $o(r)/r \downarrow 0$ as $r \downarrow 0$, this function is integrable on any bounded interval, and we may therefore define a function $\varphi : [0, \infty) \to (-\infty, 0]$ as follows:

$$\varphi(r) := -\int_r^{2r} \frac{o(s)}{s}\, ds.$$

It is readily noted that $-o(r) \geq \varphi(r) \geq -o(2r)/2$ for all positive r, whence $\varphi(r) \to 0$ as $r \downarrow 0$. We also have

$$f(y) \geq g(y) := f(x) + \langle \zeta, y - x \rangle + \varphi(\|y - x\|),$$

for all y near x. Since $\varphi(0) = 0$, it follows that $f - g$ has a local minimum at x. Clearly g is continuous, so it remains only to verify that $g'(x)$ exists and equals ζ; i.e., that the function $w(y) := \varphi(\|y - x\|)$ has $w'(x) = 0$.

For any nonzero v in $\mathbb{R}^n$ we have

$$0 \geq \frac{w(x + tv) - w(x)}{t} = \frac{\varphi(\|tv\|)}{t} = \|v\| \frac{\varphi(\|tv\|)}{\|tv\|} \geq \frac{-o(2t\|v\|)}{2t\|v\|}(\|v\|).$$

Upon taking the limit as $t \downarrow 0$ we see that the convergence to 0 is uniform on bounded sets of v, and so the Fréchet derivative of w at x exists and satisfies $w'(x) = 0$. This completes the proof. □

4.13. Exercise. Show that Proposition 4.12 remains true if in its statement the word "continuous" is replaced by "Lipschitz" or by "continuously differentiable." What if it is replaced by "C^2"?

Of course all the results of this section concerning Df and $\partial_D f$ have counterparts in terms of upper derivates and supergradients of an upper semicontinuous function f. When f is continuous, we can simply define the D-superdifferential $\partial^D f(x)$ to be $-\partial_D(-f)(x)$.

4.14. Exercise. Let f be continuous near x, and suppose that both $\partial_D f(x)$ and $\partial^D f(x)$ are nonempty. Prove that $f'(x)$ exists, and that

$$\partial_D f(x) = \partial^D f(x) = \{f'(x)\}.$$

Partial Subdifferentials

Let $f(x, y)$ be a function of two real variables, and suppose that the partial derivatives $\partial f/\partial x$ and $\partial f/\partial y$ both exist at $(0, 0)$. As we know from classical theory, it does *not* follow that f is differentiable at $(0, 0)$; some extra hypothesis is required to give this conclusion. (The classical result requires that one of the partials be defined and continuous in (x,y) near $(0,0)$.) In other words, differentiability cannot be confirmed "one direction at a time."

Similar considerations hold in the nonsmooth setting: that f has a proximal subgradient ζ relative to x and a proximal subgradient ξ relative to y does not imply that (ζ, ξ) is a proximal subgradient of f as a function of both variables (though the converse is true). In the setting of the D-calculus, the following exercise illustrates the same point.

4.15. Exercise. Let $f(x,y)$ be a Lipschitz function on $\mathbb{R}^2$.

(a) If $\partial f/\partial x(0,0)$ exists, show that

$$\partial_D(f(\cdot,0))(0) = \{\partial f/\partial x(0,0)\},$$

where $\partial_D(f(\cdot,0))(0)$ signifies the D-subdifferential of the function $g(x) := f(x,0)$ at the point $x = 0$. If $(\zeta, \xi) \in \partial_D f(0,0)$, show that ζ belongs to $\partial_D(f(\cdot,0))(0)$. Deduce that if the gradient $\nabla f(0,0)$ exists (i.e., the vector whose components are the two partial derivatives of f at $(0,0)$), then

$$\partial_D f(0,0) \subseteq \{\nabla f(0,0)\}.$$

(b) By considering the case $f(x,y) = -\min[|x|,|y|]$, show that $\partial_D f(0,0)$ can be empty even if $\nabla f(0,0)$ exists.

Despite the rather negative nature of this example, there does turn out to be a relation in finite dimensions between partial and full subdifferentials that holds "often" in the presence of Lipschitz behavior. The proof of the following will suppose known the fact that a locally Lipschitz function of a single variable is differentiable almost everywhere in the sense of Lebesgue measure on the line.

4.16. Theorem. *Let $f\colon \mathbb{R}^n \times \mathbb{R} \to \mathbb{R}$ be a Lipschitz function. Then for all $x \in \mathbb{R}^n$, for y in $\mathbb{R}$ almost everywhere, the following holds:*

$$\partial_D f(x,y) = \partial_D f(\cdot, y)(x) \times \partial_D f(x, \cdot)(y)$$
$$= \partial_D f(\cdot, y)(x) \times \left\{\frac{\partial f}{\partial y}(x,y)\right\}. \quad (6)$$

Proof. Fix any $x \in \mathbb{R}^n$, and let us limit attention to the set Y of y for which $(\partial f/\partial y)(x,y)$ exists, a set which differs from $\mathbb{R}$ by a null set, and which is therefore measurable. The left side of (6) is always contained in the right, for any $y \in Y$, by Exercise 4.15(a), so it suffices to prove that the opposite inclusion holds for almost all $y \in Y$. In turn, this would follow from the fact that for almost all $y \in Y$, we have

$$Df(x,y;v,w) \geq Df(x,y;v,0) + w\frac{\partial f}{\partial y}(x,y) \ \forall v \in \mathbb{R}^n, \ \forall w \in \mathbb{R}. \quad (7)$$

4 Derivate Calculus and Rademacher's Theorem

This is what we will prove.

The object, then, is to prove that the following subset S of $\mathbb{R}$ has measure 0 (or is contained in a set of measure zero, since Lebesgue measure is complete):

$$S := \left\{ y \in Y \colon \exists (v,w) \text{ such that } Df(x,y;v,0) \right. \\ \left. > Df(x,y;v,w) - w\frac{\partial f}{\partial y}(x,y) \right\}.$$

Note that a direction (v,w) corresponding to y as in the definition of S must have $w \neq 0$. If $y \in S$ and (v,w) is (one of) its corresponding directions, then there exist $r \in \mathbb{R}$ and $\varepsilon > 0$ such that

$$\frac{f(x+tv,y) - f(x,y)}{t} > r > Df(x,y;v,w) - w\frac{\partial f}{\partial y}(x,y)$$

$$\text{for } 0 < t < \varepsilon. \quad (8)$$

We label $C(v,w,r,\varepsilon)$ the set of $y \in Y$ for which (8) holds.

Now let $\{v_i\}$, $\{w_i\}$, $\{r_i\}$, and $\{\varepsilon_i\}$ be countable dense sets in $\mathbb{R}^n$, $\mathbb{R}$, $\mathbb{R}$, and $(0,\infty)$, respectively. Note that all the functions of (v,w) appearing in (8) are continuous (see Exercise 4.1(a)), and (crucial fact!) that

$$\left| \frac{f(x+tv,y)}{t} - \frac{f(x+tv_i,y)}{t} \right| \leq K\|v - v_i\|,$$

independently of $t > 0$, where K is the Lipschitz constant for f. It follows that if (8) holds for a given y, then it also holds (for the same y) with (v,w,r,ε) replaced by suitably close $(v_i, w_i, r_i, \varepsilon_i)$ (with $\varepsilon_i < \varepsilon$). In other words, S is contained in the countable union of the sets $C(v_i, w_i, r_i, \varepsilon_i)$, and so it suffices to prove that each such set has measure 0. We will drop the indices for this last part of the proof, and we suppose $w \neq 0$ (for otherwise $C(v,w,r,\varepsilon)$ is empty).

We will need some measurability results.

Lemma. *The function $y \mapsto Df(x,y;v,w)$ is measurable.*

Proof. For each positive integer j, let $\{t_i^j\}_{i=1}^\infty$ be a countable dense set in $(0, 1/j)$. Then, for any $\alpha \in \mathbb{R}$, we have

$$\{y \in \mathbb{R} \colon Df(x,y;v,w) > \alpha\}$$

$$= \bigcup_j \bigcap_i \left\{ y \in \mathbb{R} \colon \frac{f(x + t_i^j v, y + t_i^j w) - f(x,y)}{t_i^j} > \alpha + \frac{1}{j} \right\}.$$

3. Special Topics

The right side is a countable union/intersection of measurable sets, and so is itself measurable. The lemma follows.

We now ask the reader to contribute the following facts to the proof:

4.17. Exercise.

(a) The function $y \to \partial f/\partial y(x,y)$ defined on Y is measurable.

(b) The subset $C(v,w,r,\varepsilon)$ of Y defined by (8) is measurable.

Now let y_0 belong to $C(v,w,r,\varepsilon)$; i.e., let (8) hold. Choose $\delta > 0$ so that $\delta < 4K|w|$, and small enough so that

$$r > Df(x, y_0; v, w) - w\frac{\partial f}{\partial y}(x, y_0) + \delta, \tag{9}$$

and let $\{t_i\}$ be a sequence "realizing" $Df(x, y_0; v, w)$; i.e., such that

$$\lim_{i \to \infty} \frac{f(x + t_i v, y_0 + t_i w) - f(x, y_0)}{t_i} = Df(x, y_0; v, w).$$

Of course, we also have

$$\lim_{i \to \infty} \frac{f(x, y_0 + t_i w) - f(x, y_0)}{t_i} = w\frac{\partial f}{\partial y}(x, y_0).$$

It follows from (9) that for some i_0, for all $i \geq i_0$, the following holds:

$$r > \frac{f(x + t_i v, y_0 + t_i w) - f(x, y_0)}{t_i}$$
$$- \frac{f(x, y_0 + t_i w) - f(x, y_0)}{t_i} + \frac{\delta}{2},$$
$$= \frac{f(x + t_i v, y_0 + t_i w) - f(x, y_0 + t_i w)}{t_i} + \frac{\delta}{2}. \tag{10}$$

We now consider any point $y \in C(v, w, r, \varepsilon)$ (i.e., satisfying (8)). Then, putting $t = t_i$, we have, for i sufficiently large,

$$\frac{f(x + t_i v, y) - f(x, y)}{t_i} > r.$$

Combining this with (10) for $i \geq i_0$ and using the Lipschitz constant K for f yields

$$2K\frac{|y - y_0 - t_i w|}{t_i} > \frac{\delta}{2}.$$

This says that for all i sufficiently large, we have

$$\left\{(y_0 + t_i w) + \frac{\delta}{4K}t_i B\right\} \cap S = \emptyset.$$

Thus, letting I_i be the interval whose endpoints are y_0 and $y_0 + t_i w$, and putting $C := C(v, w, r, \varepsilon)$, we deduce (where $\mathcal{L}$ denotes Lebesgue measure on $\mathbb{R}$),

$$\mathcal{L}\{I_i \cap C\} \leq \left(1 - \frac{\delta}{4K|w|}\right) t_i |w|, \quad i \geq i_0.$$

Consequently,

$$\limsup_{i \to \infty} \frac{\mathcal{L}(I_i \cap C)}{\mathcal{L}(I_i)} \leq 1 - \frac{\delta}{4K|w|} < 1. \tag{11}$$

Let $g \colon \mathbb{R} \to \mathbb{R}$ be the function

$$g(t) := \int_0^t \chi_C(s)\, ds = \mathcal{L}\{[0,t] \cap C\},$$

where χ_C is the characteristic function of C. Then we know from integration theory on the line that for almost all t, $g'(t)$ exists and coincides with $\chi_C(t)$. But the conclusion (11), written in terms of g, says

$$\limsup_{i \to \infty} \frac{g(y_0 + t_i w) - g(y_0)}{t_i w} < 1.$$

Consequently, $g'(y_0)$ cannot exist and be equal to 1, for any $y_0 \in C$. Thus C has measure 0. $\square$

Rademacher's Theorem

Theorem 4.16 does not affirm anything in regard to the nonemptiness of $\partial_D f(x, y)$; we address that issue now. Note to begin with that the gradient of f, $\nabla f(x)$, exists for almost all x when f is a Lipschitz function on $\mathbb{R}^n$. This is because for each given $(x_1, x_2, \ldots, x_{i-1}, x_i, x_{i+1}, \ldots, x_n)$, the function

$$t \mapsto f(x_1, x_2, \ldots, x_{i-1}, t, x_{i+1}, \ldots, x_n)$$

is Lipschitz on $\mathbb{R}$, whence $\partial f / \partial x_i$ exists for almost all values of t. Thus the set Ω_i of points x in $\mathbb{R}^n$ at which the partial derivative $\partial f / \partial x_i$ fails to exist has the property that the linear measure of Ω_i intersected with any line parallel to the ith coordinate axis is zero. By Fubini's Theorem on iterated integration, it follows that Ω_i has measure 0 in $\mathbb{R}^n$. Consequently $\Omega = \bigcup_{i=1}^n \Omega_i$ has measure 0, and $\nabla f(x)$ exists for all $x \in \mathbb{R}^n \setminus \Omega$.

The mere existence of $\nabla f(x)$ does not imply that it belongs to $\partial_D f(x)$, or even that the latter is nonempty, as seen in Exercise 4.15. But we have the following:

4.18. Corollary. *With f as in the theorem, we have $\nabla f(x) \in \partial_D f(x)$ for almost all x in $\mathbb{R}^n$.*

148   3. Special Topics

Proof. The assertion is known in the case $n = 1$ (see Exercise 4.15). Let us assume it for dimension $n \geq 1$ and derive it for $n + 1$. For all $x \in \mathbb{R}^n$, property (7) holds almost everywhere y. It follows (from Fubini's Theorem again) that (7) holds for almost all (x, y) in $\mathbb{R}^n \times \mathbb{R}$. Similarly, by the induction hypothesis we have, for each $y \in \mathbb{R}$, for almost all x,

$$\nabla_x f(x, y) \in \partial_D(f(\cdot, y))(x). \tag{12}$$

It follows that (12) is valid for almost all $(x, y) \in \mathbb{R}^n \times \mathbb{R}$. Let us now take (x, y) in that set of full measure in which both (7) and (12) hold. Then, for any $(v, w) \in \mathbb{R}^n \times \mathbb{R}$, we have

$$\begin{aligned} Df(x, y; v, w) &\geq Df(x, y; v, 0) + w\frac{\partial f}{\partial y}(x, y) \quad \text{(by (7))} \\ &\geq \langle \nabla_x f(x, y), v \rangle + w\frac{\partial f}{\partial y}(x, y) \quad \text{(by (12))} \\ &= \langle \nabla f(x, y), (v, w) \rangle. \end{aligned}$$

Thus $\nabla f(x, y) \in \partial_D f(x, y)$. □

The following fact is known as *Rademacher's Theorem*.

4.19. Corollary. *A Lipschitz function on $\mathbb{R}^n$ is Fréchet differentiable almost everywhere.*

Proof. Apply Corollary 4.17 to f and $-f$; we deduce that for almost all x, both $\partial_D f(x)$ and $\partial^D f(x)$ are nonempty. But then $f'(x)$ exists for all such x, as noted in Exercise 4.14. □

We remark that Rademacher's Theorem and the theorem that implied it are essentially local: Their conclusions hold on any open set upon which f is Lipschitz. This is clear (for example) from the fact that a Lipschitz function on any bounded set can be extended to the whole space so as to be globally Lipschitz (see Problem 1.11.6).

5   Sets in L^2 and Integral Functionals

Many of the most important infinite-dimensional applications of nonlinear analysis involve Lebesgue measure and integration. In this section we will study two canonical illustrations of a set and a functional defined in such terms, namely

$$S := \{x \in L_n^2[a, b] : x(t) \in E(t) \text{ a.e.}\} \ (E(\cdot) \text{ a given multifunction}), \tag{1}$$

$$f(x) := \int_a^b \varphi(x(t)) \, dt \ (x(\cdot) \in L_n^2[a, b], \varphi \text{ a given function}). \tag{2}$$

Along the way we will develop the theory of measurable selections of multifunctions, and we will conclude by deriving the nonsmooth analogue of the Euler equation of the classical calculus of variations.

Consider first the set S defined by (1); S is called a *unilateral constraint set* in $L_n^2[a,b] =: X$, the underlying Hilbert space throughout this section.

5.1. Exercise. Let $E(t) = E\ \forall t$, where E is a given closed subset of $\mathbb{R}^n$. Prove that S is closed, and that S is convex iff E is convex.

We would like to characterize in terms of the underlying multifunction $E(\cdot)$ those vectors $\zeta(\cdot) \in L_n^2[a,b]$ which are proximal normals to S at $x \in S$. Accordingly, let $\zeta \in N_S^P(x)$; then, for some $\sigma \geq 0$ we have

$$\langle \zeta, x' - x \rangle \leq \sigma \|x' - x\|^2 \ \forall x' \in S.$$

When the L^2 inner product and norm are expressed in integral terms, this becomes:

$$\int_a^b \left\{ \langle -\zeta(t), x'(t) - x(t) \rangle + \sigma \|x'(t) - x(t)\|^2 \right\} dt \geq 0 \ \forall x'(\cdot) \in S. \qquad (3)$$

This inequality holds for any measurable and square-integrable function $x'(\cdot)$ taking values in $E(\cdot)$. It is tempting to conclude that (almost everywhere) the integrand is nonnegative on $E(t)$:

$$\langle -\zeta(t), x' - x(t) \rangle + \sigma \|x' - x(t)\|^2 \geq 0 \ \forall x' \in E(t), \text{ a.e.} \qquad (4)$$

As we will see, we are justified in yielding to this temptation under mild assumptions. The passage from an integrated to a pointwise conclusion (i.e., from (3) to (4)) can be made rigorous through the theory of *measurable selections*, which we now pause to develop. We will return later to characterizing $\zeta \in N_S^P(x)$, taking up the argument at (3).

Measurable Multifunctions

A multifunction Γ mapping $\mathbb{R}^m$ to the subsets of $\mathbb{R}^n$ is called *measurable* provided that the set

$$\Gamma^{-1}(V) := \{u \in \mathbb{R}^m : \Gamma(u) \cap V \neq \emptyset\}$$

is (Lebesgue) measurable for every open subset V of $\mathbb{R}^n$. In order to gain some familiarity with this useful concept, we recommend the following extended exercise.

5.2. Exercise.

(a) If Γ is measurable, then its *domain* $\text{dom}\,\Gamma := \{u : \Gamma(u) \neq \emptyset\}$ is a measurable set, and the multifunction $\widetilde{\Gamma}(u) := \text{cl}\,\Gamma(u)$ is also measurable.

(b) Γ is measurable iff $\Gamma^{-1}(V)$ is measurable for every closed set V (or compact set V, or (open or closed) ball V).

(c) Let $g\colon \mathbb{R}^m \times \mathbb{R}^n \to \mathbb{R}^k$ be such that for every $u \in \mathbb{R}^m$, $x \mapsto g(u,x)$ is continuous, and for every $x \in \mathbb{R}^n$, $u \mapsto g(u,x)$ is measurable. Set
$$\Gamma(u) := \{x \in \mathbb{R}^n : g(u,x) = 0\}.$$
Prove that Γ is measurable. (*Hint.* Let V be compact, and let $\{v_i\}$ be a countable dense set in V; show that we have
$$\Gamma^{-1}(V) = \bigcap_{i=1}^\infty \bigcup_{j=1}^\infty \left\{u \colon |g(u,v_j)| < \frac{1}{i}\right\}.)$$

(d) Γ is called *closed-valued* if $\Gamma(u)$ is a closed set for every $u \in \mathbb{R}^m$. Prove that when Γ is closed-valued, then Γ is measurable iff the function $u \mapsto d(x, \Gamma(u))$ mapping $\mathbb{R}^m$ to $[0,\infty]$ is measurable for each $x \in \mathbb{R}^n$.

(e) The *graph* of Γ, $\operatorname{gr}\Gamma$, is of course the set
$$\{(u,x) \in \mathbb{R}^m \times \mathbb{R}^n : x \in \Gamma(u)\}.$$
Prove that Γ is measurable if $\operatorname{gr}\Gamma$ is closed.

(f) A function $\gamma\colon \mathbb{R}^m \to \mathbb{R}^n$ is measurable iff the multifunction $\Gamma(u) := \{\gamma(u)\}$ is measurable, and when $n=1$, iff $\Gamma(u) := [\gamma(t), \infty)$ is measurable.

(g) Let Γ have closed graph, and let $\theta\colon \mathbb{R}^\ell \to \mathbb{R}^m$ be measurable. Prove that the multifunction $w \mapsto \Gamma(\theta(w))$ is measurable. (*Hint.*
$$\{w \colon \Gamma(\theta(w)) \cap V \ne \emptyset\} = \theta^{-1}(\Gamma^{-1}(V)).)$$
Use (f) to deduce that $\gamma(\theta(\cdot))$ is measurable if $\gamma\colon \mathbb{R}^m \to \mathbb{R}$ is lower semicontinuous.

(h) If Γ_1 and Γ_2 are two closed-valued measurable multifunctions, then $\Gamma(u) := \Gamma_1(u) \cap \Gamma_2(u)$ defines another closed-valued measurable multifunction. (*Hint.* Let V, $\{v_i\}$ be as in (c), and observe
$$\{u \colon \Gamma(u) \cap V \ne \emptyset\}$$
$$= \bigcap_{i=1}^\infty \bigcup_{j=1}^\infty \left\{u \colon d(v_j, \Gamma_1(u)) + d(v_j, \Gamma_2(u)) < \frac{1}{i}\right\}.)$$

(i) Let g be as in (c), with $k=1$. Prove that for any scalars c and d, the following multifunction is measurable:
$$\Gamma(u) := \{x \in \mathbb{R}^n : c \le g(u,x) \le d\}.$$

5 Sets in L^2 and Integral Functionals

5.3. Theorem (Measurable Selection). *Let Γ be closed-valued and measurable. Then there exists a measurable function γ such that*

$$\gamma(u) \in \Gamma(u) \ \forall u \in \operatorname{dom} \Gamma.$$

Proof. Let $\Delta := \operatorname{dom} \Gamma$. We begin by noting that for any ζ in $\mathbb{R}^n$ the function $s \to d_{\Gamma(s)}(\zeta)$ is measurable on Δ (where $d_{\Gamma(s)}$ is as usual the Euclidean distance function), since

$$\{s \in \Delta : d_{\Gamma(s)}(\zeta) \leq \alpha\} = \{s \in \Delta : \Gamma(s) \cap [\zeta + \alpha \overline{B}] \neq \emptyset\}.$$

Now let $\{\zeta_i\}$ be a countable dense subset of $\mathbb{R}^n$, and define a function $\gamma_0 : \Delta \to R^n$ as follows:

$$\gamma_0(s) = \text{ the first } \zeta_i \text{ such that } d_{\Gamma(s)}(\zeta_i) \leq 1.$$

Lemma. *The functions $s \to \gamma_0(s)$ and $s \to d_{\Gamma(s)}(\gamma_0(s))$ are measurable.*

To see this, observe that γ_0 assumes countably many values, and that, for each i,

$$\{s : \gamma_0(s) = \zeta_i\} = \bigcap_j \{s : d_{\Gamma(s)}(\zeta_j) > 1\} \cap \{s : d_{\Gamma(s)}(\zeta_i) \leq 1\},$$

where the intersection is over $j = 1, \ldots, i-1$. This implies that γ_0 is measurable. To complete the proof of the lemma, we need only note

$$\{s : d_{\Gamma(s)}(\gamma_0(s)) > \alpha\} = \bigcup_j [\{s : \gamma_0(s) = \zeta_j\} \cap \{s : d_{\Gamma(s)}(\zeta_j) > \alpha\}],$$

where the union is over the positive integers j.

We pursue the process begun above by defining for each integer i a function γ_{i+1} such that $\gamma_{i+1}(s)$ is the first ζ_j for which both the following hold:

$$\|\zeta_j - \gamma_i(s)\| \leq \tfrac{2}{3} d_{\Gamma(s)}(\gamma_i(s)), \quad d_{\Gamma(s)}(\zeta_j) \leq \tfrac{2}{3} d_{\Gamma(s)}(\gamma_i(s)).$$

It follows much as above that each γ_i is measurable. Furthermore, we deduce the inequalities

$$d_{\Gamma(s)}(\gamma_{i+1}(s)) \leq \left(\tfrac{2}{3}\right)^i d_{\Gamma(s)}(\gamma_0(s)) \leq \left(\tfrac{2}{3}\right)^i,$$

together with $\|\gamma_{i+1}(s) - \gamma_i(s)\| \leq \tfrac{2^{i+1}}{3}$. It follows that $\{\gamma_i(s)\}$ is a Cauchy sequence converging to a value $\gamma(s)$ for each s, and that γ is a measurable selection for Γ. □

5.4. Exercise.

(a) Let $\varphi : \mathbb{R}^n \to \mathbb{R}$ be a Lipschitz function, and let $x(t)$ be a measurable mapping from $\mathbb{R}$ to $\mathbb{R}^n$. Prove that the multifunctions $t \mapsto \partial_L \varphi(x(t))$ and $t \mapsto \partial_C \varphi(x(t))$ are measurable.

(b) Let the multifunction Γ from $\mathbb{R}$ to $\mathbb{R}^n$ be compact-valued and measurable, and let $v\colon \mathbb{R} \to \mathbb{R}^n$ be measurable. Prove the existence of a measurable selection γ for Γ such that, for all $t \in \operatorname{dom}\Gamma$, we have

$$\langle \gamma(t), v(t) \rangle = \max\{\langle \gamma', v(t)\rangle \colon \gamma' \in \Gamma(t)\}.$$

In the following, the multifunction $E(\cdot)$ and the subset S of $L_n^2[a,b]$ are related as in (1).

5.5. Corollary. *Let $g\colon [a,b] \times \mathbb{R}^n \to \mathbb{R}$ be such that $t \mapsto g(t,x)$ is measurable for each x, and $x \mapsto g(t,x)$ is continuous for t a.e. Let the multifunction $E(\cdot)$ be measurable and closed-valued. Suppose that we have*

$$\int_a^b g\big(t, x(t)\big)\,dt \geq 0$$

whenever $x(\cdot)$ belongs to S and the integral is defined, and suppose that for a certain $x_0(\cdot) \in S$ we have $g\big(t, x_0(t)\big) = 0$ a.e. Then

$$g(t,x) \geq 0 \ \forall x \in E(t),\ t \text{ a.e.}$$

Proof. It suffices to prove that for any $k > 0$, the following set has measure 0:

$$\{t\colon \text{for some } x \in E(t) \cap \overline{B}(0;k) \text{ we have } -k \leq g(t,x) \leq -1/k\}$$

(why?). If this is not the case, then the domain of the following multifunction Γ is of positive measure:

$$\Gamma(t) := \{x \in E(t) \cap \overline{B}(0,k)\colon -k \leq g(t,x) \leq -1/k\}.$$

Further, Γ is measurable, in view of Exercise 5.2(h,i). We invoke Theorem 5.3, the Measurable Selection Theorem, to deduce the existence of a selection $x(\cdot)$ for Γ on $\operatorname{dom}\Gamma$. We extend $x(\cdot)$ to all of $[a,b]$ by defining $x(t) = x_0(t)$ for $t \in [a,b]\setminus \operatorname{dom}\Gamma$. Then $x(\cdot)$ belongs to S, the function $x \mapsto g\big(t, x(t)\big)$ is integrable, and its integral over $[a,b]$ is strictly negative. This contradiction proves the corollary. $\square$

Returning now to the inequality (3), and assuming henceforth that $E(\cdot)$ is measurable and closed-valued, we are ready to pursue the analysis of $N_S^P(x)$.

5.6. Exercise. Show that if (3) holds, then (4) does, so that

$$\zeta(t) \in N_{E(t)}^P\big(x(t)\big) \text{ a.e.}$$

The condition (4) that we have obtained differs from the pointwise inclusion $\zeta(t) \in N^P_{E(t)}(x(t))$ in one important respect: The latter implies that for almost all t, there exists $\sigma(t) \geq 0$ such that

$$\langle -\zeta(t), x' - x(t)\rangle + \sigma(t)\|x' - x(t)\|^2 \geq 0 \ \forall x' \in E(t).$$

But in (4) the *same* σ independent of t appears. To take account of this, let us denote by $N^{P,\sigma}_E(x)$ those vectors ζ satisfying

$$\langle \zeta, x' - x\rangle \leq \sigma\|x' - x\|^2 \ \forall x' \in E.$$

5.7. Proposition. $\zeta \in N^P_S(x)$ *iff for some* $\sigma \geq 0$, *we have*

$$\zeta(t) \in N^{P,\sigma}_{E(t)}(x(t)) \ a.e.$$

5.8. Exercise. Prove Proposition 5.7.

An Integral Functional

We turn now to the integral functional f on $X := L^2_n[a, b]$ defined by

$$f(x) := \int_a^b \varphi(x(t))\, dt,$$

where for simplicity we will suppose that φ is a globally Lipschitz function on $\mathbb{R}^n$, with Lipschitz constant K.

5.9. Exercise. Prove that f is well defined and finite on X, and globally Lipschitz with Lipschitz constant $K(b-a)^{1/2}$.

Our first interest is to study $\partial_P f$; accordingly, let $\zeta \in \partial_P f(x)$, and let us write the corresponding proximal subgradient inequality: For some $\sigma \geq 0$ and $\eta > 0$ we have

$$\int_a^b \{\varphi(y(t)) - \varphi(x(t)) + \sigma\|y(t) - x(t)\|^2 - \langle \zeta(t), y(t) - x(t)\rangle\}\, dt \geq 0 \quad (5)$$

whenever $\|y - x\|_2 < \eta$. What can be deduced from this?

5.10. Theorem. *If* $\zeta \in \partial_P f(x)$, *then* $\zeta(t) \in \partial_P \varphi(x(t))$ *a.e.*

We remark that the theorem implies that for any $v(\cdot) \in X$, we have

$$\langle \zeta, v\rangle = \int_a^b \langle \zeta(t), v(t)\rangle\, dt,$$

where $\zeta(t) \in \partial_P \varphi(x(t))$ a.e. In abridged notation, this might be written $\partial_P \int \varphi = \int \partial_P \varphi$, which reveals the proposition as an analogue of the technique known as "differentiating under the integral."

Proof. Let any $M > 0$ be given. We will prove that for almost all t, we have
$$\varphi(y) - \varphi(x(t)) + \sigma\|y - x(t)\|^2 - \langle \zeta(t), y - x(t)\rangle \geq 0 \ \forall y \in M\overline{B}, \qquad (6)$$
which, since M is arbitrary, leads to the desired conclusion.

Pick c, d in (a, b) with $c < d$ such that
$$M(d-c)^{1/2} + \left\{\int_c^d \|x(t)\|^2 \, dt\right\}^{1/2} < \eta,$$
where η is defined in connection with inequality (5). We define
$$\hat{\varphi}(t, y) := \begin{cases} 0 & \text{if } t \notin [c, d], \\ \varphi(y) - \varphi(x(t)) + \sigma\|y - x(t)\|^2 - \langle \zeta(t), y - x(t)\rangle & \text{if } t \in [c, d]. \end{cases}$$

Let $E := M\overline{B}$, and let $y(\cdot)$ be any element of $L_n^2[a, b]$ satisfying $y(t) \in E$ a.e. Then, if $\hat{y}$ is the element of X defined by
$$\hat{y}(t) := \begin{cases} x(t) & \text{if } t \notin [c, d], \\ y(t) & \text{if } t \in [c, d], \end{cases}$$
we have
$$\|\hat{y} - x\| = \left\{\int_c^d \|y(t) - x(t)\|^2 \, dt\right\}^{1/2}$$
$$\leq \left\{\int_c^d \|y(t)\|^2 \, dt\right\}^{1/2} + \left\{\int_c^d \|x(t)\|^2 \, dt\right\}^{1/2}$$
$$\leq M(d-c)^{1/2} + \left\{\int_c^d \|x(t)\|^2 \, dt\right\}^{1/2} < \eta.$$

Thus (5) may be invoked to yield
$$\int_a^b \hat{\varphi}(t, y(t)) \, dt = \int_c^d \hat{\varphi}(t, y(t)) \, dt$$
$$= \int_c^d \{\varphi(y(t)) - \varphi(x(t)) + \sigma\|y(t) - x(t)\|^2 - \langle\zeta(t), y(t) - x(t)\rangle\} \, dt$$
$$= \int_a^b \{\varphi(\hat{y}(t)) - \varphi(x(t)) + \sigma\|\hat{y}(t) - x(t)\|^2 - \langle\zeta(t), \hat{y}(t) - x(t)\rangle\} \, dt$$
$$\geq 0.$$

Note also that $\hat{\varphi}(t, x(t)) = 0$ a.e. Since $\hat{\varphi}$ is measurable in t and continuous in y, it follows now from Corollary 5.5 that for t a.e., we have
$$\hat{\varphi}(t, y) \geq 0 \ \forall y \in E = M\overline{B}.$$

Consequently, (6) holds for $t \in [c, d]$ a.e. But since $[c, d]$ is an arbitrary (small) subinterval of $[a, b]$, (6) in fact holds a.e. in $[a, b]$, as claimed. □

5.11. Exercise.

(a) Let $\partial_P^\sigma \varphi(x)$ denote the "global proximal subgradients of rank σ"; i.e., those ζ satisfying
$$\varphi(y) - \varphi(x) + \sigma \|y-x\|^2 \geq \langle \zeta, y-x \rangle \ \forall y \in \mathbb{R}^n.$$
Show that in Theorem 5.10, the conclusion may be strengthened to $\zeta(t) \in \partial_P^\sigma \varphi(x(t))$ a.e., and that conversely this last property (for a given $\sigma \geq 0$) implies that ζ belongs to $\partial_P f(x)$.

(b) If $\zeta \in \partial_P \varphi(x)$, then show that for some $\sigma > 0$ we actually have $\zeta \in \partial_P^\sigma \varphi(x)$. (This uses the global Lipschitz hypothesis on φ.)

The exercise shows that a certain *global* and *uniform* character of the pointwise condition $\zeta(t) \in \partial_P \varphi(x(t))$ is required to imply $\zeta \in \partial_P f(x)$. Nonetheless, an approximate converse to Theorem 5.10 is provided by the following.

5.12. Proposition. *Let $\zeta \in L_n^2[0,1]$ be such that $\zeta(t) \in \partial_P \varphi(x(t))$ a.e. Then for any $\varepsilon > 0$ there exist $x' \in X$ and $\zeta' \in \partial_P f(x')$ with $\|\zeta' - \zeta\| < \varepsilon$, $\|x' - x\| < \varepsilon$.*

Proof. According to Exercise 5.11(b), there exists for almost each t a number $\sigma(t)$ such that
$$\zeta(t) \in \partial_P^{\sigma(t)} \varphi(x(t)) \text{ a.e.}$$
In fact, let us take $\sigma(t)$ equal to the first positive integer k such that $\zeta(t) \in \partial_P^k \varphi(x(t))$. Then, for any positive integer k, we have

$\{t \in [a,b] \colon \sigma(t) = k\}$
$= \{t \colon \varphi(y) - \varphi(x(t)) + k\|y - x(t)\|^2 \geq \langle \zeta(t), y - x(t) \rangle \ \forall y \in \mathbb{R}^n\}$
$\bigcap_{j=1}^{k-1} \{t \colon \varphi(y) - \varphi(x(t)) + j\|y - x(t)\|^2 < \langle \zeta(t), y - x(t) \rangle \text{ for some } y \in \mathbb{R}^n\}.$

This implies that the function $\sigma(\cdot)$ is measurable. Now let $(\bar{x}, \bar{\zeta}, \bar{\sigma})$ be any triple such that $\bar{\zeta} \in \partial_P^{\bar{\sigma}} \varphi(\bar{x})$ (Why is there such a triple?). For every $M > \bar{\sigma}$, let Ω_M be the set $\{t \colon \sigma(t) < M\}$, and define

$$x'(t) = \begin{cases} x(t) & \text{if } t \in \Omega_M, \\ \bar{x} & \text{otherwise,} \end{cases} \quad \zeta'(t) = \begin{cases} \zeta(t) & \text{if } t \in \Omega_M, \\ \bar{\zeta} & \text{otherwise.} \end{cases}$$

Then $\zeta'(t) \in \partial_P^M \varphi(x'(t))$ for t a.e., whence $\zeta' \in \partial_P f(x')$ by Exercise 5.11(a). Since $\text{meas}(\Omega_M) \to (b-a)$ as $M \to \infty$, it follows that $x' \to x$ and $\zeta' \to \zeta$ in X as $M \to \infty$, a final step in the proof that we ask the reader to verify. □

A Convexity Property

As witnessed above, the correspondence between the proximal subgradients of the integral functional f and those of the generating integrand φ is quite a close one. We turn now to consideration of the limiting subdifferential $\partial_L f(x)$. We will discover in Theorem 5.18 below an interesting convexification property of the weak-closure operation that generates $\partial_L f$ from $\partial_P f$. To begin understanding this property, consider the natural conjecture that if $\zeta \in \partial_L f(x)$, then $\zeta(t) \in \partial_L \varphi\bigl(x(t)\bigr)$ a.e. (We call this "natural" in view of the fact that we have established this fact when ∂_L is replaced by ∂_P.) It turns out that this conjecture is wrong, and the following example will lend insight as to why.

5.13. Exercise. We set $\varphi(x) = -|x|$, with $n = 1$. We will construct a sequence x_i in X converging to 0, with elements $\zeta_i \in \partial_P f(x_i)$ such that $\{\zeta_i\}$ converges weakly to 0. It follows that $0 \in \partial_L f(0)$, yet it is *not* the case that 0 belongs to $\partial_L \varphi(0) = \{-1, 1\}$.

We define a function $x_i(t)$ on $[0, 1]$ by setting $x_i(t) = 1/i$ on each subinterval of the form $(k/i, (k+1)/i)$, where k is an even integer in $[0, i-1]$, and $x_i(t) = -1/i$ elsewhere. Then $x_i \to 0$ in X. Let $\zeta_i(t) = -x_i(t)/|x_i(t)|$.

(a) Prove that ζ_i converges weakly to 0 in X.

(b) Prove that $\zeta_i \in \partial_P f(x_i)$ (use Exercise 5.11).

In the foregoing example, observe that although $0 \notin \partial_L \varphi(0) = \{-1, 1\}$, we do have $0 \in \operatorname{co} \partial_L \varphi(0) = \partial_C \varphi(0)$. We will obtain a general conclusion along these lines below.

The key to the analysis is a certain convexification property of integration relative to (nonatomic) measures, a phenomenon first observed by A. M. Lyapounov but reflected especially clearly below in a more recent theorem due to Aumann.

A multifunction F mapping $[a, b]$ to the subsets of $\mathbb{R}^n$ is said to be L^2-*bounded* if there is a function $k \in L_1^2[a, b]$ such that

$$\|v\| \leq k(t) \ \forall v \in F(t), \text{ a.e.}$$

By $\operatorname{co} F$ we mean the multifunction whose value at t is $\operatorname{co} F(t)$. Finally, $\int F$ signifies the set of all points of the form $\int_a^b f(t)\, dt$, where $f(\cdot)$ is an integrable selection of F.

5.14. Exercise. Let the multifunction $E(\cdot)$ of (1) be L^2-bounded, and have convex closed values. Prove that the set S of its measurable selections is weakly compact, and that $\int E$ is a compact convex subset of $\mathbb{R}^n$.

5 Sets in L^2 and Integral Functionals 157

5.15. Theorem (Aumann). *Let F be a measurable, L^2-bounded multifunction from $[a,b]$ to $\mathbb{R}^n$ whose values are closed and nonempty. Then*

$$\int F = \int \mathrm{co}\, F.$$

Step 1. It suffices to show that any point ξ lying in $\int \mathrm{co}\, F$ also lies in $\int F$; we may take $\xi = 0$ without loss of generality. Exercise 5.14 shows that $\int \mathrm{co}\, F$ is convex compact. Its dimension K is defined to be the dimension of the minimal subspace L of $\mathbb{R}^n$ containing it. If $K = 0$, $\int \mathrm{co}\, F$ is a point and must coincide with $\int F$, since the latter is nonempty as a consequence of the Measurable Selection Theorem 5.3. So we may assume $K \geq 1$, and that the theorem is true for multifunctions G such that $\dim \int \mathrm{co}\, G \leq K-1$. The proof will be by induction.

Step 2. Suppose now that 0 does not lie in the relative interior of $\int \mathrm{co}\, F$ (i.e., interior relative to L). Then there is a nonzero vector d in L normal to $\int \mathrm{co}\, F$ at 0; that is, such that

$$\langle d, w \rangle \leq 0 \; \forall w \in \int \mathrm{co}\, F.$$

Let S be the set of measurable (and necessarily, square integrable) selections of $\mathrm{co}\, F$ on $[a,b]$, and let s_0 be an element of S satisfying $\int_a^b s_0(t)\, dt = 0$. Then

$$\int_a^b \langle d, s(t) \rangle\, dt \leq 0 = \int_a^b \langle d, s_0(t) \rangle\, dt \; \forall s \in S.$$

Let $H(\cdot; F(t))$ designate the upper support function of $F(t)$, or equivalently, of $\mathrm{co}\, F(t)$:

$$H(x; F(t)) := \max\{\langle x, f \rangle : f \in F(t)\}.$$

It follows from Exercise 5.4(b) that we have

$$\max_{s \in S} \int_a^b \langle d, s(t) \rangle\, dt = \int_a^b H(d; F(t))\, dt.$$

We deduce

$$\int_a^b \langle d, s_0(t) \rangle \leq \int_b^b H(d; F(t))\, dt \leq 0 = \int_a^b \langle d, s_0(t) \rangle\, dt,$$

whence

$$H(d; F(t)) = \langle d, s_0(t) \rangle \text{ a.e.}$$

Let us define a new multifunction $\widetilde{F}$ via

$$\widetilde{F}(t) := \{f \in F(t) : \langle f, d \rangle = H(d; F(t))\}.$$

5.16. Exercise. Show that $\widetilde{F}$ is nonempty and closed-valued, measurable on $[a,b]$, and that $s_0(t) \in \operatorname{co} \widetilde{F}(t)$ a.e.

It follows that $\widetilde{F}$ satisfies all the hypotheses of the theorem, and that $0 \in \int \operatorname{co} \widetilde{F}$. Further $\langle d, \int \operatorname{co} \widetilde{F} \rangle = \int_a^b H(d; F(t)) \, dt = 0$, and so $\dim \int \operatorname{co} \widetilde{F} \leq K - 1$. By the inductive hypothesis, $0 \in \int \widetilde{F} \subset \int F$, which is the required conclusion.

Step 3. We saw in Step 2 that if 0 is not in the relative interior of $\int \operatorname{co} F$, then a certain construction allows us to reduce the dimension and invoke the inductive hypothesis to conclude. The remaining case is that in which a δ-neighborhood of 0 (in L) lies in $\int \operatorname{co} F$, for some $\delta > 0$.

Choose any nonzero vector d_1 and define Φ_1 on $\mathbb{R}^n \times X$ as follows:

$$\Phi_1(x, s) := \int_a^b \langle d_1 t + x, s(t) \rangle \, dt.$$

Given any x in $\mathbb{R}^n$, express x as $y + c$, where y lies in L and c in its orthogonal competent $L^\perp$. Then

$$\max_{s \in S} \Phi_1(x, s) = \max_{s \in S} \Phi_1(y, s) \quad \left(\text{since } \int \operatorname{co} F \subset L \text{ and } c \in L^\perp\right)$$

$$= \max_{s \in S} \int_a^b \langle d_1 t + y, s(t) \rangle \, dt$$

$$\geq \delta \|y\| + k$$

for some k independent of x. It follows that the function

$$x \mapsto \max_{s \in S} \Phi_1(x, s)$$

has a minimum over $\mathbb{R}^n$, say at x_1.

The fact that S is weakly compact (Exercise 5.14) permits us to invoke the Minimax Theorem (see, for instance, Aubin (1993)) to deduce the existence of $s_0 \in S$ such that

$$\min_{x \in \mathbb{R}^n} \Phi_1(x, s_0) = \max_{s \in S} \min_{x \in \mathbb{R}^n} \Phi_1(x, s) = \min_{x \in \mathbb{R}^n} \max_{s \in S} \Phi_1(x, s)$$

$$= \max_{s \in S} \Phi_1(x_1, s).$$

It follows that (x_1, s_0) is a saddlepoint of Φ_1 relative to $\mathbb{R}^n \times S$, so that the function $\Phi_1(\cdot, s_0)$ is minimized over $\mathbb{R}^n$ at x_1. Setting the derivative equal to zero gives $\int_a^b s_0(t) \, dt = 0$. The other saddlepoint inequality asserts that $\Phi_1(x_1, \cdot)$ is maximized over S at s_0. But

$$\max_{s \in S} \Phi_1(x_1, \cdot) = \int_a^b H(d_1 t + x_1; F(t)) \, dt = \Phi_1(x_1, s_0)$$

(by Exercise 5.4(b)), whence

$$H(d_1t + x_1; F(t)) = \langle d_1t + x_1, s_0(t)\rangle \text{ a.e.}$$

Let us now define

$$F_1(t) := \{f \in F(t) : H(d_1t + x_1; F(t)) = \langle d_1t + x_1, f\rangle\}.$$

We may confirm that F_1 satisfies the same hypotheses as F, and $s_0(t) \in \operatorname{co} F_1(t)$ a.e., whence $0 \in \int \operatorname{co} F_1$.

We now recommence either Step 2 or Step 3, with F_1 in place of F. If Step 2 applies, we are finished, since $\int F_1 \subset \int F$. If not, we perform Step 3 again, this time with a vector d_2 linearly independent of d_1. This generates a new function Φ_2 admitting a saddlepoint at (x_2, s_1), and a corresponding multifunction $F_2 \subset F_1$ all of whose points f satisfy

$$\langle d_2t + x_2, f\rangle = H(d_2t + x_2; F_1(t)) \text{ a.e.}$$

and such that $0 \in \int \operatorname{co} F_2$.

The process continues until either Step 2 has applied or else Step 3 has been performed n times. In the latter case, we will have defined a multifunction $F_n \subset F$ such that $0 \in \int \operatorname{co} F_n$ and such that for almost all t, every f in $F_n(t)$ satisfies (we let F_0 stand for F):

$$\langle d_it + x_i, f\rangle = H(d_it + x_i; F_{i-1}(t)) \quad (i = 1, 2, \ldots, n).$$

In matrix terms, this may be written in the form

$$(Dt + M)f = \Sigma(t),$$

where the $n \times n$ matrix D is invertible (since its rows d_i are independent). But since $Dt + M$ is invertible except for those (finitely many) t which are eigenvalues of $-D^{-1}M$, it follows that $F_n(t)$ is a singleton a.e. Consequently we have $0 \in \int \operatorname{co} F_n = \int F_n \subset \int F$. □

5.17. Exercise. Under the hypotheses of the theorem, we will prove the following corollary: Let $\gamma(\cdot)$ be a measurable selection on $[a, b]$ of $\operatorname{co} F$. Then there exists a sequence $\{f_i\}$ of measurable selections of F which converges weakly to γ.

(a) For each positive integer N, let $a =: t_0 < t_1 < \cdots < t_N := b$ be a uniform partition of $[a, b]$; invoke the theorem to deduce the existence of a selection f_N for F on $[a, b]$ such that

$$\int_{t_i}^{t_{i+1}} f_N(t)\, dt = \int_{t_i}^{t_{i+1}} \gamma(t)\, dt \quad (i = 0, 1, \ldots, N-1).$$

(b) Prove that the sequence $\{f_N\}$ admits a subsequence $\{f_{N_i}\}$ converging weakly to a limit ζ (see Exercise 5.14). Then show that ζ must be γ. (*Hint.* Study the convergence of $\langle f_{N_i}, g \rangle$ when g is a smooth function, using integration by parts.)

We now return to the analysis of $\partial_L f$, where

$$f(x) = \int_a^b \varphi(x(t)) \, dt.$$

The hypotheses remain those of Theorem 5.10.

5.18. Theorem. *The limiting subdifferential and the generalized gradient of f coincide, and we have*

$$\partial_L f(x) = \partial_C f(x) = \{\zeta \in L_n^2[a,b]: \zeta(t) \in \partial_C \varphi(x(t)) \text{ a.e.}\}.$$

Proof. We set

$$\Lambda := \{\zeta = \underset{i \to \infty}{\text{w-lim}}\, \zeta_i : \zeta_i(t) \in \partial_L \varphi(x(t)) \text{ a.e.}\}.$$

The following three relations will be established in order:

$$\partial_C f(x) \subset \{\zeta : \zeta(t) \in \partial_C \varphi(x(t)) \text{ a.e.}\} \subset \Lambda \subset \partial_L f(x).$$

This evidently implies the theorem, since $\partial_L f(x) \subset \partial_C f(x)$. The set

$$\{\zeta : \zeta(t) \in \partial_C \varphi(x(t)) \text{ a.e.}\}$$

is convex and weakly compact in X (see Exercise 5.14), as is $\partial_C f(x)$. The first relation can therefore be proven via support functions; it amounts to showing that for any $v \in X$, we have

$$f^\circ(x;v) \leq \max\{\langle v, \zeta \rangle : \zeta(t) \in \partial_C \varphi(x(t)) \text{ a.e.}\}.$$

Let $\{x_i\}$ be a sequence in X converging to x and $\{\lambda_i\}$ a positive sequence converging to 0 such that

$$f^\circ(x;v) = \lim_{i \to \infty} \frac{f(x_i + \lambda_i v) - f(x_i)}{\lambda_i}.$$

The limit can be written

$$\lim_{i \to \infty} \int_a^b \frac{\varphi(x_i(t) + \lambda_i v(t)) - \varphi(x_i(t))}{\lambda_i}$$

$$\leq \int_a^b \limsup_{i \to \infty} \frac{\varphi(x_i(t) + \lambda_i v(t)) - \varphi(x_i(t))}{\lambda_i} \, dt$$

(Why has Fatou's Lemma been applicable?)

$$\leq \int_a^b \varphi^\circ\big(x(t); v(t)\big)\, dt$$

(since $x_i(t) \to x(t)$ a.e.). By Exercise 5.4(b) there exists a measurable selection $\zeta(\cdot)$ of $\partial_C \varphi(x(\cdot))$ such that $\langle \zeta(t), v(t) \rangle = \varphi^\circ\big(x(t); v(t)\big)$ a.e. Then

$$\int_a^b \varphi^\circ\big(x(t); v(t)\big)\, dt = \langle \zeta, v \rangle,$$

and the first of the three relations follows.

The second relation follows immediately from Exercise 5.17, since

$$\partial_C \varphi\big(x(t)\big) = \operatorname{co} \partial_L \varphi\big(x(t)\big).$$

There remains the third. Let $\zeta = \text{w-}\lim_{i \to \infty} \zeta_i$, where $\zeta_i(t) \in \partial_L \varphi(x(t))$ a.e. We wish to prove that $\zeta \in \partial_L f(x)$. As observed in Problem 1.11.28, there is a countable set C in $\mathbb{R}^n \times \mathbb{R}^n$ (not depending on t) of points (ξ, y) having $\xi \in \partial_P \varphi(y)$ such that for each t we have

$$\partial_L \varphi\big(x(t)\big) = \Big\{ \lim_{k \to \infty} \xi_k : (\xi_k, y_k) \in C,\ \lim_{k \to \infty} y_k = x(t) \Big\}.$$

For a given enumeration $\{(\xi_j, y_j)\}$ of C, let $j_i(t)$ be the first j such that

$$\|\xi_j - \zeta_i(t)\| < i^{-1} \quad \text{and} \quad \|y_j - x(t)\| < i^{-1}.$$

Let us set

$$\tilde{x}_i(t) := y_{j_i(t)}, \quad \tilde{\zeta}_i(t) := \xi_{j_i(t)}.$$

Then $\tilde{x}_i(\cdot)$ and $\tilde{\zeta}_i(\cdot)$ are measurable, and

$$\tilde{\zeta}_i(t) \in \partial_P \varphi\big(\tilde{x}_i(t)\big) \text{ a.e.}$$

By Proposition 5.12, there exist x_i' and ζ_i' in X such that

$$\|x_i' - \tilde{x}_i\| < i^{-1}, \quad \|\zeta_i' - \tilde{\zeta}_i\| < i^{-1}, \quad \zeta_i' \in \partial_P f(x_i').$$

It follows that $\{\zeta_i'\}$ converges weakly to ζ, and $\{x_i'\}$ strongly to x. Of course, $f(x_i') \to f(x)$ since f is Lipschitz. But then $\zeta \in \partial_L f(x)$ by definition, as required. □

5.19. Exercise. It is not the case that $\partial_L f(x)$ is convex for every Lipschitz functional f on $X := L_n^2[a, b]$. Verify this statement with $n = 1$ for the functional

$$f(x) := -\left| \int_0^1 x(t)\, dt \right|.$$

A Problem in the Calculus of Variations

Recall that a function $x\colon [a,b] \to \mathbb{R}^n$ is said to be *absolutely continuous* if it can be expressed in the form

$$x(t) = x_0 + \int_a^t v(s)\,ds$$

for some integrable function v; we then have $\dot{x}(t) := d/dt\, x(t) = v(t)$ a.e.

We now consider the *variational problem* of minimizing the functional

$$\ell\bigl(x(b)\bigr) + \int_a^b \varphi\bigl(x(t), \dot{x}(t)\bigr)\,dt$$

over those absolutely continuous functions $x\colon [a,b] \to \mathbb{R}^n$ which satisfy $x(0) = x_0$. Here the "givens" of the problem are the point x_0, the interval $[a,b]$, the function $\ell\colon \mathbb{R}^n \to \mathbb{R}$ (assumed locally Lipschitz) and the function $\varphi\colon \mathbb{R}^n \times \mathbb{R}^n \to \mathbb{R}$ (assumed globally Lipschitz).

Our goal is to derive necessary conditions for optimality. Let us rephrase the problem so as to make the relevance of earlier results more apparent. We define the following subset A of $L_{2n}^2[a,b]$:

$$A := \left\{ (u,v) \in L_n^2[a,b] \times L_n^2[a,b] \colon u(t) = x_0 + \int_a^t v(s)\,ds,\, t \in [a,b] \right\}.$$

We note that A is closed and convex. Now define f_1, f_2 on $L_{2n}^2[a,b]$ as follows:

$$f_1(u,v) := \ell\left(x_0 + \int_a^b v(t)\,dt\right), \quad f_2(u,v) := \int_a^b \varphi\bigl(u(t), v(t)\bigr)\,dt.$$

In these terms, our problem becomes:

$$\text{minimize}\{f_1(u,v) + f_2(u,v) \colon (u,v) \in A.\}$$

We have characterized $\partial_L f_2$ in Theorem 5.18; as for $\partial_L f_1$, we have:

> **5.20. Exercise.** The functional f_1 is locally Lipschitz on $L_{2n}^2[a,b]$. If $(\theta, \zeta) \in \partial_L f_1(u,v)$, then $\theta = 0$, and for some $\zeta_0 \in \partial_L \ell\bigl(x_0 + \int_a^b v(t)\,dt\bigr)$, we have $\zeta(t) = \zeta_0$ a.e.

The final component we will need is the following version of a celebrated result of the nineteenth century.

5.21. Proposition (Dubois–Reymond Lemma). *Let $(\zeta, \xi) \in N_A^P(u,v)$. Then*

$$\xi(t) = -\int_t^b \zeta(s)\,ds, \quad a \le t \le b,$$

so that ξ is absolutely continuous and satisfies $\xi(b) = 0$.

Proof. We remark that since A is convex, the normal cones N_A^P, N_A^L, and N_A^C all coincide. The normal vector (ζ, ξ) satisfies

$$\langle (\zeta, \xi), (u', v') \rangle \leq \langle (\zeta, \xi), (u, v) \rangle \ \forall (u', v') \in A,$$

which translates as

$$\int_a^b \{\zeta(t) \cdot (u'(t) - u(t)) + \xi(t) \cdot (v'(t) - v(t))\} \, dt \leq 0.$$

Bearing in mind that $u'(t) - u(t) = \int_a^t (v'(s) - v(s)) \, ds$, we use integration by parts to derive from this last inequality the following one:

$$\int_a^b \left[\xi(t) + \int_t^b \zeta(s) \, ds \right] \cdot [v'(t) - v(t)] \, dt \leq 0 \ \forall v' \in L_n^2[a, b].$$

Since $v' - v$ is an arbitrary integrable function, the required conclusion follows immediately. □

Combining all the ingredients, we arrive at a nonsmooth version of the famous Euler equation in the calculus of variations:

5.22. Theorem (Euler Inclusion). *If x solves the variational problem, then there exists an absolutely continuous function p satisfying*

$$(\dot{p}(t), p(t)) \in \partial_C \varphi(x(t), \dot{x}(t)) \text{ a.e.,}$$
$$-p(b) \in \partial_L \ell(x(b)).$$

Proof. The pair $(x, \dot{x})$ minimizes the function

$$f_1(u, v) + f_2(u, v) + I_A(u, v)$$

over $L_n^2[a, b] \times L_n^2[a, b]$. Thus

$$(0, 0) \in \partial_L \{f_1 + f_2 + I_A\}(x, \dot{x})$$
$$\subset \partial_L f_1(x, \dot{x}) + \partial_L f_2(x, \dot{x}) + N_A^C(x, \dot{x}),$$

by Proposition 1.10.1. Combining Exercise 5.20, Theorem 5.18, and Proposition 5.21, we deduce the existence of a point $\zeta_0 \in \partial_L \ell(x(b))$ and an element ζ of $L_n^2[a, b]$ such that

$$\left(-\zeta(t), -\zeta_0 + \int_t^b \zeta(s) \, ds \right) \in \partial_C \varphi(x(t), \dot{x}(t)) \text{ a.e.}$$

The statement of the theorem follows upon setting $p(t) := -\zeta_0 + \int_t^b \zeta(s) \, ds$. □

5.23. Exercise. When φ is C^1, a solution x to the variational problem satisfies the classical Euler equation

$$\frac{d}{dt}\{\varphi_v(x(t),\dot{x}(t))\} = \varphi_u(x(t),\dot{x}(t)) \text{ a.e.},$$

and the function $t \mapsto \varphi_v(x(t),\dot{x}(t))$ is continuous.

We remark that the last conclusion of this exercise is known as the *first Erdmann condition* in the classical calculus of variations. The relation $-p(b) = \nabla \ell(x(b))$ is the *transversality condition*.

A Weak Sequential Compactness Theorem

We conclude this section with a technical result of some importance in the next chapter. It concerns a multifunction E mapping $\mathbb{R} \times \mathbb{R}^n$ to the closed convex subsets of a given compact set E_0 in $\mathbb{R}^n$. We assume that E is graph-closed.

5.24. Theorem. *Let $\{v_i\}$ be a sequence in $L_n^2[a,b]$ such that*

$$v_i(t) \in E(\tau_i(t), u_i(t)) + r_i(t)\overline{B} \text{ a.e.}, \quad t \in [a,b],$$

where the sequence of measurable functions $\{\tau_i(\cdot), u_i(\cdot)\}$ converges a.e. to $(t, u_0(t))$, and where the nonnegative measurable functions $\{r_i\}$ converge to 0 in $L_1^2[a,b]$. Then there exists a subsequence $\{v_{i_j}\}$ of $\{v_i\}$ which converges weakly in $L_n^2[a,b]$ to a limit $v_0(\cdot)$ which satisfies

$$v_0(t) \in E(t, u_0(t)) \text{ a.e.}, \quad t \in [a,b].$$

Proof. The hypotheses evidently imply that the sequence $\{v_i\}$ is bounded in $L_n^2[a,b]$. Invoking weak compactness, we know that a subsequence $\{v_{i_j}\}$ converges weakly to some limit v_0; there remains only to prove that $v_0(t)$ belongs to $E(t, u_0(t))$ a.e. for $t \in [a,b]$.

We define $h: \mathbb{R} \times \mathbb{R}^n \times \mathbb{R}^n \to (-\infty, \infty]$ as follows:

$$h(t, u, p) := \min\{\langle p, v \rangle : v \in E(t, u)\}.$$

Since E is convex-valued, a point v lies in $E(t, u)$ iff $\langle p, v \rangle \geq h(t, u, p)$ $\forall p \in \mathbb{R}^n$ (by the Separation Theorem). In view of the inclusion satisfied by v_i, we also have, for any fixed $p \in \mathbb{R}^n$, the following inequality:

$$\langle p, v_i(t) \rangle + r_i(t)\|p\| \geq h(\tau_i(t), u_i(t), p) \text{ a.e.}$$

Because the multifunction E has closed graph, it follows readily that the function $(t, u) \mapsto h(t, u, p)$ is lower semicontinuous, and hence that the function

$$t \mapsto h(\tau_i(t), u_i(t), p)$$

is measurable (Exercise 5.2(g)).

Now let A be any measurable subset of $[a,b]$. We have

$$\int_A \{\langle p, v_{i_j}(t)\rangle + \|p\|r_{i_j}(t) - h(\tau_{i_j}(t), u_{i_j}(t), p)\} \, dt \geq 0,$$

since the integrand is nonnegative almost everywhere. As $j \to \infty$, we know

$$\int_A \langle p, v_{i_j}(t)\rangle \, dt \to \int_A \langle p, v_0(t)\rangle \, dt$$

by weak convergence, and of course $\int_A r_{i_j}(t) \, dt \to 0$. In light of this, and calling upon Fatou's Lemma (Why does it apply?), we deduce

$$\int_A \langle p, v_0(t)\rangle \, dt - \int_A \liminf_{j\to\infty} h(\tau_{i_j}(t), u_{i_j}(t), p) \, dt \geq 0.$$

Because $h(t, u, p)$ is lower semicontinuous in (t, u), this implies

$$\int_A \{\langle p, v_0(t)\rangle - h(t, u_0(t), p)\} \, dt \geq 0.$$

Since A is arbitrary, we must then have

$$\langle p, v_0(t)\rangle \geq h(t, u_0(t), p) \quad \text{a.e.,} \quad t \in [a, b]. \tag{7}$$

Now let $\{p_i\}$ be a countable dense subset of $\mathbb{R}^n$. Then (7) holds for each $p = p_i$, with the exception of t in an exceptional set Ω_i of measure 0 in $[a,b]$. Let $\Omega := \bigcup_i \Omega_i$. Then for all $t \notin \Omega$, (7) holds for all $p \in \{p_i\}$. But both sides of (7) define continuous functions of p, for given t. (To see that h is continuous in p, it suffices to observe that it is concave and finite as a function of p.) Therefore for any $t \notin \Omega$, (7) must in fact hold for all $p \in \mathbb{R}^n$, which is equivalent to

$$v_0(t) \in E(t, u_0(t)) \; \forall t \in \Omega.$$

Since Ω has measure 0, the proof is complete. □

We remark that the convexity of the values of E plays a crucial role in the theorem; Problem 7.19 will make this clear.

6 Tangents and Interiors

In this section we study certain properties of sets related to tangency and interiors. The discussion will be restricted to subsets of $\mathbb{R}^n$, and one of the central issues is the following: If at a point $x \in S$ the set S admits a "large" set of tangents, does it follow that S is "substantial" near x? For example, can we assert that the interior of S is nonempty, and that $x \in \text{cl int } S$?

In answering such questions, the tangent cone $T_S^C(x)$ turns out to be more useful than $T_S^B(x)$. Furthermore, the property that int $T_S^C(x) \neq \emptyset$ will serve to identify a class of sets having important properties which play a role, for example, in the theory of equilibria and in constructing feedback controls.

Let S be a closed nonempty subset of $\mathbb{R}^n$, and let $x \in S$. We will say that S is *wedged* at x provided that int $T_S^C(x) \neq \emptyset$. Since $T_S^C(x) = \mathbb{R}^n$ for any $x \in$ int S, this property is at issue only for points x lying in bdry S.

6.1. Exercise. Which of the sets described in Exercise 2.5.6 is wedged at 0?

We have already observed in Problem 2.9.10 the following fact:

6.2. Proposition. *Let $S = \{x' : f(x') \leq 0\}$, where $f : \mathbb{R}^n \to \mathbb{R}$ is Lipschitz near x and $0 \notin \partial f(x)$. Then S is wedged at x.*

6.3. Exercise.

(a) If S is given as in Proposition 6.2, then int $S \neq \emptyset$, and $x \in$ cl(int S). (*Hint.* Consider Problem 2.9.6.)

(b) By taking $n = 2$ and $f(x, y) = |x| - |y|$, show that Proposition 6.2 fails when ∂f is replaced by $\partial_L f$.

6.4. Theorem. *A vector $v \in \mathbb{R}^n$ belongs to int $T_S^C(x)$ iff there exists $\varepsilon > 0$ such that*

$$y \in x + \varepsilon B, \quad w \in v + \varepsilon B, \quad t \in [0, \varepsilon) \implies d_S(y + tw) \leq d_S(y). \quad (1)$$

Proof. Let $v \in$ int $T_S^C(x)$. If $v = 0$, then $T_S^C(x) = \mathbb{R}^n$, so that by polarity (see Proposition 2.5.4) we have $N_S^C(x) = \{0\}$. But $N_S^L(x) \subset N_S^C(x)$ is nontrivial when $x \in$ bdry S (Exercise 2.8.5), so that $x \in$ int S necessarily. But in that case property (1) evidently holds for a suitably small $\varepsilon > 0$. So let us suppose $0 \neq v \in$ int $T_S^C(x)$. Then by polarity there exists $\delta > 0$ such that

$$\langle v, \zeta \rangle \leq -\delta \|\zeta\| \ \forall \zeta \in N_S^C(x).$$

If (1) fails to hold for any $\varepsilon > 0$, then there exist sequences $\{y_i\}$, $\{w_i\}$ converging to x and v, respectively, and a positive sequence $\{t_i\}$ decreasing to 0 such that

$$d_S(y_i + t_i w_i) - d_S(y_i) > 0.$$

We invoke the Proximal Mean Value Theorem to deduce the existence of z_i converging to x and $\zeta_i \in \partial_P d_S(z_i)$ such that $\langle \zeta_i, w_i \rangle > 0$. Take a subsequence if necessary to have $\zeta_i / \|\zeta_i\|$ converge to a limit $\zeta \in \partial_L d_S(x) \subset N_S^C(x)$. We have

$$-\delta = -\delta \|\zeta\| \geq \langle v, \zeta \rangle = \lim_{i \to \infty} \langle w_i, \zeta_i \rangle / \|\zeta_i\| \geq 0,$$

a contradiction which establishes the necessity of (1).

As for the sufficiency, it is evident that (1) implies

$$d_S^\circ(x;w) \leq 0 \ \forall w \in v + \varepsilon B,$$

whence $B(v;\varepsilon) \subset T_S^C(x)$ and $v \in \operatorname{int} T_S^C(x)$. □

Let us agree to call the following set $W(v;\varepsilon)$ a *wedge* (of axis v and radius ε):

$$W(v;\varepsilon) := \{tw : t \in [0, \varepsilon), w \in v + \varepsilon B\}.$$

The first of the several immediate consequences of the theorem given below explains our use of the term "wedged" for the condition $0 \in \operatorname{int} T_S^C(x)$, and reveals it to be of a local and uniform nature.

6.5. Exercise.

(a) S is wedged at x iff there exists a wedge $W(v;\varepsilon)$ such that $y + W(v;\varepsilon) \subset S \ \forall y \in S \cap B(x;\varepsilon)$.

(b) If S is wedged at x, then $\operatorname{int} S \neq \emptyset$ and $x \in \operatorname{cl}(\operatorname{int} S)$. If S is wedged at each of its points, then $S = \operatorname{cl}(\operatorname{int} S)$.

(c) If $v \in \operatorname{int} T_S^C(x)$, then $v \in T_S^C(x')$ for all x' near x.

(d) If $T_S^C(x) = \mathbb{R}^n$, then $x \in \operatorname{int} S$.

(e) Let $n = 2$ and set S equal to

$$\{(x,y) : \|(x,y) - (1,0)\| \leq 1\} \cup \{(x,y) : \|(x,y) - (-1,0)\| \leq 1\}.$$

Show that $T_S^B(0,0) = \mathbb{R}^2$, yet $(0,0) \notin \operatorname{int} S$. What is $T_S^C(0,0)$?

Lower Semicontinuity of $T_S^C(\cdot)$

A multifunction $\Gamma : X \to X$ is said to be *lower semicontinuous* at x if for any given $v \in \Gamma(x)$ and $\varepsilon > 0$, there exists $\delta > 0$ such that

$$x' \in \operatorname{dom} \Gamma, \quad x' \in x + \delta B \implies v \in \Gamma(x') + \varepsilon B.$$

6.6. Exercise.

(a) Let $\Delta := \operatorname{dom} \Gamma$. Then Γ is lower semicontinuous at $x \in \Delta$ iff for every $v \in \Gamma(x)$ we have

$$\limsup_{x' \xrightarrow{\Delta} x} d(v, \Gamma(x')) = 0.$$

(b) In Exercise 2.5.6, determine in which cases $T_S^C(\cdot)$ or $T_S^B(\cdot)$ is lower semicontinuous at the origin.

The lower semicontinuity of the tangent cone is a very useful property in certain contexts, and it is notable that wedged sets automatically possess it.

6.7. Proposition. *If S is wedged at x, then $T_S^C(\cdot)$ is lower semicontinuous at x.*

Proof. Given $v \in T_S^C(x)$, we must verify (see Exercise 6.6)
$$\limsup_{x' \xrightarrow{S} x} d(v, T_S^C(x')) = 0.$$

Since $T_S^C(x)$ is closed convex with nonempty interior, it is the closure of its interior. In view of this fact, it clearly suffices to verify the preceding condition for $v \in \operatorname{int} T_S^C(x)$. But for such v, it is immediate from Exercise 6.5(c). □

The following reflects a general fact about closed convex cone-valued multifunctions which are in polarity.

6.8. Proposition. *$T_S^C(\cdot)$ is lower semicontinuous at x iff $N_S^C(\cdot)$ is graph-closed at x.*

Proof. Let $T_S^C(\cdot)$ be lower semicontinuous at x, and let $\zeta_i \in N_S^C(x_i)$, where $x_i \to x$, $\zeta_i \to \zeta$. We wish to deduce $\zeta \in N_S^C(x)$. Without loss of generality, we can suppose $\|\zeta\| = 1$. If $\zeta \notin N_S^C(x)$, then there exists a vector v separating ζ from $N_S^C(x)$: For some $\delta > 0$ we have
$$\langle v, \xi \rangle \leq 0 < \delta = \langle v, \zeta \rangle \ \forall \xi \in N_S^C(x).$$

It follows that $v \in N_S^C(x)^\circ = T_S^C(x)$, and that (by lower semicontinuity) $v \in T_S^C(x_i) + (\delta/2)B$ for all i sufficiently large. Then, for some vector $u_i \in B$, and for all i large, $v + (\delta/2)u_i \in T_S^C(x_i)$, whence
$$\left\langle \zeta_i, v + \left(\frac{\delta}{2}\right) u_i \right\rangle \leq 0.$$

This implies $\langle \zeta_i, v \rangle \leq (\delta/2)\|\zeta_i\|$, and in the limit $\langle \zeta, v \rangle \leq \delta/2$, the desired contradiction.

The sufficiency part of the proof is no harder than the necessity; it constitutes one of the end-of-chapter problems. □

A cone K in $\mathbb{R}^n$ is called *pointed* if it contains no two nonzero elements whose sum is zero.

> **6.9. Exercise.** A convex cone K in $\mathbb{R}^n$ has nonempty interior iff its polar K° is pointed.

We can summarize some previous results in terms of pointedness of the normal cone, as follows:

6.10. Corollary. *If $N_S^C(x)$ is pointed, then $N_S^C(\cdot)$ is graph-closed at x, $T_S^C(\cdot)$ is lower semicontinuous at x, and S is wedged at x.*

6.11. Exercise.

(a) When $n = 1$, $N_S^C(\cdot)$ is always graph-closed at each point of S.

(b) We construct an example in $\mathbb{R}^2$ of a set S such that $N_S^C(\cdot)$ fails to be graph-closed at the origin; S will be the graph of a certain continuous function $f\colon [0,1] \to \mathbb{R}$. We set $f(x) = 0$ at every point x of the form 2^{-n} for $n = 0, 1, 2, \ldots$ and we set $f(0) = 0$. In between any two points of the form 2^{-n-1} and 2^{-n}, the graph of f describes an isosceles triangle whose apex is located at $\big((2^{-n-1} + 2^{-n})/2, 2^{-2n}\big)$. Using the Proximal Normal Formula of Theorem 2.6.1 for N_S^C, show that $N_S^C(0,0)$ is a half-space. Show also that $N_S^C = \mathbb{R}^2$ at each point of the form $(2^{-n}, 0)$, $n > 1$. Conclude that N_S^C is not graph-closed at $(0,0)$.

A General Relation Between T_S^C and T_S^B

We now establish a result which clarifies the relationship between the two notions of tangency. It affirms that asymptotically, a vector v which lies in $T_S^C(x)$ is one which lies in $T_S^B(x')$ for all x' near x.

6.12. Theorem. $v \in T_S^C(x)$ iff
$$\limsup_{x' \xrightarrow{S} x} d\big(v, T_S^B(x')\big) = 0.$$

Proof. To begin with, let v satisfy the given limit condition, and let us show that $v \in T_S^C(x)$. It suffices to show that $\langle v, \zeta \rangle \leq 0$ for any element $\zeta \in N_S^L(x)$, since $N_S^L(x)^\circ = N_S^C(x)^\circ = T_S^C(x)$. Such a vector ζ is of the form
$$\lim_{i \to \infty} \zeta_i, \quad \zeta_i \in N_S^P(x_i), \quad x_i \xrightarrow{S} x.$$

By hypothesis, there exists $v_i \in T_S^B(x_i)$ such that $v_i \to v$. Recall now that $T_S^B(x_i) \subset N_S^P(x_i)^\circ$ (Exercise 2.7.1); in consequence we have $\langle v_i, \zeta_i \rangle \leq 0$. In the limit we get $\langle \zeta, v \rangle \leq 0$, as required.

Suppose now that $v \in T_S^C(x)$, but that the limit condition of the theorem fails. Then, for some $\varepsilon > 0$ and sequence $x_i \to x$, $x_i \in S$, we have $d\big(v, T_S^B(x_i)\big) > \varepsilon$. Since a vector u lies in $T_S^B(y)$ iff $Dd_S(y; u) \leq 0$ (by Exercise 2.7.1) we have $Dd_S(x_i; v + w) > 0$ for all $w \in \varepsilon\overline{B}$. Subbotin's Theorem 4.2 applies to give $z_i \in x_i + (1/i)B$ admitting $\zeta_i \in \partial_P d_S(z_i)$ such that
$$\langle \zeta_i, v + w \rangle > 0 \; \forall w \in \varepsilon \overline{B}.$$

We deduce $\langle \zeta_i/\|\zeta_i\|, v \rangle \geq \varepsilon$. Taking a subsequence if necessary to arrange the convergence of $\zeta_i/\|\zeta_i\|$ to a limit ζ, we have $\zeta \in N_S^L(x)$ (since $\zeta_i \in \partial_P d_S(z_i) \subset N_S^P(z_i)$) and $\langle \zeta, v \rangle \geq \varepsilon$. This contradicts $v \in N_S^L(x)^\circ = T_S^C(x)$, and completes the proof. $\square$

170   3. Special Topics

Recall that S is said to be regular at x if $T_S^B(x) = T_S^C(x)$.

6.13. Corollary. *S is regular at x iff $T_S^B(\cdot)$ is lower semicontinuous at x.*

6.14. Exercise.

(a) Prove Corollary 6.12.

(b) Show that the set S of Exercise 6.11(b) is regular at $(0,0)$, but that $T_S^C(\cdot)$ is not lower semicontinuous there.

(c) A continuous function f is such that $f(x) \in T_S^B(x)$ for every x iff it satisfies $f(x) \in T_S^C(x)$ for every x.

7 Problems on Chapter 3

7.1. Let f and g be locally Lipschitz functions on $\mathbb{R}^n$ satisfying $f \geq g$. Suppose that at a point x_0 we have $f(x_0) = g(x_0)$. Prove that

$$\partial_C f(x_0) \cap \partial_C g(x_0) \neq \emptyset.$$

Show that the corresponding result is false if ∂_C is replaced by either ∂_L or ∂_D.

7.2. We adopt the notation and the hypotheses of Theorem 1.7.

(a) Obtain the following "inner estimate" for $\partial_C V(0)$: For any $x \in \Sigma(0)$, we have $\partial_C V(0) \cap M(x) \neq \emptyset$. (*Hint.* Consider the device used in Exercise 1.5, in light of the preceding problem.)

(b) We have

$$DV(0;u) \geq \inf_{x \in \Sigma(0)} \inf_{\zeta \in M(x)} \langle \zeta, u \rangle,$$

$$D(-V)(0;u) \geq \sup_{x \in \Sigma(0)} \inf_{\zeta \in M(x)} \langle -\zeta, u \rangle.$$

(c) If $M(x)$ is a singleton $\{\zeta(x)\}$ for each $x \in \Sigma(0)$, then $V'(0;u)$ exists for each u and equals $\inf_{x \in \Sigma(0)} \langle \zeta(x), u \rangle$.

(d) If $\Sigma(0)$ is a singleton $\{x\}$ and $M(x)$ is a singleton $\{\zeta\}$, then $V'(0)$ exists, and we have $\nabla V(0) = \zeta$.

7.3. A function $f \in \mathcal{F}(\mathbb{R}^n)$ is called *calm* at $x \in \text{dom } f$ if it satisfies

$$\liminf_{x' \to x} \frac{f(x') - f(x)}{\|x' - x\|} > -\infty.$$

Prove that if f is calm at x, then for some $K > 0$ and for any $\varepsilon > 0$, there exist $z \in x + \varepsilon B$ and $\zeta \in \partial_P f(z)$ with $|f(z) - f(x)| < \varepsilon$ and $\|\zeta\| \leq K$ (i.e., f admits a priori bounded proximal subgradients arbitrarily near x).

7.4. Prove the following addendum to Exercise 1.8. If $V(\cdot)$ is calm at 0, then we may take $\lambda_0 = 1$.

7.5. We adopt the notation and hypotheses of Theorem 1.14, and we assume in addition that the functions f, g_i ($i = 1, 2, \ldots, p$) are convex, and that the functions h_j ($j = 1, 2, \ldots, m$) are affine.

(a) Prove that V is convex.

(b) Prove that if every $x \in \Sigma(0,0)$ is normal, then
$$\partial_L V(0,0) = \partial_C V(0,0) = \bigcup_{x \in \Sigma(0,0)} M(x).$$

(c) Prove that any x feasible for $P(0,0)$ for which $M(x) \neq \emptyset$ is a solution to $P(0,0)$. (This is a general principle in optimization: In the convex case of the problem, the (normal) necessary conditions become sufficient as well.)

7.6. Consider the smooth, finite-dimensional case of the Mean Value Inequality, Theorem 2.3, when $n = 2$, $x = (0,0)$,
$$Y = \{(1,t) : 0 \le t \le 1\}, \quad f(u,v) = u + (1-u)v^2.$$

Find the points z satisfying the conclusion of the theorem. Observe that none of them lie in $\text{int}[0, Y]$.

7.7. Consider the general case of the Mean Value Inequality, Theorem 2.6, when $X = \mathbb{R}$, $x = 0$, $Y = \{1\}$, and
$$f(u) = \begin{cases} -\sqrt{|u|} & \text{if } u \le 0, \\ 1 & \text{if } u > 0. \end{cases}$$

Show that for all $\bar{r}$ and ε sufficiently near $\hat{r}$ and 0, respectively, the point z whose existence is asserted by the theorem must lie outside $[x, Y]$.

7.8. Let S and E be two bounded, closed, nonempty subsets of the Hilbert space X, with $0 \notin E$. Prove that for some $s \in S$ and $\zeta \in N_S^P(s)$, we have $\langle \zeta, e \rangle > 0$ for all $e \in E$.

7.9. Show that the Decrease Principle, Theorem 2.8, remains true under the following weakened hypothesis: For some $\Delta > 0$, $\rho > 0$, and $\delta > 0$, we have
$$z \in x_0 + \rho B, \quad \zeta \in \partial_P f(z), \quad f(z) < f(x_0) + \Delta \implies \|\zeta\| \ge \delta.$$

7.10. Let $f \colon \mathbb{R}^2 \to \mathbb{R}$ be differentiable, with $f(-1, v) < 0$ and $f(1, v) > 0$ for all $v \in [-1, 1]$. Prove the existence of a point (x, y) satisfying
$$|y| \le |x| \le 1 \quad \text{and} \quad f_x(x,y) > |f_y(x,y)|.$$

172   3. Special Topics

7.11. Show that for some $\delta > 0$, for all (u, v) in $\mathbb{R}^2$ sufficiently near $(0,0)$, there exists a solution (x, y, z) in $\mathbb{R}^3$ to the system
$$|x - y| + 2z = u, \quad x - |z| = v,$$
satisfying
$$\|(x, y, z)\| \leq \frac{1}{\delta}\|(u, v)\|.$$

7.12. Show that for some $\delta > 0$, for all (x, y, z) in $\mathbb{R}^3$ sufficiently near $(0, 0, 0)$, there exists (x', y', z') satisfying
$$|x' - y'| + 2z' = 0, \quad x' - |z'| = 0,$$
such that
$$\|(x', y', z') - (x, y, z)\| \leq \frac{1}{\delta}\|(|x - y| + 2z, x - |z|)\|.$$

7.13.

(a) Let $G\colon \mathbb{R}^m \to \mathbb{R}^n$ be Lipschitz near x, and let $\partial_C G(x)$ be of maximal rank. Then we have
$$0 \in \partial_L \langle \zeta, G(\cdot) \rangle(x) \implies \zeta = 0.$$

(b) Verify that $\partial_C G(0)$ is of maximal rank when $m = 3$, $n = 2$, and
$$G(x, y, z) := \big(|x - y| + 2z, x - |z|\big).$$

(What is the relevance of this to the two preceding problems?)

7.14. Let $x_0 \in S_1 \cap S_2$, where S_1 and S_2 are closed subsets of $\mathbb{R}^n$ satisfying
$$N^L_{S_1}(x_0) \cap \big(-N^L_{S_2}(x_0)\big) = \{0\}.$$
Prove that for some $\delta > 0$, for all x sufficiently near x_0, we have
$$d_{S_1 \cap S_2}(x) \leq \frac{1}{\delta} \max\{d_{S_1}(x), d_{S_2}(x)\}.$$

7.15. Let the functions g, h be as in Theorem 3.8, and in particular continue to assume the constaint qualification at x_0. Let S be the set
$$\{x \in X \colon g(x) \leq 0, h(x) = 0\}.$$
We assume that $g(x_0) = 0$, and that h is C^1 near x_0. We set
$$I := \{1, 2, \ldots, p\}, \quad J := \{1, 2, \ldots, m\}.$$

(a) $T^C_S(x_0)$ contains every vector $v \in X$ satisfying
$$g^\circ_i(x_0; v) \leq 0 \quad (i \in I), \quad \langle h'_j(x_0), v \rangle = 0 \quad (j \in J).$$

(b) If each of the functions g_i admits directional derivatives at x_0, then any vector $v \in T^B_S(x_0)$ satisfies
$$g'_i(x_0; v) \leq 0 \quad (i \in I), \quad h'_j(x_0; v) = 0 \quad (j \in J).$$

(c) If each of the functions g_i is regular at x_0, then S is regular at x_0, and we have
$$T^C_S(x_0) = T^B_S(x_0)$$
$$= \{v : g'_i(x_0; v) \leq 0 \ (i \in I), \langle h'_j(x_0), v \rangle = 0 \ (j \in J)\},$$
$$N^C_S(x_0) = \mathrm{cone}\{\zeta \in \partial_C\{\langle \gamma, g(\cdot) \rangle + \langle \lambda, h(\cdot) \rangle\}(x_0)$$
$$: \gamma \in \mathbb{R}^p, \lambda \in \mathbb{R}^m, \gamma \geq 0\}.$$

(*Hint.* Part (a) can be proven via the characterization provided by Proposition 2.5.2, together with an appeal to Theorem 3.8.)

7.16.

(a) Let $f : \mathbb{R}^2 \to \mathbb{R}$ be defined as follows:
$$f(x, y) = \begin{cases} -|x|^{3/2} & \text{if } x \leq 0, \\ 0 & \text{if } x \geq 0, \ y \leq 0, \\ \min(x, y) & \text{if } x \geq 0, \ y \geq 0. \end{cases}$$

Prove that f is Lipschitz near $(0, 0)$. Calculate the sets $\partial_P f(0,0)$, $\partial_D f(0,0)$, $\partial_L f(0,0)$, and $\partial_C f(0,0)$, and note that they are all different.

(b) Calculate $\partial_D f(0)$ where f is the function that appears in Problem 2.9.16.

7.17. Let $x \in S$, where S is a closed subset of $\mathbb{R}^n$.

(a) Prove that $\partial_D d_S(x) \subset N^D_S(x) \cap \bar{B}$.

(b) Prove now that equality holds. (*Hint.* Use Exercise 4.9(c) and Proposition 4.10.)

(c) Deduce from (b) the corresponding fact for limiting constructs:
$$\partial_L d_S(x) = N^L_S(x) \cap \bar{B}.$$

(d) Show that $\partial_C d_S(x)$ and $N_S^C(x) \cap \overline{B}$ differ in general.

7.18. Let $x \in S$, where S is a closed subset of $\mathbb{R}^n$. We will prove that S is regular at x iff d_S is regular at x.

(a) Let S be regular at x. Use the preceding exercise to justify the following steps:

$$\begin{aligned} d_S^\circ(x; v) &= \max\{\langle \zeta, v \rangle : \zeta \in \partial_L d_S(x)\} \\ &= \max\{\langle \zeta, v \rangle : \zeta \in N_S^L(x) \cap \overline{B}\} \\ &= \max\{\langle \zeta, v \rangle : \zeta \in N_S^D(x) \cap \overline{B}\} \\ &= \max\{\langle \zeta, v \rangle : \zeta \in \partial_D d_S(x)\} \\ &\leq D d_S(x; v). \end{aligned}$$

Deduce that d_S is regular at x.

(b) If $d_S(\cdot)$ is regular at x, show that S is regular at x.

7.19. Let E be a compact subset of $\mathbb{R}^n$, and let S consist of those functions $x(\cdot) \in L_n^2[a, b]$ such that $\dot{x}(t) \in E$ a.e. Prove that S is weakly compact iff E is convex. What is the weak closure of S if E is not convex?

7.20. Let $\varphi \colon \mathbb{R}^n \times \mathbb{R}^m \to \mathbb{R}^k$ be such that for every $x \in \mathbb{R}^n$, the function $u \mapsto \varphi(x, u)$ is continuous, and for every $u \in \mathbb{R}^m$, the function $x \mapsto \varphi(x, u)$ is measurable. Suppose that Γ is a closed-valued measurable multifunction from $\mathbb{R}^n$ to $\mathbb{R}^m$ such that

$$\{u \in \Gamma(x) \colon \varphi(x, u) = 0\} \neq \emptyset \ \forall x \in \mathbb{R}^n.$$

Prove the existence of a measurable selection γ for Γ such that

$$\varphi(x, \gamma(x)) = 0 \ \forall x \in \mathbb{R}^n.$$

(This is known as *Filippov's Lemma*.)

7.21. Let Γ be a multifunction from $\mathbb{R}^m$ to $\mathbb{R}^n$ whose images are compact and convex. Prove that Γ is measurable iff the function $H(x, p)$ is measurable in x for each p, where

$$H(x, p) := \sup\{\langle p, v \rangle \colon v \in \Gamma(x)\}.$$

7.22. Consider $f \colon L_1^2[0, 1] \to \mathbb{R}$ defined by

$$f(x) := \int_0^1 \varphi(x(t)) \, dt, \text{ with } \varphi(x) := -|x|.$$

(a) Prove that $\partial_P f(x) \neq \emptyset$ iff there exists $\delta > 0$ such that
$$\text{meas}\{t \in [0,1]: -\delta \leq x(t) \leq \delta\} = 0.$$

(b) Exhibit ζ and $x \in L_1^2[0,1]$ such that $\zeta(t) \in \partial_P \varphi(x(t))$ a.e., yet $\zeta \notin \partial_P f(x)$.

7.23. Show that any absolutely continuous function x satisfying the Euler inclusion and transversality condition of Theorem 5.22 is a solution of the variational problem when φ and ℓ are convex.

7.24. In the context of Theorem 5.22, let $\varphi(u,v)$ have the form $\left(1 + \|v\|^2\right)^{1/2} + g(u)$, where g is Lipschitz. Prove that any solution x of the variational problem is continuously differentiable.

7.25. Prove the sufficiency part of Proposition 6.8: If $N_S^C(\cdot)$ is graph-closed at x, then $T_S^C(\cdot)$ is lower semicontinuous at x.

7.26. Let S be a closed convex subset of $\mathbb{R}^n$, and let $x \in S$. Then S is wedged at x iff $\text{int } S \neq \emptyset$.

7.27. Let S be a closed subset of $\mathbb{R}^n$ which is wedged at x, where $x \in \text{bdry } S$. Then
$$N_S^C(x) = -N_{\hat{S}}^C(x),$$
where $\hat{S}$ is the set $\text{cl}(\text{comp } S)$. Find a counterexample to this formula when S is not wedged at x.

7.28. Let $f: \mathbb{R}^n \to \mathbb{R}$ be locally Lipschitz. Prove that epi f is wedged at each of its points. (Conversely, a set which is wedged can be viewed locally as the epigraph of a Lipschitz function, for a suitable choice of coordinate axes, as can be shown.)

7.29. The Hausdorff distance $\rho(C,S)$ between two sets C and S in $\mathbb{R}^n$ is defined via
$$\rho(C,S) = \max\left\{\sup_{c \in C} d_S(c), \sup_{s \in S} d_C(s)\right\}.$$
A multifunction Γ from $\mathbb{R}^m$ to $\mathbb{R}^n$ is (Hausdorff) *continuous* at x if
$$\rho(\Gamma(x'), \Gamma(x)) \to 0 \text{ as } x' \to x.$$
Prove that when Γ is closed-valued, then the continuity of Γ at x implies that Γ is both upper and lower semicontinuous at x.

7.30. It is known that there exists a subset S of $\mathbb{R}$ with the following property: for every $a, b \in \mathbb{R}$ with $a < b$, the Lebesgue measure $\mathcal{L}$ of the set $S \cap [a,b]$ satisfies
$$\frac{(b-a)}{3} < \mathcal{L}(S \cap [a,b]) < \frac{2(b-a)}{3}.$$

Let $f: \mathbb{R} \to \mathbb{R}$ be defined as follows:

$$f(x) = \int_0^x \chi_S(t)\, dt,$$

where χ_S is the characteristic function of S. Prove the following facts:

(a) f is Lipschitz, and we have $\partial_C f(x) = [0, 1]$ for all $x \in \mathbb{R}$.

(b) The functions $f(x)$ and $g(x) := x - f(x)$ have the same generalized gradient at every point, yet do not differ by a constant.

We remark that $\partial_P f$ and $\partial_L f$ are not known, nor is it known whether the phenomenon that occurs in (b) can take place in terms of these subdifferentials.

4
A Short Course in Control Theory

> We are guided by the beauty of our weapons.
> —Leonard Cohen, *First We Take Manhattan*

Mathematics, as well as several areas of application, abounds with situations where it is desired to control the behavior of the trajectories of a given dynamical system. The goal can be either geometric (keep the state of the system in a given set, or bring it toward the set), or functional (find the trajectory that is optimal relative to a given criterion). More specific issues arise subsequently, such as the construction of feedback control mechanisms achieving the aims we have in mind. In this chapter we will identify a complex of such fundamental and related issues, as they arise in connection with the control of ordinary differential equations in a deterministic setting. The first section sets the scene and develops a technical base for the entire chapter.

1 Trajectories of Differential Inclusions

We are given a multifunction F mapping $\mathbb{R} \times \mathbb{R}^n$ to the subsets of $\mathbb{R}^n$, and a time interval $[a, b]$. The central object of study in this chapter will be the *differential inclusion*

$$\dot{x}(t) \in F(t, x(t)) \text{ a.e.,} \quad t \in [a, b]. \tag{1}$$

A *solution* $x(\cdot)$ of (1) is taken to mean an absolutely continuous function $x \colon [a, b] \to \mathbb{R}^n$ which, together with $\dot{x}$, its derivative with respect to t,

178   4. A Short Course in Control Theory

satisfies (1). For brevity, we will refer to any absolutely continuous x from $[a, b]$ to $\mathbb{R}^n$ as an *arc* on $[a, b]$. We also refer to an arc x satisfying (1) as a *trajectory* of F.

The concept of differential inclusion subsumes that of a *standard control system*

$$\dot{x}(t) = f(t, x(t), u(t)), \qquad (2)$$

where $f\colon \mathbb{R} \times \mathbb{R}^n \times \mathbb{R}^m \to \mathbb{R}^n$, and where the control function u takes values in some prescribed subset U of $\mathbb{R}^m$; simply consider $F(t, x) := f(t, x, U)$. Filippov's Lemma (Problem 3.7.20) implies that under mild hypotheses on f, an arc x satisfies (1) iff there is a measurable function $u(\cdot)$ with values in U such that (2) holds.

A more special case of (1) is the one in which $F(t, x)$ is a singleton for all (t, x); i.e., $F(t, x) = \{f(t, x)\}$. In the classical study of the ordinary differential equation

$$\dot{x}(t) = f(t, x(t)) \qquad (3)$$

the properties of the function f play an important role. Let us recall some basic facts on this subject:

1.1. Theorem. *Suppose that f is continuous, and let $(t_0, x_0) \in \mathbb{R} \times \mathbb{R}^n$ be given. Then the following hold:*

(a) *There exists a solution of (3) on an open interval $(t_0 - \delta, t_0 + \delta)$, for some $\delta > 0$, satisfying $x(t_0) = x_0$.*

(b) *If in addition we assume linear growth, that is that there exist nonnegative constants γ and c such that*

$$\|f(t, x)\| \leq \gamma \|x\| + c \; \forall (t, x),$$

then there exists a solution of (3) on $(-\infty, \infty)$ such that $x(t_0) = x_0$.

(c) *Let us now add the hypothesis that f is locally Lipschitz. Then there exists a unique solution of (3) on $(-\infty, \infty)$ such that $x(t_0) = x_0$.*

Many readers will know that the hypotheses on f in the preceding theorem can be relaxed—for example, joint continuity of f in (t, x) can be replaced with measurability in t and continuity in x. However, the form given above will suffice for our purposes.

In developing the basic theory of differential inclusions, two properties of F turn out to be particularly important: upper semicontinuity and the Lipschitz condition. We will not see the latter intervene until the next section, but the former property is part of the *Standing Hypotheses* on F that will be in force throughout the rest of this chapter, whether explicitly mentioned or not.

1 Trajectories of Differential Inclusions

1.2. Standing Hypotheses.

(a) For every (t,x), $F(t,x)$ is a nonempty compact convex set.

(b) F is upper semicontinuous.

(c) For some positive constants γ and c, and for all (t,x),
$$v \in F(t,x) \implies \|v\| \leq \gamma\|x\| + c.$$

Parts (b) and (c) of the Standing Hypotheses are of course analogues of the continuity and linear growth conditions, respectively, familiar in the classical case. We recall that F is upper semicontinuous at x if, given any $\varepsilon > 0$, there exists $\delta > 0$ such that
$$\|x' - x\| < \delta \implies F(x') \subset F(x) + \varepsilon B.$$

1.3. Exercise.

(a) When $F(t,x) = \{f(t,x)\}$, show that Standing Hypotheses 1.2 (b,c) hold iff f is continuous and satisfies $\|f(t,x)\| \leq \gamma\|x\| + c$.

(b) In the presence of Standing Hypotheses 1.2(a,c), property (b) is equivalent to the graph of F being closed.

(c) Let Ω be a bounded subset of $\mathbb{R} \times \mathbb{R}^n$. Prove that F is uniformly bounded on Ω: there exists M such that
$$(t,x) \in \Omega, \quad v \in F(t,x) \implies \|v\| \leq M.$$

(d) Let Ω be a bounded subset of $\mathbb{R} \times \mathbb{R}^n$, and let $r > 0$ satisfy $\|(t,x)\| \leq r \ \forall (t,x) \in \Omega$. Let $\tilde{r} > r$, and define $\widetilde{F}$ as follows:
$$\widetilde{F}(t,x) = \begin{cases} F(t,x) & \text{if } \|(t,x)\| \leq \tilde{r}, \\ F\left(\dfrac{(t,x)}{\|(t,x)\|}\tilde{r}\right) & \text{if } \|(t,x)\| \geq \tilde{r}. \end{cases}$$

Prove that $\widetilde{F}$ satisfies the Standing Hypotheses, agrees with F on a neighborhood of Ω, and is globally bounded.

The role of the *linear growth* condition in the classical theory of differential equations is predicated on the a priori bounds on solutions to which it gives rise. We will benefit from it in precisely the same way. The following is known as *Gronwall's Lemma*.

1.4. Proposition. *Let x be an arc on $[a,b]$ which satisfies*
$$\|\dot{x}(t)\| \leq \gamma\|x(t)\| + c(t) \ a.e., \quad t \in [a,b],$$
where γ is a nonnegative constant and where $c(\cdot) \in L_1^1[a,b]$. Then, for all $t \in [a,b]$, we have
$$\|x(t) - x(a)\| \leq (e^{\gamma(t-a)} - 1)\|x(a)\| + \int_a^t e^{\gamma(t-s)} c(s)\, ds.$$

If the function c is constant and $\gamma > 0$, this becomes

$$\|x(t) - x(a)\| \leq (e^{\gamma(t-a)} - 1)(\|x(a)\| + c/\gamma).$$

Proof. Let $r(t) := \|x(t) - x(a)\|$, a function which is absolutely continuous on $[a, b]$, as the composition of a Lipschitz function and an absolutely continuous one. Let t be in that set of full measure in which both $\dot{x}(t)$ and $\dot{r}(t)$ exist. If $x(t) \neq x(a)$, we have

$$\dot{r}(t) = \frac{\langle \dot{x}(t), x(t) - x(a) \rangle}{\|x(t) - x(a)\|},$$

and otherwise $\dot{r}(t) = 0$ (since r attains a minimum at t). Thus, a.e. $t \in [a, b]$ we have

$$\begin{aligned}\dot{r}(t) &\leq \|\dot{x}(t)\| \leq \gamma \|x(t)\| + c(t) \\ &\leq \gamma \|x(t) - x(a)\| + \gamma \|x(a)\| + c(t) \\ &= \gamma r(t) + \gamma \|x(a)\| + c(t).\end{aligned}$$

We rewrite this inequality in the form

$$(\dot{r}(t) - \gamma r(t))e^{-\gamma t} \leq \gamma e^{-\gamma t}\|x(a)\| + e^{-\gamma t}c(t)$$

and note that the left side is the derivative of the function $t \mapsto r(t)e^{-\gamma t}$. Integrating both sides now gives the result. □

1.5. Exercise. Let C be a bounded subset of $\mathbb{R}^n$ and let $[a, b]$ be given. Show that there exists $K > 0$ and $M > 0$ with the property that any trajectory x of F on $[a, b]$ having $x(a) \in C$ is Lipschitz on $[a, b]$ of rank K, and has $\|x(t)\| \leq M$ for every $t \in [a, b]$.

Euler Solutions

Many of us will have seen methods of calculating solutions of ordinary differential equations; how would we study in concrete terms the calculation of trajectories of the differential inclusion (1)? The most straightforward approach to calculating a trajectory is to first find a *selection* f of F; i.e., a function f such that $f(t, x) \in F(t, x)$ for all (t, x). Then, we consider the differential equation $\dot{x} = f(t, x)$; any solution will presumably satisfy (1).

The problem with this approach lies in finding selections f with the regularity properties (e.g., continuity) required by the usual theory of differential equations. This selection issue is an interesting and well-studied one, but not one that we intend nor need to dwell upon. Instead, we will consider a generalized concept of solution to $\dot{x} = f(t, x)$, one which requires no particular regularity of f.

Let us now consider the so-called Cauchy or *initial-value problem*

$$\dot{x}(t) = f(t, x(t)), \quad x(a) = x_0, \qquad (4)$$

where f is simply any function from $[a, b] \times \mathbb{R}^n$ to $\mathbb{R}^n$. (We will return presently to our differential inclusion.) How would we begin to calculate numerically a solution of (4)? Recalling the classical Euler iterative scheme from ordinary differential equations, we suspect that a reasonable answer is obtained by discretizing in time. So let

$$\pi = \{t_0, t_1, \ldots, t_{N-1}, t_N\}$$

be a partition of $[a, b]$, where $t_0 = a$ and $t_N = b$. (We do not require uniform partitions; thus the interval lengths $t_i - t_{i-1}$ may differ.)

We proceed by considering, on the interval $[t_0, t_1]$, the differential equation with *constant* right-hand side

$$\dot{x}(t) = f(t_0, x_0), \quad x(t_0) = x_0.$$

Of course this has a unique solution $x(t)$ on $[t_0, t_1]$, since the right side is constant; we define $x_1 := x(t_1)$. Next we iterate, by considering on $[t_1, t_2]$ the initial-value problem

$$\dot{x}(t) = f(t_1, x_1), \quad x(t_1) = x_1.$$

The next so-called *node* of the scheme is $x_2 := x(t_2)$. We proceed in this manner until an arc x_π (which is in fact piecewise affine) has been defined on all of $[a, b]$. We use the notation x_π to emphasize the role played by the particular partition π in determining x_π, which has been called in the past (and in our present) the *Euler polygonal arc* corresponding to the partition π, or similar words to that effect.

The *diameter* (or *mesh size*) μ_π of the partition π is given by

$$\mu_\pi := \max\{t_i - t_{i-1} : 1 \leq i \leq N\}.$$

An *Euler solution* to the initial-value problem (4) means any arc x which is the uniform limit of Euler polygonal arcs x_{π_j}, corresponding as above to some sequence π_j such that $\pi_j \to 0$, where this connotes convergence of the diameters $\mu_{\pi_j} \to 0$ (evidently, the corresponding number N_j of partition points in π_j must then go to infinity). We will also say that an arc x on $[a, b]$ is an *Euler arc* for f when x is an Euler solution on $[a, b]$ as above to the initial-value problem (4) for the "right" initial condition, namely $x_0 = x(a)$.

There are potential pathologies associated with these Euler solutions when f is discontinuous, one of which is that the equality $\dot{x}(t) = f(t, x(t))$ may fail completely. The following exercise serves to illustrate this, among other perhaps counterintuitive and unexpected features.

1.6. Exercise.

(a) Define $f: [0,1] \times \mathbb{R} \to \mathbb{R}$ as follows:

$$f(t,x) = \begin{cases} e^t & \text{if } x = e^t, \\ 1 & \text{if } x < e^t, \\ e & \text{if } x > e^t. \end{cases}$$

Let $x_0 = 1$, $a = 0$, and $b = 1$. Show that problem (4) admits a unique Euler solution but that the arc $\hat{x}(t) = e^t$ is *not* an Euler solution, even though it satisfies $\dot{x}(t) = f(t, x(t))$ at every point.

(b) Let f be the function

$$f(x) = \begin{cases} 1 & \text{if } x \leq 0, \\ -1 & \text{if } x > 0. \end{cases}$$

Set $x_0 = 0$, $a = 0$, and $b = 1$. Show that $x(t) \equiv 0$ is the unique Euler solution of (4), although it fails to satisfy $\dot{x}(t) = f(t, x(t))$ at any t.

(c) Now let f be the function

$$f(x) = \begin{cases} 1 & \text{if } x = 0, \\ -1 & \text{otherwise}, \end{cases}$$

where again $x_0 = 0$, $a = 0$, and $b = 1$. Show that $x(t) \equiv 0$ and $x(t) = -t$ are both Euler solutions of the initial-value problem. (*Hint.* The first Euler solution is arrived at by considering only uniform partitions.)

(d) Use part (c) in order to illustrate the fact that if we are restricted to exclusively uniform partitions of a compact interval $[a,b]$, then the set of Euler solutions obtained thereby can be a proper subset of the general class of Euler solutions (even when f is dependent only upon x).

(e) Provide an example of a function $f: \mathbb{R} \to \mathbb{R}$ such that there exist two distinct Euler solutions of (4) on $[0,1]$, each corresponding to a sequence of uniform partitions.

On the other hand, here are some important positive things that can be said about Euler solutions:

1.7. Theorem. *Suppose that for positive constants γ and c and for all $(t,x) \in [a,b] \times \mathbb{R}^n$, we have the linear growth condition*

$$\|f(t,x)\| \leq \gamma \|x\| + c,$$

where f is otherwise arbitrary. Then:

(a) *At least one Euler solution x to the initial-value problem (4) exists on $[a,b]$, and any Euler solution is Lipschitz.*

(b) *Any Euler arc x for f on $[a,b]$ satisfies*
$$\|x(t) - x(a)\| \leq (t-a)e^{\gamma(t-a)}(c + \gamma\|x(a)\|), \quad a \leq t \leq b.$$

(c) *If f is continuous, then any Euler arc x for f on $[a,b]$ is continuously differentiable on (a,b) and satisfies $\dot{x}(t) = f(t, x(t))\ \forall t \in (a,b)$.*

Proof. Let $\pi := \{t_0, t_1, \ldots, t_N\}$ be a partition of $[a,b]$, and let x_π be the corresponding Euler polygonal arc, with the nodes of x_π being denoted $x_0, x_1, \ldots, x_N$ as usual. On the interval (t_i, t_{i+1}) we have
$$\|\dot{x}_\pi(t)\| = \|f(t_i, x_i)\| \leq \gamma\|x_i\| + c,$$
whence
$$\begin{aligned}
\|x_{i+1} - x_0\| &\leq \|x_{i+1} - x_i\| + \|x_i - x_0\| \\
&\leq (t_{i+1} - t_i)(\gamma\|x_i\| + c) + \|x_i - x_0\| \\
&\leq [(t_{i+1} - t_i)\gamma + 1]\|x_i - x_0\| + (t_{i+1} - t_i)(\gamma\|x_0\| + c).
\end{aligned}$$

We now require the following exercise in induction:

1.8. Exercise. Let $r_0, r_1, \ldots, r_N$ be nonnegative numbers satisfying
$$r_{i+1} \leq (1 + \delta_i)r_i + \Delta_i, \quad i = 0, 1, \ldots, N-1,$$
where $\delta_i \geq 0$ and $\Delta_i \geq 0$, $r_0 = 0$. Then
$$r_N \leq \left(\exp\left(\sum_{i=0}^{N-1} \delta_i\right)\right) \sum_{i=0}^{N-1} \Delta_i.$$

Returning now to the authors' part of the task, we apply this result to derive, for $i = 1, 2, \ldots, N$,
$$\|x_i - x_0\| \leq M,$$
where
$$M := (b-a)e^{\gamma(b-a)}(\gamma\|x_0\| + c).$$

Thus all the nodes x_i lie in the ball $\overline{B}(x_0; M)$; by convexity this is true of all the values $x_\pi(t)$, $a \leq t \leq b$. Since the derivative along any linear portion of x_π is determined by the values of f at the nodes, we obtain as well the following uniform bound on $[a,b]$:
$$\|\dot{x}_\pi\|_\infty \leq k := \gamma M + c.$$

Therefore x_π is Lipschitz of rank k on $[a, b]$.

Now let π_j be a sequence of partitions such that $\pi_j \to 0$; i.e., such that μ_{π_j} goes to zero, and (necessarily) $N_j \to \infty$. Then the corresponding polygonal arcs x_{π_j} on $[a, b]$ all satisfy

$$x_{\pi_j}(a) = x_0, \quad \|x_{\pi_j} - x_0\|_\infty \leq M, \quad \|\dot{x}_{\pi_j}\|_\infty \leq k.$$

It follows that the family $\{x_{\pi_j}\}$ is equicontinuous and uniformly bounded; then, by the well-known theorem of Arzela and Ascoli, some subsequence of it converges uniformly to a continuous function x. The limiting function inherits the Lipschitz rank k on $[a, b]$, and in consequence is absolutely continuous (i.e., x is an arc). Thus by definition x is an Euler solution of the initial-value problem (4) on $[a, b]$, and assertion (a) of the theorem is proved.

The inequality in part (b) of the theorem is inherited by x from the sequence of polygonal arcs generating it (we identify t with b). There remains to prove part (c) of the theorem.

To this end, let x_{π_j} denote a sequence of polygonal arcs for problem (4) converging uniformly to an Euler solution x. As shown above, the arcs x_{π_i} all lie in a certain ball $\overline{B}(x_0; M)$ and they all satisfy a Lipschitz condition of the same rank k. Since a continuous function on $\mathbb{R}^n$ is uniformly continuous on compact sets, for any $\varepsilon > 0$, we can find $\delta > 0$ such that

$$t, \tilde{t} \in [a, b], \quad x, \tilde{x} \in \overline{B}(x_0; M), \quad |t - \tilde{t}| < \delta,$$
$$\|x - \tilde{x}\| < \delta \implies \|f(t, x) - f(\tilde{t}, \tilde{x})\| < \varepsilon.$$

Now let j be large enough so that the partition diameter μ_{π_j} satisfies $\mu_{\pi_j} < \delta$ and $k\mu_{\pi_j} < \delta$. For any point t which is not one of the finitely many points at which $x_{\pi_j}(t)$ is a node, we have $\dot{x}_{\pi_j}(t) = f(\tilde{t}, x_{\pi_j}(\tilde{t}))$ for some $\tilde{t}$ within $\mu_{\pi_j} < \delta$ of t. Thus, since

$$\|x_{\pi_j}(t) - x_{\pi_j}(\tilde{t})\| \leq k\mu_{\pi_j} < \delta,$$

we deduce

$$\|\dot{x}_{\pi_j}(t) - f(t, x_{\pi_j}(t))\| = \|f(t, x_{\pi_j}(t)) - f(\tilde{t}, x_{\pi_j}(\tilde{t}))\| < \varepsilon.$$

It follows that for any t in $[a, b]$, we have

$$\left\| x_{\pi_j}(t) - x_{\pi_j}(a) - \int_a^t f(t, x_{\pi_j}(\tau))\, d\tau \right\|$$
$$= \left\| \int_a^t \{\dot{x}_{\pi_j}(\tau) - f(\tau, x_{\pi_j}(\tau))\}\, d\tau \right\| < \varepsilon(t - a) \leq \varepsilon(b - a).$$

Letting $j \to \infty$, we obtain from this

$$\left\| x(t) - x_0 - \int_a^t f(\tau, x(\tau))\, d\tau \right\| \le \varepsilon(b - a).$$

Since ε is arbitrary, it follows that

$$x(t) = x_0 + \int_a^t f(\tau, x(\tau))\, d\tau,$$

which implies (since the integrand is continuous) that x is C^1 and $\dot{x}(t) = f(t, x(t))$ for all $t \in (a, b)$. □

Euler arcs possess a uniform Lipschitz property akin to the one established in Exercise 1.5 for trajectories:

1.9. Exercise. Let C be a bounded subset of $\mathbb{R}^n$ and let $[a, b]$ be given. Assume that f is given as in Theorem 1.7. Show that there exists $K > 0$ with the property that any Euler arc x for f on $[a, b]$ having $x(a) \in C$ is Lipschitz on $[a, b]$ of rank K.

We are reassured by part (c) of the theorem that when f is continuous (which is the minimal assumption under which the classical study of differential equations operates), Euler arcs satisfy the usual pointwise definition of solution. Part (a) of the following exercise will provide a converse, and in part (b), a further counterintuitive feature is noted:

1.10. Exercise.

(a) Let f be locally Lipschitz, and let the arc x on $[a, b]$ satisfy $\dot{x}(t) = f(t, x(t))$ $\forall t \in (a, b)$. Prove that x is an Euler arc for f on $[a, b]$.

(b) Let $f(t, x) := (3/2)x^{1/3}$ (for $n = 1$), $x_0 = 0$, $a = 0$, $b = 1$. Prove that the initial-value problem (4) has three distinct classical solutions of the form $x(t) = \alpha t^\beta$, but only one Euler solution.

Compactness of Approximate Trajectories

We now return to considering the trajectories of a multifunction F satisfying the Standing Hypotheses, and specifically the existence issue for differential inclusions.

Let f be any *selection* for F; that is,

$$f(t, x) \in F(t, x)\ \forall (t, x).$$

Then f evidently inherits the linear growth condition from F. By Theorem 1.7, an Euler solution x to the initial-value problem (4) exists. We are tempted to conclude that $\dot{x}(t)$, being equal to $f(t, x(t))$ a.e., must lie in

$F(t, x(t))$ a.e.; i.e., that x is a trajectory for F. This reasoning is fallacious, however, since an Euler solution to (4) may not satisfy the pointwise condition $\dot{x} = f(t, x)$ for any t; witness Exercise 1.6(b). The key to the correct argument is a sequential compactness property of approximate trajectories, one which we will have frequent occasion to invoke, and which we establish in some generality for future purposes.

1.11. Theorem. *Let $\{x_i\}$ be a sequence of arcs on $[a, b]$ such that the set $\{x_i(a)\}$ is bounded, and satisfying*

$$\dot{x}_i(t) \in F(\tau_i(t), x_i(t) + y_i(t)) + r_i(t)B \text{ a.e.,}$$

where $\{y_i\}$, $\{r_i\}$, $\{\tau_i\}$ are sequences of measurable functions on $[a, b]$ such that y_i converges to 0 in L^2, $r_i \geq 0$ converges to 0 in L^2, τ_i converges a.e. to t. Then there is a subsequence of $\{x_i\}$ which converges uniformly to an arc x which is a trajectory of F, and whose derivatives converge weakly to $\dot{x}$.

Proof. From the differential inclusion and the linear growth condition we have

$$\|\dot{x}_i(t)\| \leq \gamma \|x_i(t) + y_i(t)\| + |r_i(t)|.$$

Using Gronwall's Lemma (Proposition 1.4), this implies a uniform bound on $\|x_i\|_\infty$ and hence on $\|\dot{x}_i\|_2$. Invoking weak compactness in $L_n^2[a, b]$ allows the extraction of a subsequence $\dot{x}_{i_j}$ converging weakly to a limit v_0; we may also suppose (by Arzela and Ascoli) that x_{i_j} converges uniformly to a continuous function x. Passing to the limit in

$$x_{i_j}(t) = x_{i_j}(a) + \int_a^t \dot{x}_{i_j}(s)\, ds$$

shows that $x(t) = x(a) + \int_a^t v_0(s)\, ds$, whence x is an arc and $\dot{x} = v_0$ a.e. The fact that x is a trajectory for F is an immediate consequence of Theorem 3.5.24. □

The following important consequence of the theorem vindicates our attempt to calculate trajectories of F by way of selections, and proves that trajectories do exist, given any x_0 and $[a, b]$.

1.12. Corollary. *Let f be any selection of F, and let x be an Euler solution on $[a, b]$ of $\dot{x} = f(t, x)$, $x(a) = x_0$. Then x is a trajectory of F on $[a, b]$.*

Proof. Let x_{π_j} be the polygonal arcs whose uniform limit is x, as in the proof of Theorem 1.7. Let $t \in (a, b)$ be a nonpartition point, and let $\tau_j(t)$ designate the partition point t_i immediately before t. Then

$$\dot{x}_{\pi_j}(t) = f(t_i, x_i) \in F(t_i, x_i) = F(\tau_j(t), x_{\pi_j}(t) + y_j(t)),$$

where $y_j(t) := x_i - x_{\pi_j}(t) = x_{\pi_j}(\tau_j(t)) - x_{\pi_j}(t)$. Since the functions x_{π_j} admit a common Lipschitz constant k, we have

$$\|y_j(t)\|_\infty \le k \sup_{t \in [a,b]} |\tau_j(t) - t| \le k\mu_{\pi_j}.$$

It follows that τ_j and y_j are (measurable) functions converging uniformly to t and 0, respectively. It is now a consequence of the theorem that x, the uniform limit of x_{π_j}, is a trajectory of F. □

1.13. Exercise.

(a) *Trajectory continuation.* Let z be a trajectory of F on $[a,b]$. Prove that there exists a trajectory of F on $[a,\infty)$ which coincides with z on $[a,b]$. What is more, show that there exists a trajectory of F on $(-\infty,\infty)$ which coincides with z on $[a,b]$.

(b) *Variable intervals.* Let x_i be a sequence of trajectories of F on $[a_i, b_i]$, where $a_i \to a$, $b_i \to b$, $a < b$, and where the sequence $x_i(a_i)$ is bounded. Let the trajectories x_i be continued to $(-\infty, \infty)$ as in part (a). Prove that a subsequence $\{x_{i_j}\}$ of $\{x_i\}$ has the property that, for some trajectory $\bar{x}$ of F on $[a,b]$, x_{i_j} converges uniformly to $\bar{x}$ on $[a,b]$.

(c) *Uniform convergence on bounded intervals.* Let x_i be a sequence of trajectories of F on $[a, \infty)$ such that the sequence $x_i(a)$ is bounded. Prove that there is a trajectory $\bar{x}$ of F on $[a, \infty)$ and a subsequence x_{i_j} having the property that for any $b > a$, x_{i_j} converges uniformly to $\bar{x}$ on $[a,b]$.

(d) Let $f: \mathbb{R}^k \to \mathbb{R}^\ell$ only be required to satisfy a linear growth condition. Is there a *minimal* multifunction F from $\mathbb{R}^k$ to $\mathbb{R}^\ell$ which satisfies the Standing Hypotheses such that $f(x) \in F(x)$ for all x?

(e) Let $f: \mathbb{R}^n \times \mathbb{R}^m \to \mathbb{R}^n$ and $g: \mathbb{R}^n \times \mathbb{R}^m \to \mathbb{R}^m$ satisfy a linear growth condition, and let (x, y) be an Euler arc for (f, g) on $[a, b]$. Suppose that f is continuous. Prove that x is C^1 on (a, b) and satisfies $\dot{x}(t) = f(x(t), y(t))$ on (a, b).

We can extend the notion of Euler arcs for f from finite intervals to semi-infinite ones of the form $[a, \infty)$ as follows: we say that the arc $x(\cdot)$ defined on $[a, \infty)$ is an Euler arc on $[a, \infty)$ provided that for any $b \in (a, \infty)$, the arc x restricted to $[a, b]$ is an Euler arc for f on $[a, b]$ (or, equivalently, an Euler solution on $[a, b]$ of the initial-value problem $\dot{y} = f(t, y), y(a) = x(a)$).

In general, it is not the case that the concatenation of two Euler arcs for f is an Euler arc. That is, if x on $[a, b]$ and y on $[b, c]$ are Euler arcs, the arc consisting of x followed by y may not be one (on $[a, c]$). Nonetheless, we have

188   4. A Short Course in Control Theory

1.14. Exercise.

(a) Let $f\colon [a,\infty)\times\mathbb{R}^n \to \mathbb{R}^n$ have linear growth. For any x_0, show that there is an Euler arc x for f on $[a,\infty)$ such that $x(a) = x_0$. (*Hint.* Construct an appropriate family of polygonal arcs defined on $[a,\infty)$, then a subsequence converging on $[a, a+1]$, a further subsequence converging on $[a, a+2]$, and so on.)

(b) Prove Corollary 1.12 when $[a,\infty)$ replaces $[a,b]$.

The *lower Hamiltonian* h and *upper Hamiltonian* H corresponding to F will play an important role in what follows. These are functions from $\mathbb{R}\times\mathbb{R}^n\times\mathbb{R}^n$ to $\mathbb{R}$ defined as follows:

$$h(t,x,p) := \min_{v\in F(t,x)} \langle p, v\rangle, \quad H(t,x,p) := \max_{v\in F(t,x)} \langle p, v\rangle.$$

We gather some basic properties of h and H in the following, under the Standing Hypotheses on F. (Proposition 2.1.3 on support functions is relevant here.)

1.15. Exercise.

(a) h is lower semicontinuous in (x,p), and concave and continuous in p.

(b) h is superadditive in p:
$$h(t,x,p+q) \geq h(t,x,p) + h(t,x,q),$$
and $h(t,x,0) = 0$.

(c) A given vector v belongs to $F(t,x)$ iff
$$h(t,x,p) \leq \langle p, v\rangle \ \forall p \in \mathbb{R}^n;$$
v belongs to $F(t,x) + r\overline{B}$ (where $r \geq 0$) iff
$$h(t,x,p) \leq r\|p\| + \langle p, v\rangle \ \forall p \in \mathbb{R}^n.$$

(d) What are the counterparts of (a)–(c) in terms of H?

2 Weak Invariance

A venerable notion from the classical theory of dynamical systems is that of *invariance*. When the basic model consists of an autonomous ordinary differential equation $\dot{x}(t) = f(x(t))$ with locally Lipschitz right-hand side and a set $S \subseteq \mathbb{R}^n$, then flow invariance of the pair (S, f) is the property that for every initial point x_0 in S, the (unique) trajectory emanating from $x(0) = x_0$ is defined on $[0,\infty)$ and satisfies $x(t) \in S$ for all $t \geq 0$. In this section, we will study a generalization of this concept to situations wherein the differential equation is replaced by a differential inclusion (possibly nonautonomous). The largely geometric analysis undertaken here will have important ramifications later in the chapter. Once again, we will find it advantageous to consider first Euler arcs.

Proximal Aiming

Suppose that we are studying the flow of an ordinary differential equation

$$\dot{x}(t) = f(t, x(t)), \quad x(0) = x_0,$$

with the issue being to determine whether the resulting trajectory $x(t)$ approaches a given closed set S in $\mathbb{R}^n$. One natural way to test whether this is the case is to pick a point $s \in \text{proj}_S(x(t))$ (for a given t), and check the sign of the quantity

$$\langle f(t, x(t)), x - s \rangle.$$

If it is negative, $\dot{x}(t)$ will "point toward s," and hence certainly toward S. If this prevails at every point $(t, x(t))$, then the state $x(t)$ should indeed "move toward S." The next result, a simple and important one, confirms this heuristic observation in the general framework of Euler arcs.

2.1. Proposition. *Let f satisfy the linear growth condition*

$$\|f(t,x)\| \leq \gamma\|x\| + c \; \forall (t,x),$$

and let $x(\cdot)$ be an Euler arc for f on $[a,b]$. Let Ω be an open set containing $x(t)$ $\forall t \in [a,b]$, and suppose that every $(t,z) \in [a,b] \times \Omega$ satisfies the following "proximal aiming" condition: there exists $s \in \text{proj}_S(z)$ such that $\langle f(t,z), z - s \rangle \leq 0$. Then we have

$$d_S(x(t)) \leq d_S(x(a)) \; \forall t \in [a, b].$$

Proof. Let x_π be one polygonal arc in the sequence converging uniformly to x, as per the definition of the Euler solution. As usual, we denote its node at t_i by x_i ($i = 0, 1, \ldots, N$), and so $x_0 = x(a)$. We may suppose that $x_\pi(t)$ lies in Ω for all $t \in [a, b]$. Accordingly, there exists for each i a point $s_i \in \text{proj}_S(x_i)$ such that $\langle f(t_i, x_i), x_i - s_i \rangle \leq 0$. Letting k be the a priori bound on $\|\dot{x}_\pi\|_\infty$ as in the proof of Theorem 1.7, we calculate

$$\begin{aligned}
d_S^2(x_1) &\leq \|x_1 - s_0\|^2 \quad \text{(since } s_0 \in S\text{)} \\
&= \|x_1 - x_0\|^2 + \|x_0 - s_0\|^2 + 2\langle x_1 - x_0, x_0 - s_0 \rangle \\
&\leq k^2(t_1 - t_0)^2 + d_S^2(x_0) + 2\int_{t_0}^{t_1} \langle \dot{x}_\pi(t), x_0 - s_0 \rangle \, dt \\
&= k^2(t_1 - t_0)^2 + d_S^2(x_0) + 2\int_{t_0}^{t_1} \langle f(t_0, x_0), x_0 - s_0 \rangle \, dt \\
&\leq k^2(t_1 - t_0)^2 + d_S^2(x_0).
\end{aligned}$$

190   4. A Short Course in Control Theory

The same estimates at any node apply to give
$$d_S^2(x_i) \leq d_S^2(x_{i-1}) + k^2(t_i - t_{i-1})^2,$$
whence
$$d_S^2(x_i) \leq d_S^2(x_0) + k^2 \sum_{\ell=1}^{i}(t_\ell - t_{\ell-1})^2$$
$$\leq d_S^2(x_0) + k^2 \mu_\pi \sum_{\ell=1}^{i}(t_\ell - t_{\ell-1})$$
$$\leq d_S^2(x_0) + k^2 \mu_\pi (b - a).$$

Now consider the sequence x_{π_j} of polygonal arcs converging to x. Since the last estimate holds at every node and since $\mu_{\pi_j} \to 0$ (and the same k applies to each x_{π_j}), we deduce in the limit $d_S(x(t)) \leq d_S(x(a))$ $\forall t \in [a, b]$, as claimed. □

2.2. Exercise. Suppose that in Proposition 2.1, the proximal aiming condition is changed to
$$\langle f(t, z), z - s \rangle \leq \theta(t, z) d_S(z)$$
for a continuous function θ, other things being equal.

(a) Show that the conclusion becomes
$$d_S^2(x(t)) - d_S^2(x(\tau)) \leq 2 \int_\tau^t \theta(r, x(r)) \, d_S(x(r)) \, dr$$
for any $a \leq \tau < t \leq b$.

(b) Prove that this implies
$$\frac{d}{dt} d_S(x(t)) \leq \theta(t, x(t)) \text{ a.e.}$$
on any interval on which $d_S(x(t)) > 0$, or on any interval in which $\theta(t, x(t)) \geq 0$.

Weak Invariance (Autonomous Case)

Our real goal is not to work with a given $f(t, x)$, which allows no scope for control considerations, but rather with the possible trajectories of a differential inclusion $\dot{x}(t) \in F(x(t))$. Note that we are taking F to be independent of t (the so-called *autonomous* case for the moment; we will return to the nonautonomous case later. Our attention is still focused on approaching a given closed set S. More precisely, if the initial state x_0 actually lies in S already, is there a trajectory starting at x_0 that remains in S thereafter?

This concept is important enough to merit some terminology:

2 Weak Invariance

2.3. Definition. The system (S, F) is called *weakly invariant* provided that for all $x_0 \in S$, there exists a trajectory x on $[0, \infty)$ such that

$$x(0) = x_0, \quad x(t) \in S \ \forall t \geq 0.$$

Note that we speak of this property (which is also called *viability*) as being one of the *pair* (S, F), and not just of S. Here is a key proximal sufficient condition for weak invariance, in terms of the lower Hamiltonian defined earlier (see Exercise 1.15):

2.4. Theorem. *Suppose that for every $x \in S$, we have*

$$h(x, N_S^P(x)) \leq 0. \tag{1}$$

Then (S, F) is weakly invariant.

Let us dispel any doubt as to the meaning of (1): this Hamiltonian inequality says that for every $\zeta \in N_S^P(x)$, we have $h(x, \zeta) \leq 0$. Observe that (1) is automatically true if $N_S^P(x)$ reduces to $\{0\}$, since $h(x, 0) = 0$.

Proof. Let us define a function f_P as follows: for each x in $\mathbb{R}^n$, choose any $s = s(x)$ in $\text{proj}_S(x)$, and let v in $F(s)$ minimize over $F(s)$ the function

$$v \mapsto \langle v, x - s \rangle.$$

We set $f_P(x) = v$. Note that f_P is autonomous; i.e., t-independent. Since $x - s \in N_S^P(s)$, this minimum is nonpositive by (1); i.e., $\langle f_P(x), x - s \rangle \leq 0$, which is the main hypothesis of Proposition 2.1. If s_0 is any given point in S, then

$$\begin{aligned}
\|f_P(x)\| = \|v\| &\leq \gamma \|s\| + c \ \text{(since F has linear growth)} \\
&\leq \gamma \|s - x\| + \gamma \|x\| + c \\
&= \gamma d_S(x) + \gamma \|x\| + c \ \text{(since $s \in \text{proj}_S(x)$)} \\
&\leq \gamma \|x - s_0\| + \gamma \|x\| + c \\
&\leq 2\gamma \|x\| + \|\gamma s_0\| + c,
\end{aligned}$$

and so f_P satisfies the linear growth condition. Now set $[a, b] = [0, 1]$ and apply Proposition 2.1 for any $x_0 \in S$. We conclude that the resulting Euler solutions to $\dot{x} = f_P(x)$, $x(0) = x_0$ on $[0, 1]$ necessarily lie in S. We can extend x to $[0, \infty)$ in the evident way by considering next the interval $[1, 2]$, etc.

The proof will be complete if we can show that x is a trajectory for F. (Note that f is *not* a selection for F, so Corollary 1.12 is not available.) Let us define another multifunction F_S as follows:

$$F_S(x) := \text{co}\{F(s) \colon s \in \text{proj}_S(x)\}.$$

We ask the reader to prove some facts concerning F_S.

2.5. Exercise. F_S satisfies the Standing Hypotheses, and $F_S(x) = F(x)$ if $x \in S$.

Returning to the proof, observe that by construction, f_P is a selection for the multifunction F_S. Therefore, by Corollary 1.12, the arc x defined above is a trajectory for F_S; i.e., $\dot{x}(t) \in F_S(x(t))$ a.e. on $[0,1]$. But since $x(t) \in S$ on $[0,1]$, and since $F = F_S$ on S, it follows that x is a trajectory for F. □

Note that the proof, a paradigm for other more complicated ones later, is constructive in nature, and actually produces more than is asserted by the theorem, as exemplified by the next corollary.

2.6. Corollary. *There is an autonomous function f_P with linear growth such that, for any Euler arc x for f_P, we have*

$$d_S(x(t)) \leq d_S(x(a)) \ \forall t \geq a.$$

In particular, S is invariant under f_P, in the sense that any Euler arc for f_P which begins in S must remain in S thereafter.

The term *feedback* is used for any function $f(x)$ designed or constructed for the purpose of generating Euler solutions. As attractive as the "proximal aiming" feedback f_P may be, it does have some drawbacks. It is discontinuous in general, of course, since the metric projection multifunction proj_S does not generally admit continuous selections. In addition, observe that $f_P(x)$ is completely arbitrary within $F(x)$ when $x \in S$ (for then $s = x$). Finally, note that $f_P(x) \in F(x)$ for $x \in S$, but not necessarily otherwise: f_P is *not* a selection of F. The issue of finding feedback *selections* that lead to invariance will be explored later. First we look at an example of the construction above.

2.7. Exercise. For $n = 2$, let $S = \{(x,y): \max\{x,y\} \geq 0\}$ (the closed complement of the third quadrant), and let

$$F(x,y) = \{(|y| - u - 1/2, u): -1 \leq u \leq 1\}.$$

(a) Show that (S, F) is weakly invariant via Theorem 2.4.

(b) Let the function f_P constructed above be expressed in the form $f_P(x,y) = (|y| - u(x,y) - 1/2, u(x,y))$. Show that $u(x,y) = 1$ for $x < y < 0$, -1 for $y < x < 0$, so that f_P is discontinuous. What are the possible values of f_P when $x = y < 0$? When $(x,y) \in S$?

(c) Prove that no *continuous* function f exists having the global invariance property of f_P cited in Corollary 2.6.

Tangential and Other Conditions for Weak Invariance

It turns out that weak invariance can be characterized in a number of different ways, one of which involves a multifunction that plays an important role in control theory.

2.8. Definition. The *attainable set* $\mathcal{A}(x_0;t)$ for $t \geq 0$ is the set of all points of the form $x(t)$, where $x(\cdot)$ is any trajectory on $[0,t]$ satisfying $x(0) = x_0$.

2.9. Exercise. Prove the following:

(a) $\mathcal{A}(x_0;t)$ is compact and nonempty.

(b) (S,F) is weakly invariant $\implies \forall x_0 \in S, \forall t > 0, \mathcal{A}(x_0;t) \cap S \neq \emptyset$.

(c) *Semigroup property.* For $\Delta > 0$, $\mathcal{A}(x_0;t+\Delta) = \mathcal{A}(\mathcal{A}(x_0;t);\Delta)$.

(d) For given $t > 0$, the multifunction $x \mapsto \mathcal{A}(x;t)$ satisfies the Standing Hypotheses, except that $\mathcal{A}(x;t)$ can fail to be a convex set.

2.10. Theorem. *The following are equivalent:*

(a) $F(x) \cap T_S^B(x) \neq \emptyset \ \forall x \in S$.

(b) $F(x) \cap \operatorname{co} T_S^B(x) \neq \emptyset \ \forall x \in S$.

(c) $h(x, N_S^P(x)) \leq 0 \ \forall x \in S$.

(d) (S,F) *is weakly invariant.*

(e) $\forall x_0 \in S, \forall \varepsilon > 0, \exists \delta \in (0, \varepsilon)$ *such that* $\mathcal{A}(x_0;\delta) \cap S \neq \emptyset$.

Proof. Condition (a) evidently implies (b). Recall that (Exercise 2.7.1)

$$\operatorname{co} T_S^B(x) \subseteq \left[N_S^P(x)\right]^\circ,$$

from which it follows that (b) implies (c). That (c) implies (d) is Theorem 2.4, and the implication (d) $\implies$ (e) was noted in Exercise 2.9. Thus only the implication (e) $\implies$ (a) remains to be proven.

When (e) holds, for a sequence $\delta_i \downarrow 0$, there exist trajectories x_i on $[0, \delta_i]$ with $x_i(0) = x_0$, $x_i(\delta_i) \in S$. Since the trajectories have a common Lipschitz constant, there exists $K > 0$ such that $|x_i(\delta_i) - x_0| \leq K\delta_i$ for all large i. Then, upon taking a subsequence (we eschew relabeling), we can assume that

$$\frac{x_i(\delta_i) - x_0}{\delta_i} \to v$$

for some v. Note that $v \in T_S^B(x)$ by definition of the Bouligand tangent cone. Therefore we need only to show that v lies in $F(x)$ in order to deduce (a).

We have

$$x_i(\delta_i) - x_0 = \int_0^{\delta_i} \dot{x}_i(t)\, dt, \qquad (2)$$

where $\dot{x}_i(t) \in F(x_i(t))$, $0 \leq t \leq \delta_i$. Now for any given $\Delta > 0$, for all i sufficiently large, the set $\{x_i(t): 0 \leq t \leq \delta_i\}$ lies in $x_0 + \Delta B$. Further, given any $\varepsilon > 0$, taking Δ small enough will ensure that $x \in x_0 + \Delta B$ implies $F(x) \subset F(x_0) + \varepsilon B$. The upshot of all this, in view of (2), is that for all i large enough we have

$$\frac{x_i(\delta_i) - x_0}{\delta_i} \in F(x_0) + \varepsilon B.$$

Thus $v \in F(x_0) + \varepsilon \overline{B}$. Since ε is arbitrary, $v \in F(x_0)$ as required. □

2.11. Exercise.

(a) (S, F) is weakly invariant iff given any $x_0 \in S$ there exists $\delta > 0$ (depending on x_0) and a trajectory x on $[0, \delta]$ such that $x(0) = x_0$, $x(t) \in S$ ($0 \leq t \leq \delta$). (Thus, weak invariance can also be characterized in terms of the *local* existence of a trajectory.)

(b) Let x be a trajectory for F defined on $[a, \infty)$. Then the set

$$S_x := \{x(t): t \geq a\}$$

has the property that (S_x, F) is weakly invariant. Provide an example in $\mathbb{R}^2$ where S_x is not closed.

(c) The system (S, F) is termed *weakly preinvariant* if, given any $x_0 \in S$, there exists a trajectory x for F defined on $(-\infty, 0]$ such that $x(0) = x_0$. Show that the system (S, F) of Exercise 2.7 is weakly invariant but not weakly preinvariant.

(d) Prove that (S, F) is weakly preinvariant iff $(S, -F)$ is weakly invariant, and that this is in turn equivalent to

$$H(x, N_S^P(x)) \geq 0 \ \forall x \in S.$$

(*Hint.* Reverse time.)

(e) Let x be a trajectory for F defined on $[a, \infty)$, and let

$$G := \{(t, x(t)): t \geq a\}.$$

Prove that G is closed, and that $(G, \{1\} \times F)$ is weakly invariant, where $\{1\} \times F$ signifies the multifunction whose value at (t, x) is the set $\{1\} \times F(x)$ in $\mathbb{R} \times \mathbb{R}^n$.

(f) Show that S is weakly invariant iff it is the union of trajectories of F on $[0, \infty)$.

(g) Show that Theorem 2.10 holds true if the linear growth condition is deleted from the Standing Hypotheses, but S is assumed to be compact. (*Hint.* Redefine F outside a neighborhood of S so that the new multifunction has uniformly bounded images, as was done in Exercise 1.3(d).)

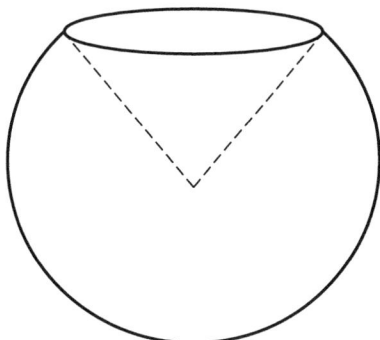

FIGURE 4.1. The set S of Exercise 2.12.

Since $T_S^B(x)$ always includes $T_S^C(x)$, the tangential or "inwardness" condition that $F(x) \cap T_X^C(x) \neq \emptyset$ for all $x \in S$ is certainly sufficient for weak invariance of (S, F). The next exercise, however, demonstrates that the tangent cone $T_S^C(x)$ cannot replace the Bouligand cone in Theorem 2.10, since necessity can fail (possibly only at a single point x).

2.12. Exercise. Let $n = 3$, and let $S := \text{cl}(\sqrt{2}B \backslash K)$, where K (see Figure 4.1) is the cone

$$\{(x_1, x_2, x_3) \in \mathbb{R}^3 : \|(x_1, x_2)\| \leq x_3\}.$$

Define F as follows: let

$$W := \{(x_1, x_2, 1) \in \mathbb{R}^3 : \|(x_1, x_2)\| = 1\},$$
$$Q := \{(x_1, x_2, 1) \in \mathbb{R}^3 : \|(x_1, x_2)\| \leq 1\},$$

and set

$$F(x_1, x_2, x_3) = F(x) = \begin{cases} Q & \text{if } x \in S \backslash W, \\ \text{co}\{Q, (-x_2, x_1, 0)\} & \text{otherwise.} \end{cases}$$

Prove that F satisfies the Standing Hypotheses, that

$$F(x) \cap T_S^B(x) \neq \emptyset \;\forall x \in S.$$

but that

$$F(0) \cap T_S^C(0) = \emptyset.$$

3 Lipschitz Dependence and Strong Invariance

We now wish to focus upon designing feedbacks f that are actual selections for F. For example, it can be useful to know whether a given trajectory of F arises as an Euler solution corresponding to some selection. That this is not generally the case under merely the hypotheses invoked thus far is illustrated by the following.

3.1. Exercise. We define F on $\mathbb{R}^2$ as follows:

$$F(x,y) := \begin{cases} \{(1,e)\} & \text{if } y > e^x, \\ \{(1,r): 1 \leq r \leq e\} & \text{if } y = e^x, \\ \{(1,1)\} & \text{if } y < e^x. \end{cases}$$

(a) Confirm that F satisfies the Standing Hypotheses, and that $(\bar{x},\bar{y})(t) = (t, e^t)$ is a trajectory on $[0,1]$. Show that there is *no* selection f for F (autonomous or not) admitting $(\bar{x},\bar{y})$ as an Euler solution of $(\dot{x},\dot{y}) = f(t,x,y)$, $(x(0),y(0)) = (0,1)$.

(b) The set $\{(t,e^t): t \geq 0\}$ is weakly invariant with respect to F, by part (b) of Exercise 2.11. Deduce from this that there exist weakly invariant systems which fail to admit a feedback selection f whose Euler arcs leave S invariant.

A better behavior on the part of F that will preclude such pathology is provided by a multivalued version of local Lipschitz behavior:

3.2. Definition. F is said to be *locally Lipschitz* provided that every point x admits a neighborhood $U = U(x)$ and a positive constant $K = K(x)$ such that

$$x_1, x_2 \in U \implies F(x_2) \subseteq F(x_1) + K\|x_1 - x_2\|\overline{B}. \tag{1}$$

We then say that F is Lipschitz of rank K on the set U.

3.3. Exercise. Let F be locally Lipschitz.

(a) Let C be any bounded subset of $\mathbb{R}^n$. Show that there exists K such that (1) holds on C; thus F is "Lipschitz on bounded sets."

(b) Fix $v_0 \in \mathbb{R}^n$, and let $f(x)$ be the point v in $F(x)$ closest to v_0. Prove that f is a continuous selection for F. (In general, f need not be locally Lipschitz.)

(c) Prove that F is locally Lipschitz iff $x \mapsto h(x,p)$ is locally Lipschitz for each p.

We now confirm that when F is locally Lipschitz, a weakly invariant system (S,F) admits a feedback selection whose Euler arcs all leave S invariant.

3.4. Theorem. *Let (S,F) be weakly invariant, where F is locally Lipschitz. Then there exists a feedback selection g_P for F under which S is invariant; that is, such that for any Euler arc x for g_P on $[a,b]$ having $x(a) \in S$, we have $x(t) \in S \; \forall t \in [a,b]$.*

Proof. Let f_P be defined precisely as in the proof of Theorem 2.4. Then $f_P(x)$ lies not necessarily in $F(x)$, but in $F(s)$, where $s = s(x) \in \text{proj}_S(x)$. We define $g_P(x)$ to be the point in $F(x)$ closest to $f_P(x)$, so that g_P is a selection for F.

Now let $x_0 \in S$ and $b > a$ be given. We will complete the proof by showing that any Euler solution $\bar{x}$ on $[a, b]$ from x_0 generated by g_P is such that $\bar{x}(t) \in S$ $\forall t \in [a, b]$. (We already know that $\bar{x}$ is a trajectory from Corollary 1.12.)

It follows from Exercise 1.9 that there is some a priori bound for $\bar{x}(t)$ on $[a, b]$, let us say $\|\bar{x}(t) - x_0\| < M$. Let K be a Lipschitz constant for F on the set $x_0 + 2MB$. Note now that if $\|x - x_0\| < M$, then

$$\|s - x_0\| \leq \|s - x\| + \|x - x_0\| = d_S(x) + \|x - x_0\|$$
$$\leq 2\|x - x_0\| < 2M.$$

This allows us to estimate as follows:

$$\langle g_P(x), x - s \rangle = \langle f_P(x), x - s \rangle + \langle g_P(x) - f_P(x), x - s \rangle$$
$$\leq \|g_P(x) - f_P(x)\| \|x - s\|$$
$$\leq K\|x - s\|^2 = K d_S^2(x).$$

This shows that the extension of Proposition 2.1 provided by Exercise 2.2 is applicable, with $\theta(t, x) := K d_S(x)$. Letting $\bar{x}$ be an Euler solution corresponding to g_P, we conclude that

$$\tfrac{d}{dt} d_S(\bar{x}(t)) \leq K d_S(\bar{x}(t)) \text{ a.e. in } [a, b].$$

The Gronwall inequality then implies that

$$d_S(\bar{x}(t)) \leq d_S(\bar{x}(a)) e^{K(t-a)} = 0,$$

since $\bar{x}(a) = x_0 \in S$. Thus $\bar{x}(t) \in S$, $a \leq t \leq b$. $\square$

The proof yields the following estimate for solutions not necessarily beginning in S:

3.5. Corollary. *Let K be a Lipschitz constant for F on the set $x_0 + 2MB$, and let $b > a$ satisfy*

$$(b - a) e^{\gamma(b-a)} (c + \gamma\|x_0\|) < M.$$

Then any Euler solution on $[a, b]$ of $\dot{x} = g_P(x)$, $x(a) = x_0$ is a trajectory satisfying

$$d_S(x(t)) \leq d_S(x_0) e^{K(t-a)}, \quad a \leq t \leq b.$$

Proof. The inequalities in the proof hold for $\|x - x_0\| < M$, for the given K; on the other hand, the solutions $x(t)$ continue to satisfy $\|x(t) - x_0\| < M$ for $t \in [a, b]$ when $b - a$ is small enough (as stated), according to Theorem 1.7. Thus, at least on $[a, b]$, the upper bound on $d_S(x(t))$ ensues. $\square$

3.6. Exercise. Observe that the multifunction F of Exercise 2.7 is locally Lipschitz. What is the feedback selection g_P of Theorem 3.4 in this case?

The following consequence affirms that for Lipschitz F, any trajectory can be generated by a feedback selection, which resolves an earlier question.

3.7. Corollary. *When F is locally Lipschitz, a given arc $\bar{x}$ on $[a,b]$ is a trajectory for F iff there exists a (not necessarily autonomous) feedback selection f for F such that $\bar{x}$ is an Euler solution on $[a,b]$ for the initial-value problem $\dot{x} = f(t,x)$, $x(a) = \bar{x}(a)$. In fact, given a trajectory $\bar{x}$, there is a feedback selection for F for which $\bar{x}$ is the unique Euler solution of the associated initial-value problem.*

Proof. That selections give rise to trajectories was noted in Corollary 1.12, so it remains to exhibit, for a given trajectory $\bar{x}$, a feedback selection generating it. The set G of Exercise 2.11(e) is weakly invariant under the trajectories of $\{1\} \times F$, allowing us to invoke Theorem 3.4. This provides a selection for $\{1\} \times F$, necessarily of the form $(1, f(t,x))$, where f is a selection for F, and having the property that any Euler solution x of the initial-value problem $\dot{x} = f(t,x)$, $x(a) = \bar{x}(a)$ is such that $(t, x(t)) \in G$ on $[a,b]$. But then $x(t) = \bar{x}(t)$ for $t \in [a,b]$, proving that $\bar{x}$ is the (unique) Euler solution. □

Strong Invariance

The system (S, F) is said to be *strongly invariant* if *every* trajectory x on $[0, \infty)$ for which $x(0) \in S$ is such that $x(t) \in S$ for all $t \geq 0$.

Remaining still in the case in which F is autonomous, let us characterize strong invariance in terms resembling those of Theorem 2.10, but with a Lipschitz hypothesis imposed upon the multifunction F. Note that now T_S^C, the tangent cone of §2.5, joins the cast by playing a role in one of the criteria for strong invariance; in this regard recall Exercise 2.12, which showed that T_S^C could *not* be included in the characterizations of weak invariance. Note that the *upper* Hamiltonian H figures in (d) below, so that (d) is a strong counterpart of the weak proximal normal condition that figured in Theorem 2.10.

3.8. Theorem. *Let F be locally Lipschitz. Then the following are equivalent:*

(a) $F(x) \subseteq T_S^C(x) \ \forall x \in S$.

(b) $F(x) \subseteq T_S^B(x) \ \forall x \in S$.

(c) $F(x) \subseteq \operatorname{co} T_S^B(x) \ \forall x \in S$.

(d) $H(x, N_S^P(x)) \leq 0 \ \forall x \in S$.

(e) (S, F) is strongly invariant.

(f) $\forall x_0 \in S$, $\exists \varepsilon > 0 \ni \mathcal{A}(x_0; t) \subseteq S \ \forall t \in [0, \varepsilon]$.

Proof. That (a) $\implies$ (b) $\implies$ (c) is a tautology, since $T_S^C(x) \subseteq T_S^B(x)$. Since every element of $\operatorname{co} T_S^B(x)$ belongs to $N_S^P(x)^\circ$, we know (c) $\implies$ (d). Let us now address the implication (d) $\implies$ (e).

Let $\bar{x}$ be any trajectory for F on $[a, b]$, with $\bar{x}(a) \in S$. According to Corollary 3.7, there exists a feedback selection f such that $\bar{x}$ is an Euler solution of the initial-value problem $\dot{x} = f(t, x)$, $x(a) = \bar{x}(a) =: x_0$.

Let $M > 0$ be such that all Euler solutions x of this initial-value problem above satisfy $\|x(t) - x_0\| < M$, $a \leq t \leq b$. If x lies in $x_0 + MB$ and $s \in \operatorname{proj}_S(x)$, then

$$\|s - x_0\| \leq \|s - x\| + \|x - x_0\| \leq 2\|x - x_0\|,$$

so that $s \in x_0 + 2MB$.

Now let K be a Lipschitz constant for F on $x_0 + 2MB$, and consider any $x \in x_0 + MB$ and $s \in \operatorname{proj}_S(x)$. Then $x - s \in N_S^P(s)$. Since $f(t, x) \in F(x)$, there exists $v \in F(s)$ such that $\|v - f(t, x)\| \leq K\|s - x\| = Kd_S(x)$. Further, by condition (d), we have $\langle v, x - s \rangle \leq 0$. We deduce

$$\langle f(t, x), x - s \rangle \leq Kd_S(x)^2.$$

Thus a special case of Exercise 2.2 holds, for $\theta(t, x) := Kd_S(x)$, and we obtain

$$\frac{d}{dt} d_S(\bar{x}(t)) \leq K d_S(\bar{x}(t)), \quad a \leq t \leq b, \quad d_S(\bar{x}(0)) = 0,$$

which implies that $d_S(\bar{x}(t)) = 0$ in $[a, b]$ by the now familiar Gronwall inequality. Thus $\bar{x}(t) \in S \ \forall t \in [a, b]$, and (e) holds.

That (e) and (f) are equivalent is easy to see; let us now show (e) $\implies$ (d). Consider any $\tilde{x} \in S$, let any $\tilde{v}$ in $F(\tilde{x})$ be given, and set $\widetilde{F}(x) = \{\tilde{f}(x)\}$, where $\tilde{f}(x)$ is the closest point in $F(x)$ to $\tilde{v}$. Note that $\tilde{f}(\tilde{x}) = \tilde{v}$, and that $\tilde{f}$ is a continuous selection of F (Exercise 3.3(c)). Thus $\widetilde{F}$ satisfies the Standing Hypotheses, and clearly $(S, \widetilde{F})$ is strongly, and hence weakly, invariant. By Theorem 2.10, we must have

$$\tilde{h}(\tilde{x}, N_S^P(\tilde{x})) \leq 0;$$

here, $\tilde{h}$ of course denotes the lower Hamiltonian associated with $\widetilde{F}$. This is the same as asserting that $\langle \tilde{v}, \zeta \rangle$ is nonpositive for any $\zeta \in N_S^P(\tilde{x})$. Since $\tilde{v}$ is arbitrary in $F(\tilde{x})$, (d) follows.

200   4. A Short Course in Control Theory

To complete the proof of the theorem, it suffices to prove that (d) $\Longrightarrow$ (a). Let $\tilde{v}$ be any element of $F(\tilde{x})$; we need to show that $\tilde{v}$ belongs to $N_S^L(\tilde{x})^\circ = T_S^C(\tilde{x})$. Now any element ζ of $N_S^L(\tilde{x})$ is of the form $\zeta = \lim_i \zeta_i$, where $\zeta_i \in N_S^P(x_i)$ and $x_i \to \tilde{x}$. For each i, there exists $v_i \in F(x_i)$ such that

$$\|v_i - \tilde{v}\| \leq K \|x_i - \tilde{x}\|,$$

and (by (d)) we have $\langle \zeta_i, v_i \rangle \leq 0$. We derive $\langle \zeta, \tilde{v} \rangle \leq 0$ as required. □

3.9. Exercise.

(a) Show that the local Lipschitz assumption on F cannot be dispensed with in the preceding theorem, by taking $F(t,x) = \{f(x)\}$, and $S = \{0\}$, where f is the function of Exercise 1.10(b).

(b) Show that the system of Exercise 2.7 is weakly but not strongly invariant.

The Nonautonomous Case

The invariance results obtained in the preceding sections can all be extended to the case in which F depends on t as well as on x (the *nonautonomous* case), and there is an easy route for getting the nonautonomous results from the autonomous versions. This technique is called *state augmentation*; here, it consists of viewing t as simply a component of the state, one whose derivative is always 1.

To be more precise, we will think of t as being the zeroth coordinate of x, and we implement this notationally through the convention wherein $\bar{x}$ denotes a vector (x^0, x) in $\mathbb{R} \times \mathbb{R}^n$. We proceed to define an augmented multifunction $\bar{F}$ which satisfies the Standing Hypotheses whenever F does:

$$\bar{F}(\bar{x}) = \bar{F}(x^0, x) = \{1\} \times F(x^0, x).$$

Then if $\bar{x}(\cdot) = (x^0(\cdot), x(\cdot))$ is a trajectory for $\bar{F}$, with $\bar{x}(a) = \bar{x}_0 = (x_0^0, x_0)$, it follows that x is a trajectory for F with $x(a) = x_0$, and that $x^0(t) = x_0^0 + t - a$. Conversely, if x is a trajectory for F, we augment it to a trajectory $\bar{x}$ for $\bar{F}$ by setting $x^0(t) = x_0^0 + t - a$ (for any choice of x_0^0).

As before, let S be a given closed subset of $\mathbb{R}^n$, and now let $F(t,x)$ possibly depend on t. We extend Definition 2.3 as follows: the system (S, F) is called *weakly invariant* provided that for all (t_0, x_0) with $x_0 \in S$, there exists a trajectory x of F on $[t_0, \infty)$ satisfying $x(t_0) = x_0$, $x(t) \in S$ $\forall t \geq t_0$.

3.10. Exercise. For the nonautonomous case described above, show that (S, F) is weakly invariant iff

$$h\big(t, x, N_S^P(x)\big) \leq 0 \; \forall t \in \mathbb{R}, \; \forall x \in S.$$

When F is locally Lipschitz, show that we obtain the corresponding characterization of strong invariance. (*Hint.* Augment the state, and consider $\bar{S} := \mathbb{R} \times S$.)

Dependence on Initial Conditions

In the general case in which $F(t,x)$ is nonautonomous, which remains our interest here, the attainable set depends of course on the initial-value of t as well. Thus, $\mathcal{A}(t_0, x_0; T)$, for $T \geq t_0$, is defined as the set of all points of the form $x(T)$, where x is a trajectory for F on $[t_0, T]$ satisfying $x(t_0) = x_0$.

Classically, the theorem by which the solution of a differential equation depends regularly on initial conditions is a well-known and useful result. Here is its counterpart for differential inclusions:

3.11. Theorem. *Let $F(t, x)$ be locally Lipschitz in (t, x). Then for any fixed $T \in \mathbb{R}$, the multifunction $(t_0, x_0) \mapsto \mathcal{A}(t_0, x_0; T)$ is locally Lipschitz on $(-\infty, T] \times \mathbb{R}^n$.*

Proof. Fix $a < T$ and $y \in \mathbb{R}^n$. We will exhibit a constant ρ with the property that for any $(t_0, x_0) \in [a, T) \times B(y; 1)$, for any trajectory z on $[t_0, T]$ having $z(t_0) = x_0$, and for any other $(\tau, \alpha) \in [a, T) \times B(y; 1)$, there exists a trajectory x on $[\tau, T]$ with $x(\tau) = \alpha$ such that

$$\|x(T) - z(T)\| \leq \rho \|(\tau - t_0, \alpha - x_0)\|.$$

This will establish that the mapping $(t, x) \mapsto \mathcal{A}(t, x; T)$ is locally Lipschitz on $(-\infty, T) \times \mathbb{R}^n$; the extension to $(-\infty, T]$ will be treated subsequently.

We augment the state as described above, setting

$$\bar{z}(t) = (t, z(t)), \quad \bar{z}_0 = (t_0, x_0), \quad G = \{(t, z(t)) : a \leq t < \infty\},$$

where we have extended z to the interval $[a, \infty)$. Then $\bar{z}$ is a trajectory on $[a, T]$ for the multifunction $\bar{F} := \{1\} \times F$, so that $(G, \bar{F})$ is weakly invariant. This allows us to invoke Corollary 3.5 to deduce the existence of an arc $\bar{x}(t) = (t, x(t))$ on $[\tau, T]$ such that $x(\tau) = \alpha$, x is a trajectory of F, and

$$d_G(t, x(t)) \leq e^{K(T-\tau)} d_G(\tau, \alpha),$$

where K is a Lipschitz constant for F on an appropriately large ball. Then x is a trajectory for F on $[\tau, T]$ and we have

$$d_G(T, x(T)) \leq e^{K(T-a)} \|(\tau - t_0, \alpha - x_0)\|,$$

since $(t_0, x_0) \in G$. Accordingly, for some point $(t', z(t')) \in G$ we have

$$\|(T - t', x(T) - z(t'))\| \leq e^{K(T-a)} \|(\tau - t_0, \alpha - x_0)\|.$$

Let $k \geq 1$ be a common Lipschitz constant for all trajectories of F on $[a, T]$ with initial-value in $B(y; 1)$. Then

$$\|z(T) - x(T)\| \leq \|z(T) - z(t')\| + \|z(t') - x(T)\|$$
$$\leq k|T - t'| + k\|z(t') - x(T)\| \leq 2k\|(T - t', x(T) - z(t'))\|$$
$$\leq 2k e^{K(T-a)} \|(\tau - t_0, \alpha - x_0)\|.$$

This reveals the required constant $\rho := 2ke^{K(T-a)}$.

To complete the proof, observe that $\mathcal{A}(T, x_0; T) = \{x_0\}$, so it suffices to show that
$$\mathcal{A}(\tau, \alpha; T) \subset \{x_0\} + \rho \|(T - \tau, x_0 - \alpha)\| \overline{B}.$$
But if x is any trajectory on $[\tau, T]$ with $x(\tau) = \alpha$, then
$$\|x(T) - x_0\| \leq \|x(T) - \alpha\| + \|\alpha - x_0\|$$
$$= \|x(T) - x(\tau)\| + \|\alpha - x_0\| \leq k|T - \tau| + k\|\alpha - x_0\|$$
$$\leq 2k\|(T - \tau, x_0 - \alpha)\| \leq \rho\|(T - \tau, x_0 - \alpha)\|,$$
which completes the proof. □

The following exercise provides useful facts regarding an optimal control problem to be studied later.

3.12. Exercise. Let $\ell \colon \mathbb{R}^n \to \mathbb{R}$ be continuous. For $\tau \leq T$ and for any $\alpha \in \mathbb{R}^n$, set
$$V(\tau, \alpha) := \inf\{\ell(x(T)) : x \text{ is a trajectory of } F \text{ on } [\tau, T] \text{ with } x(\tau) = \alpha\}.$$

(a) Prove that the infimum defining $V(\tau, \alpha)$ is attained.

(b) Now let F be locally Lipschitz. Prove that V is continuous on $(-\infty, T] \times \mathbb{R}^n$, and locally Lipschitz if ℓ is locally Lipschitz.

4 Equilibria

Consider the issue of stabilizing at a given point x^* a system described as the trajectory $x(t)$ of an autonomous differential inclusion $\dot{x} \in F(x)$. If the system is already at x^*, then it will be possible to remain there iff $0 \in F(x^*)$, a condition which is described by saying that x^* is a *zero* (or *equilibrium*, or *rest point*) of the multifunction F. Note that this is equivalent to the weak invariance of the system $(\{x^*\}, F)$, so that formally, the study of equilibria amounts to that of invariant singleton sets, an observation that motivates the dynamic approach to the issue that we will develop in this section.

The Classical Case and Brouwer's Theorem

At the heart of very many theorems on the existence of equilibria is the celebrated Brouwer Fixed Point Theorem, which can be stated as follows: *If g is a continuous function mapping $\overline{B}$ to itself, where $\overline{B}$ is the closed unit ball in $\mathbb{R}^n$, then there is a point u in $\overline{B}$ such that $g(u) = u$.* Let us note the bearing of this theorem upon the differential equation

$$\dot{x}(t) = f(x(t)), \tag{1}$$

where f is a Lipschitz mapping from $\mathbb{R}^n$ to $\mathbb{R}^n$. Suppose that the solution $x(t)$ to (1), for any initial condition $x(0) = x_0 \in \overline{B}$, has the property that $x(t) \in \overline{B}$ $\forall t \geq 0$. That is, suppose that the system $(\overline{B}, f)$ is invariant; "weakly" and "strongly" coalesce in this setting. It follows from the Brouwer Fixed Point Theorem that f must admit a zero in $\overline{B}$. This and more is taken up in the following exercise:

4.1. Exercise.

(a) Prove that in the situation described above there is a point x^* in $\overline{B}$ such that $f(x^*) = 0$. (*Hint.* For given $\tau > 0$, consider the map $f_\tau(\alpha) := x(\tau; \alpha)$; that is, the value of $x(\tau)$, where x is the (unique) solution of the initial-value problem satisfying the initial condition $x(0) = \alpha$. Invoke the Brouwer Fixed Point Theorem to deduce that for each $\tau > 0$ there exists $x_\tau \in \overline{B}$ such that $x(\tau; x_\tau) = x_\tau$, and then let $\tau \downarrow 0$.)

(b) Conversely, given the "zero point theorem" of part (a), deduce from it the Brouwer Fixed Point Theorem. (*Hint.* Given g, consider $f(x) := g(x) - x$.)

We would now like to consider sets S other than the ball, in fact sets that are not even convex. It is certainly the case, however, that some topological hypothesis will have to be made. Consider for example the annulus S in $\mathbb{R}^2$ described in polar coordinates (r, θ) via $S := \{(r, \theta): 1 \leq r \leq 2\}$, and let $g(r, \theta) := (r, \theta + \theta_0)$ for $0 < \theta_0 < 2\pi$. Then g is a smooth function mapping S to itself, yet g admits no fixed point.

One immediate extension of the Brouwer Fixed Point Theorem is obtained by taking S to be *homeomorphic* to a closed unit ball. This means that $S = h(\overline{B})$ where $\overline{B}$ is the closed unit ball in $\mathbb{R}^k$, $h: \overline{B} \to S$ is continuous, and where the inverse $h^{-1}: S \to \overline{B}$ exists and is continuous.

4.2. Exercise.

(a) Let $g: S \to S$ be continuous, and let S be homeomorphic to a closed unit ball. Prove that g has a fixed point in S.

(b) Show that part (a) of Exercise 4.1 holds true if the unit ball is replaced by a set which is homeomorphic to the ball.

A Conjecture in the Multivalued Case

The next step is to replace functions by multifunctions. A natural conjecture, in light of the above, is the following:

Conjecture. Let S in $\mathbb{R}^n$ be homeomorphic to a closed unit ball, and let F be a locally Lipschitz multifunction from $\mathbb{R}^n$ to $\mathbb{R}^n$ such that (S, F) is weakly invariant. Then F has a zero in S.

204   4. A Short Course in Control Theory

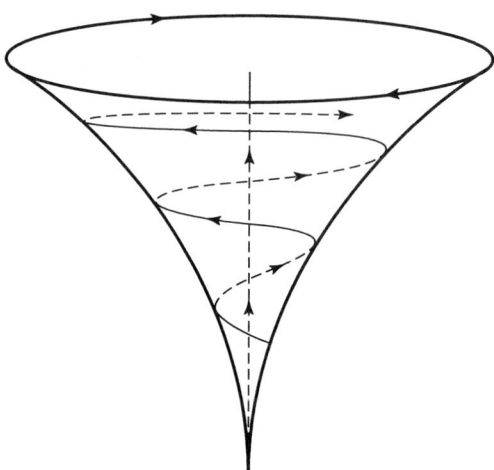

FIGURE 4.2. The set S of Exercise 4.3.

Natural, but false! Let us construct a counterexample. S is the following subset of $\mathbb{R}^3$ (a "fluted funnel"; see Figure 4.2):

$$\{(x, y, z) : x^2 + y^2 = z^4, 0 \leq z \leq 1\}.$$

We define $f: \mathbb{R}^3 \to \mathbb{R}^3$ as follows:

$$f(x, y, z) := \left[-yz + \frac{2x(1-z)}{(x^2+y^2)^{1/4}}, xz + \frac{2y(1-z)}{(x^2+y^2)^{1/4}}, 1-z \right]$$

for $x^2 + y^2 \neq 0$, and

$$f(0, 0, z) := [0, 0, 1-z].$$

4.3. Exercise.

(a) S is homeomorphic to the closed unit ball in $\mathbb{R}^2$, and S is regular.

(b) f is continuous everywhere, not Lipschitz at the origin, and f has no zero in S.

(c) For all $(x, y, z) \in S$, we have $f(x, y, z) \in T_S^B(x, y, z)$. Deduce that the system (S, f) is weakly invariant, but show that it is not strongly invariant. (Figure 4.2 indicates some trajectories of f, including two that start at 0, one which remains in the set S, while the other leaves it.)

The preceding exercise shows that our conjecture is false when the multifunction F is a continuous (single-valued) function. However, the exercise does not directly address the case of a *Lipschitz* multifunction. That gap will be bridged by a useful result which asserts that upper semicontinuous multifunctions can be approximated "from above" by Lipschitz ones.

4.4. Proposition. *Let F satisfy the Standing Hypotheses. Then there exists a sequence of locally Lipschitz multifunctions $\{F_k\}$ also satisfying the Standing Hypotheses such that:*

(i) *For each $k \in \mathbb{N}$, for every $x \in \mathbb{R}^n$,*

$$F(x) \subseteq F_{k+1}(x) \subseteq F_k(x) \subseteq \overline{\mathrm{co}}\, F\big(B(x; 3^{-k+1})\big).$$

(ii) $\bigcap_{k \geq 1} F_k(x) = F(x) \; \forall x \in \mathbb{R}^n$.

Proof. Let us fix an integer $k \geq 1$ and consider the set of all points (or grid) in $\mathbb{R}^n$ whose coordinates are all of the form $\pm m/4^k$, $m \in \{0, 1, 2, \dots\}$. Let $\{x_i\}$ be an enumeration of this countable set ($i = 1, 2, \dots$), and observe that the collection of open balls $B(x_i; 3^{-k})$ of radius 3^{-k} around x_i ($i = 1, 2, \dots$) is a covering of $\mathbb{R}^n$. This covering has the property that any bounded set intersects only finitely many members of the collection. This implies the existence of a locally Lipschitz *partition of unity* for the covering; that is, a collection of functions $p_i(x)$, each locally Lipschitz and taking values in $[0, 1]$, with the following properties:

$$x \notin B(x_i; 3^{-k}) \implies p_i(x) = 0 \; \forall i,$$
$$\sum_i p_i(x) = 1 \; \forall x.$$

Note that only finitely many terms in the last sum are nonzero, so there is no difficulty regarding its interpretation.

4.5. Exercise. Show that the desired Lipschitz partition of unity is obtained by setting

$$p_i(x) := \frac{d(x, \mathrm{comp}\, B_i^k)}{\sum_j d(x, \mathrm{comp}\, B_j^k)},$$

where $B_i^k := B(x_i; 3^{-k})$.

Returning to the proof, we define the compact convex set

$$C_i^k := \overline{\mathrm{co}}\, F\big(B(x_i; 2 \cdot 3^{-k})\big)$$

and the multifunction F_k via

$$F_k(x) := \sum_i p_i^k(x) C_i^k,$$

where the dependence of p_i on k has been reflected by the notation p_i^k.

Then F_k is locally Lipschitz, as is easily seen. In addition, $F_k(x)$ contains $F(x)$. To see this, note that for given x, those terms actually contributing in the sum defining $F_k(x)$ have $p_i^k(x) > 0$, whence $x \in B_i^k$. But for such "active indices" i, we then have $F(x) \subset C_i^k$. Since the coefficients $p_i^k(x)$ taken over the active indices form a convex combination, it follows that $F_k(x)$ contains $F(x)$. Also, when $x \in B_i^k$ we have reciprocally that $x_i \in B(x; 3^{-k})$, whence $B(x_i, 2 \cdot 3^{-k}) \subset B(x, 3^{-k+1})$. Thus, for each such i, C_i^k is contained in $\overline{\text{co}}\, F(B(x; 3^{-k+1}))$, and so is $F_k(x)$. This confirms the last estimate in (i), which clearly implies (ii) in light of the upper semicontinuity of F. There remains only the decreasing nature of the approximation to check.

Letting $F_{k+1}(x) = \sum_i p_i^{k+1}(x) C_i^{k+1}$, consider any two indices i and j which are active for $F_{k+1}(x)$ and $F_k(x)$, respectively (these are the only ones that actually matter). Then $p_i^{k+1}(x) > 0$, $p_j^k(x) > 0$, whence $x \in B(x_i, 3^{-k-1}) \cap B(x_j, 3^{-k})$, which implies $\|x_i - x_j\| < 3^{-k-1} + 3^{-k}$, which in turn implies $C_i^{k+1} \subset C_j^k$. This shows that each set contributing to the sum defining $F_{k+1}(x)$ is contained in every set in the sum defining $F_k(x)$, which completes the proof. □

Now let us return to the burial of the conjecture. Applying Proposition 4.4 to the multifunction $F(x) := \{f(x)\}$, where f is the function of Exercise 4.3, gives rise to a sequence of locally Lipschitz multifunctions F_k such that (S, F_k) is weakly invariant for each k; yet F_k cannot admit a zero in S for k arbitrarily large, otherwise f would do so as well.

To summarize, the condition

$$F(x) \cap T_S^B(x) \neq \emptyset \quad (x \in S)$$

although it characterizes weak invariance, is not adequate (even when F is Lipschitz) to yield a zero of F, even when S is homeomorphic to the closed unit ball (though it is adequate when S *is* a closed unit ball, as will follow from results below).

Existence of Equilibria in the Multivalued Case

Having seen what fails to be true, let us proceed to derive a positive result; the key is to use T_S^C instead of T_S^B.

4.6. Theorem. *Let S be homeomorphic to a closed unit ball in $\mathbb{R}^n$, and let F satisfy the Standing Hypotheses. Suppose that:*

(i) $\text{int}\, T_S^C(x) \neq \emptyset \; \forall x \in S$ *(i.e., S is wedged).*

(ii) $F(x) \cap T_S^C(x) \neq \emptyset \; \forall x \in S$.

Then F has a zero in S. Even if F is Lipschitz, this fails in general in the absence of (i) *(even when S is regular), or if $T_S^B(x)$ replaces $T_S^C(x)$ in* (ii).

Proof. We make the *temporary hypothesis* that F is locally Lipschitz and satisfies

$$F(x) \cap \text{int}\, T_S^C(x) \neq \emptyset \; \forall x \in S.$$

Let $x \in S$, and let $\varepsilon > 0$ be given. Then, because F is Lipschitz near x, and because T_S^C is lower semicontinuous at x (by Proposition 3.6.7), there exists for each $y \in F(x) \cap \text{int } T_S^C(x)$ a scalar $\delta(x, y) > 0$ such that

$$x' \in x + \delta(x,y)B \implies y \in \{F(x') + \varepsilon B\} \cap \text{int } T_S^C(x').$$

The family of open balls $x + \delta(x, y)B$ forms an open covering of the compact set S; let the family of sets $\{x_i + \delta(x_i, y_i)B\}$ ($i = 1, 2, \ldots, m$) be a finite subcover. Now associate to this subcover a Lipschitz partition of unity $\{p_i(x)\}$ subordinate to it, and define a locally Lipschitz function f_ε as follows:

$$f_\varepsilon(x) := \sum_{i=1}^{m} p_i(x) y_i.$$

When $p_i(x) \neq 0$, we have $x \in x_i + \delta(x_i, y_i)B$, whence

$$y_i \in \{F(x) + \varepsilon B\} \cap \text{int } T_S^C(x)$$

by definition of $\delta(x_i, y_i)$. Because $F(x) + \varepsilon B$ and $\text{int } T_S^C(x)$ are both convex sets, we infer

$$f_\varepsilon(x) \in \{F(x) + \varepsilon B\} \cap \text{int } T_S^C(x).$$

Thus $f_\varepsilon(x) \in T_S^B(x)$, so that the system (S, f_ε) is weakly invariant by Theorem 2.10. By Exercise 4.2, there is a point $x_\varepsilon \in S$ such that $f_\varepsilon(x_\varepsilon) = 0$, which implies

$$0 \in F(x_\varepsilon) + \varepsilon B.$$

Now let $\varepsilon \downarrow 0$. Some subsequence $\{x_{\varepsilon_i}\}$ of the corresponding points converges to a limit $x^* \in S$, and the upper semicontinuity of F yields $0 \in F(x^*)$. Thus the theorem is proven, under the temporary hypothesis.

To remove the need for this hypothesis, let us observe first that it is satisfied by each of the multifunctions

$$\widetilde{F}_k(x) := F_k(x) + \gamma \overline{B},$$

together with the hypotheses of the theorem under consideration, where F_k is the Lipschitz approximation to F provided by Proposition 4.4, and where $\gamma > 0$ is fixed.

By the case of the theorem proved above, there exists $x_k \in S$ such that $0 \in \widetilde{F}_k(x_k)$, and by passing to a subsequence (without relabeling) we may assume $x_k \to x^* \in S$. Now the monotonicity of the F_k scheme gives us:

$$0 \in F_k(x_j) + \gamma \overline{B} \ \forall j \geq k.$$

Letting $j \to \infty$ yields $0 \in F_k(x^*) + \gamma \overline{B}$, since F_k is Lipschitz and all the more upper semicontinuous. We invoke (ii) of Proposition 4.4 to deduce $0 \in F(x^*) + \gamma \overline{B}$. Since γ is an arbitrary positive number, an evident sequential argument together with the upper semicontinuity of F confirms that F has a zero in S.

There remain to prove the two negative assertions in the statement of the theorem. The first of these is confirmed by the example above that buried the conjecture, while the second follows from the example in Exercise 2.12 (see Exercise 4.8 below). □

Under the right hypotheses, nonconvex sets which are weakly invariant necessarily contain equilibria, which is the assertion of the next corollary.

4.7. Corollary. *Let S be homeomorphic to the closed unit ball in $\mathbb{R}^n$. Further assume that S is wedged and regular, and that the system (S, F) is weakly invariant, where F satisfies the Standing Hypotheses. Then S contains a zero of F.*

4.8. Exercise.

(a) Show that for a suitable locally Lipschitz approximation F_k of the multifunction F appearing in Exercise 2.12, F_k together with the set S of that exercise satisfies all the hypotheses of Theorem 4.6, but with T_S^C replaced by T_S^B. Yet F_k has no zero in S. (*Hint. S is star-shaped.*)

(b) Prove Corollary 4.7.

5 Lyapounov Theory and Stabilization

Consider the differential equation

$$\dot{x}(t) = f(x(t)),$$

where f is a smooth function from $\mathbb{R}^n$ to itself, and suppose that the point x^* is an equilibrium: $f(x^*) = 0$. Then of course the constant function $x(t) \equiv x^*$ is a solution of the differential equation. If for any $\alpha \in \mathbb{R}^n$ the solution $x(\cdot)$ of the differential equation satisfying $x(0) = \alpha$ exists on $[0, \infty)$ and has the property that $x(t) \to x^*$ as $t \to \infty$, then the equilibrium is said to be *globally asymptotically stable*. We recall that if $\alpha \neq x^*$, then x^* cannot be reached in finite time: $x(t) \neq x^* \ \forall t > 0$.

A simple but far-reaching criterion to assure this asymptotic stability can be given in terms of *Lyapounov functions*. Suppose that smooth functions Q and W exist such that:

(i) $Q(x) \geq 0$, $W(x) \geq 0 \ \forall x \in \mathbb{R}^n$, and $W(x) = 0$ iff $x = x^*$;

(ii) the sets $\{x \colon Q(x) \leq q\}$ are compact $\forall q \in \mathbb{R}$; and

(iii) $\langle \nabla Q(x), f(x) \rangle \leq -W(x) \ \forall x$.

(These properties are referred to as *positive definiteness, growth,* and *infinitesimal decrease*, respectively.) It follows that x^* must be globally asymptotically stable. For let $x(\cdot)$ be any local solution of the differential equation. We have

$$\frac{d}{dt}Q(x(t)) = \langle \nabla Q(x(t)), f(x(t))\rangle \leq -W(x(t))$$

by condition (iii). It follows that the nonnegative function

$$t \mapsto Q(x(t)) + \int_0^t W(x(\tau))\,d\tau$$

is decreasing and hence bounded on $[0, \infty)$. Since $W \geq 0$, this implies that the function $t \mapsto Q(x(t))$ is bounded and hence that $x(t)$ remains bounded. But then the solution $x(t)$ of the differential equation exists on $[0, \infty)$, and $W(x(t))$ is a globally Lipschitz function on $[0, \infty)$. That $W(x(t))$ converges to 0 now follows from:

5.1. Exercise. Let $r(t)$ be a nonnegative, globally Lipschitz function on $[0, \infty)$, and suppose that $\int_0^t r(\tau)\,d\tau$ is bounded for $t \geq 0$. Then $r(t) \to 0$ as $t \to \infty$.

Knowing that $W(x(t)) \to 0$, it follows readily that $x(t)$ converges to $x^* = W^{-1}(0)$ as $t \to \infty$, as claimed. Note the prominence of *monotonicity* in this classical argument; it will play a central role in the results of this section as well.

The method under discussion here was introduced by A. M. Lyapounov in the theory of differential equations. Its extension to control systems is conditioned by the fact that we typically consider situations in which *some* but not all trajectories from a given initial condition approach the equilibrium. That is, an element of "controllability" is involved. With this in mind, here is a natural way to proceed in the case of a differential inclusion $\dot{x} \in F(x)$ admitting an equilibrium at x^* (i.e., such that $0 \in F(x^*)$). We define a (smooth) *Lyapounov pair* (Q, W) to be one having the same properties (i) and (ii) as above, together with the following replacement for (iii):

(iii)$'$ $\min_{v \in F(x)}\langle \nabla Q(x), v\rangle \leq -W(x)\ \forall x$.

This does the trick, as we ask the reader to check.

5.2. Exercise. If Q and W are C^1 and satisfy (i), (ii), and (iii)$'$, then for any $\alpha \in \mathbb{R}^n$ there is a trajectory x on $[0, \infty)$ with $x(0) = \alpha$ such that $x(t) \to x^*$ as $t \to \infty$. (*Hint.* Consider trajectories of the multifunction

$$\widetilde{F}(x) := \{v \in F(x) : \langle \nabla Q(x), v\rangle \leq -W(x)\}.)$$

We remark that the very existence of (Q, W) implies that x^* is an equilibrium, and that in contrast to the classical case of a differential equation, $x(\cdot)$ may reach x^* in finite time.

One of the interesting questions about this approach is whether it is necessary as well as sufficient for asymptotic stability. In this connection, and for other reasons as well, it turns out to be essential to develop a theory encompassing nonsmooth Lyapounov functions. In light of the earlier chapters, our proposal in this direction is the following: take Q and W in the class $\mathcal{F}(\mathbb{R}^n)$, retain properties (i) and (ii), and add to them the following:

for every x, for every $\zeta \in \partial_P Q(x), \exists v \in F(x)$ such that $\langle \zeta, v \rangle \leq -W(x)$.

We summarize, and in so doing express this proximal form of the infinitesimal decrease property in terms of the lower Hamiltonian h. Two functions Q and W in $\mathcal{F}(\mathbb{R}^n)$ are said to be a *Lyapounov pair* for x^* if they satisfy the following properties:

- Positive definiteness: $Q, W \geq 0$; $W(x) = 0$ iff $x = x^*$.
- Growth: the level sets $\{x \in \mathbb{R}^n : Q(x) \leq q\}$ are compact $\forall q \in \mathbb{R}$.
- Infinitesimal decrease: $h(x, \partial_P Q(x)) \leq -W(x) \ \forall x \in \mathbb{R}^n$.

Note that this last property holds automatically at $x = x^*$ when x^* is an equilibrium, since then $0 \in F(x^*)$. Furthermore, infinitesimal decrease can also be expressed in an equivalent form using the subderivates of §3.4:

5.3. Proposition. *We have $h(x, \partial_P Q(x)) \leq -W(x) \ \forall x \in \mathbb{R}^n$ iff*

$$\inf_{v \in F(x)} DQ(x; v) \leq -W(x) \ \forall x \in \mathrm{dom}\, Q.$$

Proof. Suppose first that the derivate condition holds. Let $\zeta \in \partial_P Q(x)$, and pick $v \in F(x)$ such that $DQ(x; v) \leq -W(x)$ (we know from Exercise 3.4.1 that $DQ(x; \cdot)$ is lower semicontinuous, so the infimum is attained). Since $DQ(x; v) \geq \langle \zeta, v \rangle$ in general, we have $\langle \zeta, v \rangle \leq -W(x)$, whence $h(x, \partial_P Q(x)) \leq -W(x)$; i.e., the Hamiltonian inequality holds.

The necessity is a deeper result. If the derivate condition fails, then for some $\delta > 0$ we have

$$DQ(x; v) \geq -W(x) + \delta \ \forall v \in F(x).$$

Since $DQ(x; \cdot)$ is lower semicontinuous and $F(x)$ is compact, this implies that for some $\eta > 0$ we have

$$DQ(x; v) > -W(x) + \tfrac{\delta}{2} \ \forall v \in F(x) + \eta \overline{B}.$$

Applying Subbotin's Theorem 3.4.2, we deduce that for some z arbitrarily near x, for some $\zeta \in \partial_P Q(z)$, we have

$$\langle \zeta, v \rangle > -W(x) + \tfrac{\delta}{2} \ \forall v \in F(x) + \eta \overline{B}.$$

For z near enough to x, we have $F(z) \subset F(x) + \eta B$, since F is upper semicontinuous, as well as $W(z) \geq W(x) - \delta/4$, since W is lower semicontinuous. For such z we deduce

$$\langle \zeta, v \rangle > -W(z) + \tfrac{\delta}{4} \ \forall v \in F(z).$$

This implies $h(z, \zeta) \geq -W(z) + \delta/4$, which shows that the Hamiltonian condition fails at z, thereby completing the proof. □

It turns out that without any loss of generality, we can always assume that W has additional properties, in particular, being finite-valued. However, the fact that Q can be extended-valued is useful, for example, in subsuming local stability, and is to be retained.

5.4. Exercise.

(a) If Q and W in $\mathcal{F}(\mathbb{R}^n)$ form a Lyapounov pair for x^*, prove that there is a function $\widetilde{W}$ for which $(Q, \widetilde{W})$ continues to be a Lyapounov pair for x^*, where $\widetilde{W}$ is locally Lipschitz and satisfies a linear growth condition. (*Hint.*

$$\widetilde{W}(x) := \min\{W(y) + \|x - y\| : y \in \mathbb{R}^n\}.)$$

(b) Prove that $Q(x) > 0$ for any $x \neq x^*$. (*Hint.* Otherwise, $0 \in \partial_P Q(x)$.)

(c) Given $x \neq x^*$, there exist points y arbitrarily near x^* having $Q(y) < Q(x)$.

(d) Q attains a global minimum at x^* (which we can always take to be 0 by redefining Q if necessary).

5.5. Theorem. *Let $0 \in F(x^*)$, and let there exist Q and W in $\mathcal{F}(\mathbb{R}^n)$ such that (Q, W) constitutes a Lyapounov pair for x^*. Then for any $\alpha \in \text{dom}\, Q$ there is a trajectory x for F on $[0, \infty)$ having $x(0) = \alpha$ such that $x(t) \to x^*$ as $t \to \infty$.*

The proof of this important result requires a proximal characterization of a certain system monotonicity property, and is postponed until after that acquisition.

Weakly Decreasing Systems

Let $\varphi \in \mathcal{F}(\mathbb{R}^n)$. The system (φ, F) is said to be *weakly decreasing* if for any $\alpha \in \mathbb{R}^n$, there exists a trajectory x of F on $[0, \infty)$ with $x(0) = \alpha$ which satisfies

$$\varphi(x(t)) \leq \varphi(x(0)) = \varphi(\alpha) \ \forall t \geq 0.$$

Note that only points $\alpha \in \operatorname{dom} \varphi$ need be tested in this condition, since it holds automatically otherwise.

We will see that this system property is the functional counterpart of weak invariance of sets.

5.6. Exercise.

(a) Let φ be the indicator of a closed set S. Show then that (φ, F) is weakly decreasing iff (S, F) is weakly invariant.

(b) Show that (φ, F) is weakly decreasing iff $(\operatorname{epi} \varphi, F \times \{0\})$ is weakly invariant.

5.7. Theorem. (φ, F) *is weakly decreasing iff*

$$h(x, \partial_P \varphi(x)) \leq 0 \ \forall x \in \mathbb{R}^n.$$

Proof. Note the meaning of this Hamiltonian inequality: for any $x \in \mathbb{R}^n$, and $\zeta \in \partial_P \varphi(x)$, we have $h(x, \zeta) \leq 0$. Suppose first that (φ, F) is weakly decreasing. Then by Exercise 5.6, $(\operatorname{epi} \varphi, F \times \{0\})$ is weakly invariant. In light of Theorem 2.4, this is characterized by the condition that for any vector $(\zeta, \lambda) \in N^P_{\operatorname{epi} \varphi}(x, r)$, where $(x, r) \in \operatorname{epi} \varphi$, we have

$$\min\{(\zeta, \lambda) \cdot (v, 0) \colon v \in F(x)\} \leq 0.$$

Now if $\zeta \in \partial_P \varphi(x)$, then $(\zeta, -1) \in N^P_{\operatorname{epi} \varphi}(x, \varphi(x))$, and so we deduce

$$h(x, \zeta) = \min\{\langle \zeta \cdot v \rangle \colon v \in F(x)\} \leq 0,$$

which confirms the Hamiltonian inequality.

For the converse, suppose that the Hamiltonian condition holds. In order to deduce that (φ, F) is weakly decreasing, or equivalently that $(\operatorname{epi} \varphi, F \times \{0\})$ is weakly invariant, it suffices to exhibit for any $(\zeta, \lambda) \in N^P_{\operatorname{epi} \varphi}(x, r)$ an element $v \in F(x)$ such that $\langle \zeta, v \rangle \leq 0$. We know that $\lambda \leq 0$ and that $(\zeta, \lambda) \in N^P_{\operatorname{epi} \varphi}(x, \varphi(x))$, by Exercise 1.2.1(d). If $\lambda < 0$, then we have

$$(\zeta/(-\lambda), -1) \in N^P_{\operatorname{epi} \varphi}(x, r),$$

which implies $-\zeta/\lambda \in \partial_P \varphi(x)$. Then the Hamiltonian condition applies to imply the existence of $v \in F(x)$ for which $\langle (-\zeta/\lambda), v \rangle \leq 0$. But then $\langle \zeta, v \rangle \leq 0$ as required.

The remaining case to consider is that in which $\lambda = 0$. Then we have

$$(\zeta, 0) \in N^P_{\operatorname{epi} \varphi}(x, \varphi(x)),$$

and we invoke Problem 1.11.23 to deduce the existence of sequences $(\zeta_i, -\varepsilon_i)$, with $\varepsilon_i > 0$, and $(x_i, \varphi(x_i))$ such that

$$(\zeta_i, -\varepsilon_i) \to (\zeta, 0), \quad (\zeta_i, -\varepsilon_i) \in N^P_{\operatorname{epi} \varphi}(x_i, \varphi(x_i)), \quad (x_i, \varphi(x_i)) \to (x, \varphi(x)).$$

Then, as in the case $\lambda < 0$ above, there exists $v_i \in F(x_i)$ for which $\langle \zeta_i, v_i \rangle \leq 0$. Since F is locally bounded, the sequence $\{v_i\}$ is bounded. Passing to a subsequence, we can suppose that v_i converges to a limit v; then $v \in F(x)$ as a consequence of the upper semicontinuity of F. We deduce $\langle \zeta, v \rangle \leq 0$, which completes the proof. □

Proof of Theorem 5.5

Let (Q, W) be a Lyapounov pair for x^*, where W has the additional properties provided by Exercise 5.4, and let $\alpha \in \mathrm{dom}\, Q$. We will prove the existence of the trajectory x cited in the statement of the theorem.

We define $\widetilde{Q} \colon \mathbb{R}^n \times \mathbb{R} \to (-\infty, \infty]$ via $\widetilde{Q}(x, y) := Q(x) + y$, and the multifunction $\widetilde{F}$ as follows:

$$\widetilde{F}(x, y) := F(x) \times \{W(x)\}.$$

Note that $\widetilde{F}$ satisfies the Standing Hypotheses. We claim that the system $(\widetilde{Q}, \widetilde{F})$ is weakly decreasing. Indeed, with an eye to applying Theorem 5.7, let $(\zeta, r) \in \partial_P \widetilde{Q}(x, y)$. Then $\zeta \in \partial_P Q(x)$ and $r = 1$. The infinitesimal decrease property of (Q, W) provides the existence of $v \in F(x)$ such that $\langle (v, W(x)), (\zeta, 1) \rangle \leq 0$, which verifies the Hamiltonian inequality of Theorem 5.7.

We deduce the existence of a trajectory (x, y) of $\widetilde{F}$ beginning at $(\alpha, 0)$ and satisfying $\widetilde{Q}(x(t), y(t)) \leq \widetilde{Q}(\alpha, 0) = Q(\alpha)$ for $t \geq 0$. This translates to

$$Q(x(t)) + \int_0^t W(x(\tau))\, d\tau \leq Q(\alpha),$$

where x is a trajectory of F. This implies that $Q(x(t))$ (and hence $x(t)$) is bounded, as well as $\int_0^t W(x(\tau))\, d\tau$. Since F is bounded on bounded sets, we observe that $\dot{x}(t)$ is bounded too, whence x satisfies a global Lipschitz condition on $[0, \infty)$, let us say of rank k. We conclude by proving what amounts to Exercise 5.1.

Suppose that $x(t)$ fails to converge to x^* as $t \to \infty$. Then for some $\varepsilon > 0$, there exist points t_i tending to $+\infty$ such that $\|x(t_i) - x^*\| \geq \varepsilon$ $(i = 1, 2, \dots)$. We can assume $t_{i+1} - t_i > \varepsilon/(2k)$. Let $\eta > 0$ be such that

$$\|x - x^*\|_\infty \geq \|u - x^*\| \geq \tfrac{\varepsilon}{2} \implies W(u) \geq \eta$$

(such η exists because W is continuous and positive on the (nonempty) annulus in question). Then

$$|t - t_i| < \tfrac{\varepsilon}{2k} \implies \|x(t) - x(t_i)\| < \tfrac{\varepsilon}{2} \implies \|x(t) - x^*\| \geq \tfrac{\varepsilon}{2},$$

so that

$$\int_{t_{i-1}}^{t_{i+1}} W(x(\tau))\, d\tau \geq \tfrac{\eta \varepsilon}{k}.$$

This would imply that $\int_0^t W(x(\tau))\,d\tau$ diverges, a contradiction that completes the proof. □

Construction of a Stabilizing Feedback

The proof of Theorem 5.5 just given says nothing explicitly about how to construct a trajectory converging to x^*, but in fact this is implicit from the constructive nature in which we proved Theorem 2.4, which lies at the heart of Theorem 5.7 and hence of Theorem 5.5 itself. Let us now make explicit the proximal aiming method underlying these results.

We define a function $f\colon \mathbb{R}^n \times \mathbb{R} \times \mathbb{R} \to \mathbb{R}^n$ as follows: given $(x,y,r) \in \mathbb{R}^n \times \mathbb{R} \times \mathbb{R}$, select any (x',y',r') in $\text{proj}_S(x,y,r)$, where

$$S := \{(x',y',r'): Q(x') + y' \leq r'\}.$$

(This is the set used in the proof of Theorem 5.7 when $\varphi = Q(x) + y$, as is the case in the proof of Theorem 5.5.)

Now select any $v \in F(x')$ which minimizes over $F(x')$ the function $v \mapsto \langle v, x - x' \rangle$, and set $f(x,y,r) = v$. The following restates Corollary 2.6 in the present setting, where the initial condition on (x,y,r) is taken to be $(\alpha, 0, Q(\alpha)) \in S$; see also Exercise 1.14.

5.8. Proposition. *Let (Q,W) be a Lyapounov pair for x^*, where W has the additional properties of Exercise 5.4. Then, for any $\alpha \in \text{dom}\,Q$, there exists at least one Euler solution (x,y) on $[0,\infty)$ of the initial-value problem*

$$\dot{x}(t) = f(x(t), y(t), Q(\alpha)), \quad \dot{y}(t) = W(x(t)), \quad x(0) = \alpha, \ y(0) = 0,$$

and any such solution defines a trajectory $x(\cdot)$ of F on $[0,\infty)$ which converges to x^* as $t \to \infty$.

The function f above fails to satisfy $f(x,y,r) \in F(x)$ in general; it is not a feedback selection for F. When F is locally Lipschitz however, we showed in Theorem 3.4 how to define a selection g that can play the role of f. In our present setting, we define $g(x,y,r)$ as follows: given $f(x,y,r) = v \in F(x')$ as above, let w be the point in $F(x)$ closest to v; set $g(x,y,r) = w \in F(x)$. Then we have

5.9. Proposition. *In the context of Proposition 5.8, when F is locally Lipschitz, there is at least one Euler solution (x,y) to the initial-value problem*

$$\dot{x}(t) = g(x(t), y(t), Q(\alpha)), \quad \dot{y}(t) = W(x(t)), \quad x(0) = \alpha, \ y(0) = 0,$$

and any such solution defines a trajectory x of F on $[0,\infty)$ which converges to x^* as $t \to \infty$.

The construction of the "stabilizing feedback" in Proposition 5.9 is quite explicit. Note however that it requires the calculation of another variable

$y(t) = \int_0^t W(x(\tau)) \, d\tau$ involving the past history of x; this is an instance of *dynamic feedback*. It is of considerable interest to be able to define a *static* feedback selection $g(x) \in F(x)$ depending only on x, and giving rise to trajectories converging to the equilibrium, whatever the initial condition. When a Lyapounov pair (Q, W) exists having Q locally Lipschitz, this can be done by proximal aiming techniques, but doing so would take us beyond the scope of this chapter.

6 Monotonicity and Attainability

We have seen the relevance of weak decrease along trajectories in §5, in connection with asymptotic controllability. This and other monotonicity issues will arise again in other contexts, and we will require some corresponding *localized* proximal characterizations. We will derive these first in the autonomous setting.

Let Ω be an open subset of $\mathbb{R}^n$, and let $\varphi \in \mathcal{F}(\Omega)$. Extending our earlier definition (which dealt with the case $\Omega = \mathbb{R}^n$), we say that (φ, F) is *weakly decreasing* on Ω if for any $\alpha \in \text{dom}\,\varphi \subset \Omega$, there exists a trajectory x of F on $[0, \infty)$ with $x(0) = \alpha$ having the property that for any interval $[0, T]$ for which $x([0, T]) \subset \Omega$, we have

$$\varphi(x(t)) \leq \varphi(\alpha) \;\; \forall t \in [0, T].$$

Recall that trajectories x on $[0, T]$ can always be extended to $[0, \infty)$, but above we only require $\varphi(x(t)) \leq \varphi(\alpha)$ to hold until the *exit time*

$$\tau(x, \Omega) := \inf\{t \geq 0 : x(t) \in \text{comp}\,\Omega\}.$$

Of course, τ can equal $+\infty$, precisely when $x(t) \in \Omega \;\forall t \geq 0$.

6.1. Theorem. *Let $\varphi \in \mathcal{F}(\Omega)$. The system (φ, F) is weakly decreasing on Ω iff*

$$h(x, \partial_P \varphi(x)) \leq 0 \;\; \forall x \in \Omega.$$

Proof. Let (φ, F) be weakly decreasing on Ω, and let $\alpha \in \Omega$ and $\zeta \in \partial_P \varphi(\alpha)$ be given. Let $\delta > 0$ be such that $\overline{B}(\alpha; \delta) \subset \Omega$, and define

$$S := \{(x, r) \in \mathbb{R}^n \times \mathbb{R} : x \in \overline{B}(\alpha; \delta), \varphi(x) \leq r\}.$$

Then

$$(\zeta, -1) \in N_S^P(\alpha, \varphi(\alpha)).$$

Also define

$$\widetilde{F}(x, r) := F(x) \times \{0\} \;\text{ if } x \in B(\alpha; \delta),$$

and otherwise set
$$\widetilde{F}(x,r) := \mathrm{co}\left\{\bigcup_{\|y-\alpha\|=\delta} F(y) \cup \{0\}\right\} \times \{0\}.$$

Then $\widetilde{F}$ satisfies the Standing Hypotheses and $(S, \widetilde{F})$ is weakly invariant (why?).

It follows from Theorem 2.10 that for some $(v, 0) \in \widetilde{F}(\alpha, \varphi(\alpha))$ we have $\langle (v,0), (\zeta, -1)\rangle \leq 0$. This implies $h(\alpha, \zeta) \leq 0$, whence the Hamiltonian inequality holds everywhere in Ω.

For the converse, let the Hamiltonian inequality hold, and let $\alpha \in \mathrm{dom}\,\varphi \subset \Omega$ be given. Let $\Omega_k := \Omega \cap B(\alpha; k)$, and define a closed set S_k and a multifunction $\widetilde{F}_k$ by

$$S_k := \{(x, r) \in \mathbb{R}^n \times \mathbb{R} : x \in \Omega_k, \varphi(x) \leq r\} \cup (\mathrm{comp}\,\Omega_k \times \mathbb{R}),$$
$$\widetilde{F}_k(x, r) := F(x) \times \{0\} \quad \text{if } x \in \Omega_k,$$

and otherwise

$$\widetilde{F}_k(x, r) := \mathrm{co}\left\{\bigcup_{y \in \mathrm{bdry}\,\Omega_k} F(x) \cup \{0\}\right\} \times \{0\}.$$

We claim that $(S_k, \widetilde{F}_k)$ is weakly invariant. Let us verify this by the criterion of Theorem 2.4. If $x \notin \Omega_k$, it is evidently satisfied, since $0 \in \widetilde{F}_k(x, r)$. If $x \in \Omega_k$, then a proximal normal (ζ, λ) to S_k at (x, r) belongs to $N^P_{\mathrm{epi}\,\varphi}(x, r)$, and precisely the argument used to prove the "converse" part of Theorem 5.7 demonstrates that $\widetilde{h}(x, r, \zeta, \lambda) \leq 0$. This establishes the claim.

Since $(S_1, \widetilde{F}_1)$ is weakly invariant, we deduce the existence of a trajectory x for F on $[0, T]$ with $x(0) = \alpha$ such that $\varphi(x(t)) \leq \varphi(\alpha)$ for $t \in [0, \tau_1)$, where τ_1 is the exit time of x from Ω_1. If $\tau_1 = \infty$, or if $x(\tau_1) \in \mathrm{bdry}\,\Omega$, then the trajectory x satisfies the requirement of the definition whereby (φ, F) is weakly decreasing on Ω. Otherwise, $x(\tau_1) \in \Omega$ and $\|x(\tau_1) - \alpha\| = 1$. In this case, we invoke the weak invariance of $(S_2, \widetilde{F}_2)$ to construct a trajectory that extends x, beginning at $x(\tau_1)$, to the interval $[0, \tau_1 + \tau_2)$, where τ_2 is the exit time of the extension from Ω_2, and where for $t \in [\tau_1, \tau_1 + \tau_2)$ we have $\varphi(x(t)) \leq \varphi(x(\tau_1)) \leq \varphi(\alpha)$. Again, if either $\tau_2 = \infty$ or $x(\tau_1 + \tau_2) \in \mathrm{bdry}\,\Omega$, we have the required trajectory; otherwise $x(\tau_1 + \tau_2) \in \Omega$, $\|x(\tau_1+\tau_2)-\alpha\| = 2$, and we begin again. If the process fails to end after finitely many steps, the resulting trajectory x is defined on $[0, \infty)$, since

$$\left\|x\left(\sum_1^k \tau_i\right) - \alpha\right\| = k \text{ implies that } \sum_1^k \tau_i \to \infty \text{ as } k \to \infty$$

(i.e., trajectories of F do not blow up in finite time). Then x on $[0, \infty)$ is itself the required trajectory, and the proof is complete. □

Strongly Decreasing Systems on Ω

Let $\varphi \in \mathcal{F}(\Omega)$, where Ω is an open subset of $\mathbb{R}^n$. The system (φ, F) is said to be *strongly decreasing* on Ω if for any $\alpha \in \operatorname{dom} \varphi \subset \Omega$, for any trajectory x of F on an interval $[a, b]$ which lies in Ω and satisfies $x(a) = \alpha$, we have

$$\varphi(x(t)) \leq \varphi(\alpha) \ \forall t \in [a, b].$$

6.2. Exercise. Show that (φ, F) is strongly decreasing on Ω iff every trajectory x of F on an interval $[a, b]$ which lies in Ω is such that the function $t \mapsto \varphi(x(t))$ is decreasing on $[a, b]$.

6.3. Theorem. *Let F be locally Lipschitz. Then (φ, F) is strongly decreasing on Ω iff*

$$H(x, \partial_P \varphi(x)) \leq 0 \ \forall x \in \Omega.$$

Proof. Let (φ, F) be strongly decreasing on Ω, and let $\zeta \in \partial_P \varphi(\alpha)$, where $\alpha \in \Omega$. Fix any $v_0 \in F(\alpha)$. We wish to prove that $\langle v_0, \zeta \rangle \leq 0$, to allow us to deduce the Hamiltonian inequality of the theorem. Pick $\delta > 0$ such that $\overline{B}(\alpha; \delta) \subset \Omega$, and set

$$S := \{(x, r) \in \mathbb{R}^n \times \mathbb{R} : x \in \overline{B}(\alpha; \delta), \varphi(x) \leq r\}.$$

Let $f(x) := v$ in $F(x)$ closest to v_0. Then f is continuous (Exercise 3.3(c)) and $f(\alpha) = v_0$. We define the multifunction

$$\widetilde{F}(x, r) := \{(f(x), 0)\} \quad \text{if } x \in B(\alpha; \delta),$$
$$:= \operatorname{co}\{0, f(y) : \|y - \alpha\| = \delta\} \times \{0\} \quad \text{otherwise}.$$

Then S is closed, $\widetilde{F}$ satisfies the Standing Hypotheses, and $(S, \widetilde{F})$ is weakly invariant. Since $\zeta \in \partial_P \varphi(\alpha)$, we have $(\zeta, -1) \in N_S^P(\alpha, \varphi(\alpha))$, so that by Theorem 2.10,

$$\langle (f(\alpha), 0), (\zeta, -1) \rangle = \langle v_0, \zeta \rangle \leq 0,$$

proving the necessity.

For the converse, let the Hamiltonian inequality hold, and let $\hat{x}$ be an Euler arc on $[0, T]$ lying in Ω. In order to prove that $\varphi(\hat{x}(t))$ is decreasing, it suffices to prove that for any $a \in [0, T)$, for all $b \in (a, T]$ sufficiently close to a, we have $\varphi(\hat{x}(b)) \leq \varphi(\hat{x}(a))$ (we may assume $\varphi(\hat{x}(a))$ finite). To this end, set

$$S := \operatorname{cl}\{(x, r) : x \in \Omega, r \geq \varphi(x)\},$$

and pick $M > 0$ such that $\hat{x}(a) + 4M\overline{B} \subset \Omega$, and then pick $b > a$ sufficiently near a so that

$$\|\hat{x}(t) - \hat{x}(a)\| < M \ \forall t \in [a, b].$$

It follows that for any

$$(x, r) \in (\hat{x}(a) + M\overline{B}) \times (\varphi(\hat{x}(a)) + M\overline{B}),$$

for any $(x', r') \in \text{proj}_S(x, r)$, we have
$$\|(x', r') - (\hat{x}(a), \varphi(\hat{x}(a)))\| < 4M.$$

Now let f be a selection of F whose unique Euler arc beginning at $(a, \hat{x}(a))$ is $\hat{x}$ (Corollary 3.7). If
$$(x, r) \in (\hat{x}(a) + M\overline{B}) \times (\varphi(\hat{x}(a)) + M\overline{B}),$$
then
$$(\zeta, \lambda) := (x - x', r - r') \in N_S^P(x', r'),$$
where $x' \in \hat{x}(a) + 4M\overline{B} \subset \Omega$. It follows that $(\zeta, \lambda) \in N_{\text{epi}\,\varphi}^P(x', r')$. If $\lambda < 0$, then $-\zeta/\lambda \in \partial_P \varphi(x')$, so by hypothesis we have $\langle f(x'), -\zeta/\lambda \rangle \leq 0$, whence $\langle f(x'), x - x' \rangle \leq 0$. If $\lambda = 0$, then
$$(\zeta, 0) \in N_{\text{epi}\,\varphi}^P(\hat{x}(a), \varphi(\hat{x}(a))),$$
and Problem 1.11.23 implies the existence of
$$x_i \xrightarrow{\varphi} x', \quad \varepsilon_i \to 0 \text{ (with } \varepsilon_i > 0\text{)}, \quad \zeta_i \to \zeta,$$
such that $(\zeta_i, -\varepsilon_i) \in \partial_P \varphi(x_i)$. Since F is Lipschitz, there exist $v_i \in F(x_i)$ with $\|v_i - f(x')\| \leq K\|x_i - x'\|$. As above, we have $\langle \zeta_i, v_i \rangle \leq 0$, and a passage to the limit gives $\langle \zeta, f(x') \rangle \leq 0$ once more.

The above confirms that Proposition 2.1 applies to S together with the map $(x, r) \mapsto (f(x), 0)$. Since the unique Euler arc for this map beginning at $(\hat{x}(a), \varphi(\hat{x}(a)))$ is $(\hat{x}(\cdot), \varphi(\hat{x}(a)))$, we deduce
$$d_S(\hat{x}(t), \varphi(\hat{x}(a))) \leq d_S(\hat{x}(a), \varphi(\hat{x}(a))) = 0 \; \forall t \in [a, b].$$
This implies $\varphi(\hat{x}(b)) \leq \varphi(\hat{x}(a))$, as required. $\square$

A Nonautonomous Extension

Suppose now that φ and F depend on t as well as x. We can easily extend the monotonicity characterizations obtained above to this nonautonomous case, and make them relative to a given t-interval (t_0, t_1), where we allow $t_0 = -\infty$ and/or $t_1 = \infty$. We will say that (φ, F) is *weakly decreasing* on $(t_0, t_1) \times \Omega$ if for any $\tau \in (t_0, t_1)$ and $\alpha \in \Omega$ there exists a trajectory x on $[\tau, t_1)$ with $x(\tau) = \alpha$ such that
$$\varphi(t, x(t)) \leq \varphi(\tau, \alpha) \; \forall t \in [\tau, b],$$
where $[\tau, b]$ is any subinterval of $[\tau, t_1)$ upon which $x(t)$ remains in Ω.

If this holds for *all* trajectories x, then (φ, F) is said to be *strongly decreasing* on $(t_0, t_1) \times \Omega$. The state augmentation device introduced in §3, together with Theorems 6.1 and 6.3, leads easily to:

6.4. Exercise. Let $\varphi \in \mathcal{F}((t_0,t_1) \times \Omega)$.

(a) Then (φ, F) is weakly decreasing on $(t_0, t_1) \times \Omega$ iff

$$\theta + h(t,x,\zeta) \leq 0 \;\forall (\theta,\zeta) \in \partial_P \varphi(t,x),\; \forall (t,x) \in (t_0,t_1) \times \Omega.$$

(b) If F is locally Lipschitz, then (φ, F) is strongly decreasing on $(t_0, t_1) \times \Omega$ iff

$$\theta + H(t,x,\zeta) \leq 0 \;\forall (\theta,\zeta) \in \partial_P \varphi(t,x),\; \forall (t,x) \in (t_0,t_1) \times \Omega.$$

A certain number of other monotonicity properties of systems have not been discussed. We will conclude with *strong increase*, leaving to the end-of-chapter problems an exhaustive study of all the variations on this theme. In the nonautonomous case still, (φ, F) is said to be *strongly increasing* on $(t_0, t_1) \times \Omega$ provided that for any interval $[a, b]$ contained in (t_0, t_1), for any trajectory x of F on $[a,b]$ for which $x(t) \in \Omega\; \forall t \in [a,b]$, we have

$$\varphi(t, x(t)) \leq \varphi(b, x(b)) \;\forall t \in [a,b].$$

Of course, this last inequality is automatically satisfied when $(b, x(b)) \notin \mathrm{dom}\,\varphi$. As in the autonomous case, this strong increase property is equivalent to the requirement that the function $t \mapsto \varphi(t, x(t))$ be increasing on $[a, b]$ whenever x is a trajectory on some interval $[a,b] \subset (t_0, t_1)$ for which $x(t)$ remains in Ω.

6.5. Proposition. *Let F be locally Lipschitz. Then (φ, F) is strongly increasing on $(t_0, t_1) \times \Omega$ iff*

$$\theta + h(t,x,\zeta) \geq 0 \;\forall (\theta,\zeta) \in \partial_P \varphi(t,x),\; \forall (t,x) \in (t_0,t_1) \times \Omega.$$

Proof. Let x be a trajectory of F on (t_0, t_1), and define $y(t) := x(t^* - t)$, where t^* is a point in (t_0, t_1). Then the function y is defined on the interval $(t^* - t_1, t^* - t_0)$, and we have

$$\dot{y}(t) = -\dot{x}(t^* - t) \in -F(t^* - t, x(t^* - t)) = -F(t^* - t, y(t)) \text{ a.e.}$$

This shows that y is a trajectory of F^* on $(t^* - t_1, t^* - t_0)$, where $F^*(t, y) := -F(t^* - t, y)$. Clearly this correspondence between x and y is one-to-one, so that $\varphi(t, x(t))$ is increasing for all such x (while $x(t) \in \Omega$) iff $\varphi(t^* - t, y(t))$ is decreasing for all such y (in each case on the relevant interval). Letting $\varphi^*(t,y) := \varphi(t^* - t, y)$, we have proven that strong increase on $(t_0, t_1) \times \Omega$ for (φ, F) is the same as strong decrease on $(t^* - t_1, t^* - t_0) \times \Omega$ for (φ^*, F^*). Applying Exercise 6.4(b), and denoting by H^* the upper Hamiltonian of F^*, the latter is equivalent to

$$\theta + H^*(t,y,\zeta) \leq 0 \;\forall (\theta,\zeta) \in \partial_P \varphi^*(t,y),\; \forall (t,y) \in (t^* - t_1, t^* - t_0) \times \Omega.$$

But it is easy to see that

$$H^*(t, y, \zeta) = -h(t^* - t, y, \zeta),$$
$$(\theta, \zeta) \in \partial_P \varphi^*(t, y) \iff (-\theta, \zeta) \in \partial_P \varphi(t^* - t, y).$$

The proposition then follows immediately. □

Local Attainability

We have addressed the issue of *remaining* in a given set, and also that of asymptotically *approaching* an equilibrium. We now examine under what conditions an initial-value *outside* a given set can be steered to it in *finite* time.

We say that the system (S, F) is *locally attainable* if there exists $r > 0$ and $T > 0$ such that the following holds: for all α having $d_S(\alpha) < r$, there exists a trajectory x of F on $[0, \infty)$ such that

$$x(0) = \alpha \text{ and } x(t) \in S \ \forall t \geq T.$$

6.6. Theorem. *Let S be compact, and let F be locally Lipschitz. Suppose that for some $\delta > 0$ we have*

$$h(x, \zeta) \leq -\delta \|\zeta\| \ \forall \zeta \in N_S^P(x), \ \forall x \in S. \tag{1}$$

Then (S, F) is locally attainable.

Proof. Let F be Lipschitz of rank K on $S + \eta B$, where $\eta > 0$, and let $r > 0$ satisfy $r < \min(\delta/K, \eta)$. Set $\lambda := \delta - Kr$. We claim that for $y \in \mathbb{R}$, the system $\bigl(d_S(x) + \lambda y, F(x) \times \{1\}\bigr)$ is weakly decreasing on

$$\Omega := \bigl\{(S + rB) \backslash S\bigr\} \times \mathbb{R}.$$

We will verify this by means of the criterion of Theorem 6.1.

Any proximal subgradient (ζ, θ) of $d_S(x) + \lambda y$ in Ω is of the form (ζ, λ) where $\zeta \in \partial_P d_S(x)$. According to Theorem 1.6.1, $\|\zeta\| = 1$ and $\zeta \in N_S^P(s)$, where $s \in \text{proj}_S(x)$. By hypothesis (1), we have $h(s, \zeta) + \delta \leq 0$. The Lipschitz condition on F gives

$$h(x, \zeta) + \lambda \leq h(s, \zeta) + K\|x - s\| + \delta - Kr$$
$$\leq -\delta + Kd_S(x) + \delta - Kr < 0.$$

This proves the claim.

It follows that for any $\alpha \in (S + rB) \backslash S$, there is a trajectory x of F on $[0, \infty)$ such that $x(0) = \alpha$, and such that

$$d_S\bigl(x(t)\bigr) + \lambda t \leq d_S(\alpha)$$

for all $t > 0$ up to the first $T > 0$ such that $d_S(x(T)) = 0$; i.e., such that $x(T) \in S$. Note that since $\lambda > 0$, such T must exist (uniformly for α).

Once $x(T) \in S$, then there is a trajectory beginning at $x(T)$ which extends x and remains in S thereafter, since (1) implies that (S, F) is weakly invariant. □

6.7. Exercise.

(a) Prove that in Theorem 6.6, the parameter T can be taken to be $r/(\delta - Kr)$.
(b) The time required to steer $\alpha \in S + \bar{r}B$ to S is no greater than $d_S(\alpha)/(\delta - Kd_S(\alpha))$, and an approach to S at rate $\gamma := \delta - Kd_S(\alpha) > 0$ can be guaranteed, in the sense that we have

$$d_S(x(t)) - d_S(\alpha) \leq -\gamma t$$

until S is reached.

A sufficient condition for local attainability can also be given in tangential terms akin to those employed in proving the existence of equilibria in Theorem 4.6.

6.8. Proposition. *Let S be compact, and let F be locally Lipschitz. Suppose that*

$$F(x) \cap \text{int } T_S^C(x) \neq \emptyset \ \forall x \in S.$$

Then (S, F) is locally attainable.

Proof. Let $x \in S$. Then there exists $v_x \in F(x)$ and $\delta_x > 0$ such that $v_x + \delta_x \bar{B} \subset T_S^C(x)$. Given any $\zeta \in N_S^L(x)$, since

$$T_S^C(x) = \left(N_S^L(x)\right)^0,$$

we have

$$\langle v_x + \delta_x u, \zeta \rangle \leq 0 \quad \text{for any } u \in \bar{B},$$

which implies $\langle v_x, \zeta \rangle \leq -\delta_x \|\zeta\|$ and consequently $h(x, \zeta) \leq -\delta_x \|\zeta\|$. This shows that the system (S, F) satisfies a *pointwise* version of condition (1), and with N_S^P replaced by N_S^L.

If the system fails to satisfy (1) uniformly, then there exist sequences $\{x_i\}$, $\{\zeta_i\}$ with $x_i \in S$, $\|\zeta_i\| = 1$, $\zeta_i \in N_S^P(x_i)$, and such that

$$h(x_i, \zeta_i) \geq -\frac{1}{i}\|\zeta_i\| = -\frac{1}{i}.$$

By passing to subsequences, we can suppose $x_i \to x \in S$, $\zeta_i \to \zeta$, where $\zeta \in N_S^L(x)$ and $\|\zeta\| = 1$. Since $h(x, p)$ is locally Lipschitz in x (Exercise 3.3(d)) and continuous in p (as a concave real-valued function), we obtain $h(x, \zeta) \geq 0$. But this contradicts what was proven above.

Thus the system (S, F) satisfies (1) for some positive δ, and so is locally attainable by Theorem 6.6. □

7 The Hamilton–Jacobi Equation and Viscosity Solutions

For the first time, we now consider the issue of finding a trajectory which is *best* relative to a given criterion. We consider the following *optimal control problem* (P)

$$\text{minimize } \ell(x(T)) \text{ subject to } \dot{x}(t) \in F(x(t)) \text{ a.e.,} \quad t \in [0,T], \; x(0) = x_0.$$

Here $T > 0$, $x_0 \in \mathbb{R}^n$, and a continuous function $\ell \colon \mathbb{R}^n \to \mathbb{R}$ are given. We seek therefore to minimize the *endpoint cost* $\ell(x(T))$ over all trajectories x of F on $[0,T]$ originating at x_0. It follows from the Standing Hypotheses that problem (P) admits a solution (Exercise 3.12). We assume throughout this section that F is locally Lipschitz and autonomous.

Verification Functions

We now describe how a venerable idea in the calculus of variations, originating with Legendre, leads to sufficient conditions for optimality in control problems. Suppose that we have a feasible arc $\bar{x}$ that we suspect of being optimal for our problem. How can we confirm that $\bar{x}$ is a solution? Here is one way: produce a smooth (C^1) function $\varphi(t, x)$ such that

$$\varphi_t(t, x) + \langle \varphi_x(t, x), v \rangle \geq 0 \; \forall x, \; \forall t \in (0,T), \; \forall v \in F(x), \tag{1}$$

$$\varphi(T, \cdot) = \ell(\cdot),$$

$$\varphi(0, x_0) = \ell(\bar{x}(T)).$$

Let us see how the existence of φ verifies that $\bar{x}$ is optimal. Let x be any other arc feasible for (P). Then a.e. on $[0, T]$ we have

$$\frac{d}{dt}\varphi(t, x(t)) = \varphi_t(t, x(t)) + \langle \varphi_x(t, x(t)), \dot{x}(t) \rangle \geq 0 \quad \text{(by (1))}.$$

Integrating this on $[0, T]$ yields

$$\varphi(T, x(T)) = \ell(x(T)) \geq \varphi(0, x_0) = \ell(\bar{x}(T)).$$

So $\bar{x}$ gives the least possible value of $\ell(x(T))$, as required. It also follows that $\varphi(0, x_0)$ is the value V of (P); i.e., the minimum cost.

In this simple argument, the *Hamilton–Jacobi inequality* (1) was really only used to deduce that the map $t \to \varphi(t, x(t))$ is increasing whenever x is a trajectory. In the terminology of the previous section, we want (φ, F) to be strongly increasing on $(0, T) \times \mathbb{R}^n$. Our previous results allow us to characterize this system property even when φ is nondifferentiable. Here is a corresponding extension of the verification argument whose proof is

7 The Hamilton–Jacobi Equation and Viscosity Solutions

an adaptation of the classical one. In this and subsequent results, it is notationally convenient to use the *augmented Hamiltonian* $\bar{h}$ defined by

$$\bar{h}(x, \theta, \zeta) := \theta + h(x, \zeta).$$

7.1. Proposition. *Let $\bar{x}$ be feasible for* (P), *and suppose that there exists a continuous $\varphi(t, x)$ on $[0, T] \times \mathbb{R}^n$ satisfying*

$$\bar{h}(x, \partial_P \varphi(t, x)) \geq 0 \; \forall (t, x) \in (0, T) \times \mathbb{R}^n, \tag{2}$$

$$\varphi(T, \cdot) = \ell(\cdot),$$
$$\varphi(0, x_0) = \ell(\bar{x}(T)). \tag{3}$$

Then $\bar{x}$ solves (P), *and the value of* (P) *is $\varphi(0, x_0)$.*

7.2. Exercise. Prove Proposition 7.1.

This is an extension to nonsmooth φ of Legendre's approach to sufficient conditions, which in the calculus of variations has also been called the "royal road of Carathéodory." A continuous function φ satisfying the hypotheses (2), (3), and (4) of Proposition 7.1 is called a *verification function* (for $\bar{x}$).

The obvious question to ask at this point is how applicable the method turns out to be, or to rephrase this: Can we be sure that a verification function φ for $\bar{x}$ exists when $\bar{x}$ is optimal? And how do we find one?

Considerable insight into this question arises from applying the technique of *invariant embedding*. Suppose that instead of the problem (P) considered above, we consider a family of problems $P(\tau, \alpha)$ parametrized by the initial data $(\tau, \alpha) \in [0, T] \times \mathbb{R}^n$; i.e., the initial condition is

$$x(\tau) = \alpha$$

rather than $x(0) = x_0$. Let $V(\tau, \alpha)$ denote the value of $P(\tau, \alpha)$; then we observe that the classical verification argument actually gives not only $V(0, x_0) = \varphi(0, x_0)$ as noted earlier, but also

$$V(\tau, \alpha) \geq \varphi(\tau, \alpha) \; \forall (\tau, \alpha) \in [0, T) \times \mathbb{R}^n.$$

We are quite naturally led to consider whether we could take V itself as the function φ in Proposition 7.1. We know that V is continuous on $(-\infty, T] \times \mathbb{R}^n$ (see Exercise 3.12).

We can easily see that V indeed satisfies (2). The reason is that $V(t, x(t))$ is always increasing when x is a trajectory; the minimum of $\ell(x(T))$ can only be greater or equal starting from an intermediate point $(t', x(t'))$ than it was from an earlier "less committed" point $(\tau, x(\tau))$. (This is an instance of the logic known as the *principle of optimality*.) This strong increase property of the system (V, F), by Proposition 6.5, implies the Hamilton–Jacobi inequality (2) on all $(-\infty, T) \times \mathbb{R}^n$.

Finally, it is clear that V satisfies (3), and satisfies (4) iff $\bar{x}$ is optimal. We obtain therefore the following satisfying justification of the verification method:

7.3. Proposition. *A feasible arc $\bar{x}$ is optimal iff there exists a continuous verification function for $\bar{x}$; the value function V is one such verification function for any optimal arc.*

7.4. Exercise. We set $n = 1$, $F(x) = \bigl[-|x|, |x|\bigr]$, $\ell(x) = x$, $T = 1$.

(a) Calculate $V(\tau, \alpha)$ for any $\tau \leq 1$, and verify that it satisfies the properties of a verification function. Note that V is nondifferentiable.

(b) Find a different verification function confirming the optimality of the arc that solves $P(0, 0)$.

There are in general many possible verification functions for a given optimal $\bar{x}$. We have seen in this section, however, how naturally the value function V arises in connection with the verification method, and hence the associated Hamilton–Jacobi inequality. Might it be possible to establish an even closer relationship, perhaps even a characterization of V, in Hamilton–Jacobi terms?

The Proximal Hamilton–Jacobi Equation

The following theorem shows that the value function is the *unique continuous solution* of a suitable generalization of the classical Hamilton–Jacobi, whose general form is

$$\varphi_t + H(x, \varphi_x) = 0,$$

with boundary condition. Recall that $\bar{h}(x, \theta, \zeta)$ is defined as $\theta + h(x, \zeta)$. We call a function φ that satisfies (5) below a *proximal solution* of the Hamilton–Jacobi equation (for h).

7.5. Theorem. *There is a unique continuous function $\varphi: (-\infty, T] \times \mathbb{R}^n \to \mathbb{R}$ satisfying*

$$\bar{h}\bigl(x, \partial_P \varphi(t, x)\bigr) = 0 \ \forall (t, x) \in (-\infty, T) \times \mathbb{R}^n, \tag{4}$$

$$\ell(x) = \varphi(T, x) \ \forall x \in \mathbb{R}^n. \tag{5}$$

That function is the value function V.

Proof. That V satisfies (6) was noted earlier, as well as "half" of (5); there remains to show

$$\bar{h}\bigl(x, \partial_P V(t, x)\bigr) \leq 0 \ \forall (t, x) \in (-\infty, T) \times \mathbb{R}^n. \tag{6}$$

But whenever $V(\tau,\alpha)$ is finite, there is an optimal arc $\bar{x}$ for the problem $P(\tau,\alpha)$, and along $\bar{x}$, V is constant (i.e., $t \to V(t,\bar{x}(t))$ is constant on $[\tau,T]$). Thus the system (V,F) is weakly decreasing relative to $t \in (-\infty,T)$, so that (7) holds by Exercise 6.4. We have shown that V satisfies (5) and (6).

Now let φ be any other function as described in the theorem. Let us show first that $V \leq \varphi$. To this end, let (τ,α) be any point with $\tau < T$. Then the system (φ,F) is weakly decreasing relative to $t < T$, so that there is a trajectory x on $[\tau,T]$ with $x(\tau) = \alpha$ such that

$$\varphi(t,x(t)) \leq \varphi(\tau,\alpha) \ \forall t \in [\tau,T).$$

Letting $t \uparrow T$, we derive $\ell(x(T)) = \varphi(T,x(T)) \leq \varphi(\tau,\alpha)$, which implies $V(\tau,\alpha) \leq \varphi(\tau,\alpha)$.

We now proceed to show $V \geq \varphi$. Let (τ,α) be any point with $\tau < T$. Then there exists a trajectory $\bar{x}$ optimal for $P(\tau,\alpha)$. Because (φ,F) is strongly increasing, we derive

$$\varphi(T,\bar{x}(T)) \geq \varphi(\tau,\alpha).$$

But $\varphi(T,\bar{x}(T)) = \ell(\bar{x}(T)) = V(\tau,\alpha)$, which completes the proof. □

The proof actually establishes two *comparison theorems* that we proceed to note formally:

7.6. Corollary. *Let $\varphi \colon (-\infty,T] \times \mathbb{R}^n \to \mathbb{R}$ be continuous and satisfy:*

(a) $\bar{h}(x,\partial_P\varphi(t,x)) \leq 0 \ \forall (t,x) \in (-\infty,T) \times \mathbb{R}^n$; and

(b) $\ell(x) \leq \varphi(T,x) \ \forall x \in \mathbb{R}^n$.

Then $\varphi \geq V$.

7.7. Corollary. *Let $\varphi \colon (-\infty,T] \times \mathbb{R}^n \to \mathbb{R}$ be continuous and satisfy:*

(a) $\bar{h}(x,\partial_P\varphi(t,x)) \geq 0 \ \forall (t,x) \in (-\infty,T) \times \mathbb{R}^n$; and

(b) $\ell(x) \geq \varphi(T,x) \ \forall x \in \mathbb{R}^n$.

Then $\varphi \leq V$.

We remark that Corollary 7.6 is valid without the Lipschitz hypothesis on F, in contrast to its companion.

Minimax Solutions

It is possible to express the extended Hamilton–Jacobi equation in terms of other constructs of nonsmooth analysis besides proximal subgradients, for example, via subderivates. Subbotin has called the concept of solution that appears below a *minimax solution*.

7.8. Proposition. *V is the unique continuous function $\varphi \colon (-\infty,T] \times \mathbb{R}^n \to \mathbb{R}$ satisfying:*

(a) $\inf_{v \in F(x)} D\varphi(t, x; 1, v) \leq 0 \ \forall (t, x) \in (-\infty, T) \times \mathbb{R}^n$;

(b) $\sup_{v \in F(x)} D\varphi(t, x; -1, -v) \leq 0 \ \forall (t, x) \in (-\infty, T) \times \mathbb{R}^n$; and

(c) $\varphi(T, \cdot) = \ell(\cdot)$.

Proof. It suffices to prove that the two conditions (a) of Corollary 7.6 and of Proposition 7.8 are equivalent, and that conditions (a) of Corollary 7.7 and (b) of the proposition are equivalent.

First, let (a) of Proposition 7.8 hold, and let (θ, ζ) belong to $\partial_P \varphi(t, x)$. Let $\varepsilon > 0$ be given, and let v in $F(x)$ satisfy $D\varphi(t, x; 1, v) < \varepsilon$. Then

$$\varepsilon > D\varphi(t, x; 1, v) \geq \langle (\theta, \zeta), (1, v) \rangle \geq \bar{h}(x, \theta, \zeta).$$

We deduce (a) of Corollary 7.6, since ε is arbitrary.

Now let (a) of Corollary 7.6 hold, and suppose that for some $\varepsilon > 0$, we had

$$D\varphi(t, x; 1, v) > \varepsilon \ \forall v \in F(x).$$

Since $D\varphi$ is lower semicontinuous in v (Exercise 3.4.1(e)), for some $\delta > 0$ we have

$$D\varphi(t, x; 1, v) > \frac{\varepsilon}{2} \ \forall v \in F(x) + \delta B.$$

By Subbotin's Theorem 3.4.2 this implies the existence, for any $r > 0$, of $(\theta, \zeta) \in \partial_P \varphi(t', x')$, where (t', x') is of distance less than r from (t, x), such that

$$\langle (\theta, \zeta), (1, v) \rangle > \frac{\varepsilon}{3} \ \forall v \in F(x) + \delta B.$$

Thus, as soon as $F(x')$ lies within $F(x) + \delta B$ we deduce

$$\bar{h}(x', \theta, \zeta) \geq \frac{\varepsilon}{3} \text{ for some } (\theta, \zeta) \in \partial_P \varphi(t', x'),$$

contradicting (a) of Corollary 7.6 as required. This completes the proof that the corresponding (a) parts are equivalent. The remainder of the proof is similar to the above, and is left as an exercise. □

7.9. Exercise. Prove that condition (b) of Proposition 7.8 and condition (a) of Corollary 7.7 are equivalent.

Viscosity Solutions

We now establish that the value function is also the unique *viscosity solution* of the Hamilton–Jacobi boundary-value problem. This celebrated solution concept, developed by M. Crandall and P. L. Lions, is bilateral like minimax solutions, but even more so, in that it uses both subdifferentials and superdifferentials. Of course, given the uniqueness of the solution, it is evident that the proximal, minimax, and viscosity solution concepts all coincide in our present setting.

Let us recall the D-subdifferential $\partial_D f$ of §3.4, and the characterization provided by Proposition 3.4.12: ζ belongs to $\partial_D f(x)$ iff there is a function g differentiable at x, with $g'(x) = \zeta$, such that $f - g$ has a local minimum at x. The D-superdifferential $\partial^D f(x)$ is defined analogously, with $f - g$ having a local maximum at x.

7.10. Proposition. *V is the unique continuous function* $\varphi\colon (-\infty, T] \times \mathbb{R}^n \to \mathbb{R}$ *satisfying*:

(a) $\bar{h}(x, \partial_D \varphi(t,x)) \leq 0 \; \forall (t,x) \in (-\infty, T) \times \mathbb{R}^n$;

(b) $\bar{h}(x, \partial^D \varphi(t,x)) \geq 0 \; \forall (t,x) \in (-\infty, T) \times \mathbb{R}^n$; *and*

(c) $\varphi(T, \cdot) = \ell(\cdot)$.

Proof. It is an immediate consequence of Proposition 3.4.5 that (a) of this proposition and (a) of Corollary 7.6 are equivalent. It suffices then to prove the equivalence of (b) of the proposition to (a) of Corollary 7.7. Now the latter condition is equivalent to the strong increase of the system (φ, F) on $(-\infty, T)$, as we know. This in turn is equivalent to the strong decrease of the system $(-\varphi, F)$ on $(-\infty, T)$, which in turn is equivalent to

$$\theta + H(x, \theta, \zeta) \leq 0 \; \forall (\theta, \zeta) \in \partial_P(-\varphi)(t,x), \; \forall (t,x) \in (-\infty, T) \times \mathbb{R}^n,$$

by Exercise 6.4. Invoking Proposition 3.4.5 again, this is equivalent to

$$\theta + H(x, \theta, \zeta) \leq 0 \; \forall (\theta, \zeta) \in \partial_D(-\varphi)(t,x), \; \forall (t,x) \in (-\infty, T) \times \mathbb{R}^n.$$

Since $\partial^D \varphi$ is precisely $-\partial_D(-\varphi)$, this last condition coincides with (b) of Proposition 7.10, and we are done. □

We remark that in terms of the usual conventions in the literature of viscosity solutions, Proposition 7.10 asserts that V is the viscosity solution of the equation $-\bar{h}(x, \nabla\varphi) = 0$ rather than $\bar{h}(x, \nabla\varphi) = 0$. The minus sign here makes a difference, as it clearly must in any bilateral mode of definition. In the unilateral proximal setting, the difference resides in the fact that it is $\partial_P \varphi$ figuring in (5), and not $\partial^P \varphi$.

7.11. Exercise.

(a) Let φ be a continuous function satisfying (5). Prove that at any point (t,x) at which φ is differentiable, we have

$$\varphi_t(t,x) + h(x, \varphi_x(t,x)) = 0. \tag{7}$$

Deduce that if φ is locally Lipschitz, then this holds a.e. on $(-\infty, T) \times \mathbb{R}^n$. This defines an earlier notion of generalized solution to the Hamilton–Jacobi equation, one that could be called an *almost everywhere solution*. We proceed to show that almost everywhere solutions are not necessarily unique, and hence differ from the proximal ones.

(b) We set $n = 1$, $F(x) = [-1, 1]$, $\ell(x) = |x|$, and $T = 1$. Show that the value function V is given by

$$V(t, x) = \max\{|x| + t - 1, 0\}.$$

(c) Confirm that V is a Lipschitz almost everywhere solution of the Hamilton–Jacobi equation with boundary condition $V(1, x) = |x|$. Show that *another* is $\varphi(t, x) := |x| + t - 1$, but that φ fails to satisfy (5), in contrast to V.

(d) Show that both φ and V satisfy (5) if ∂_P is replaced by ∂^P.

(e) If φ is differentiable at every point and satisfies (8) on $(-\infty, T) \times \mathbb{R}^n$ *without exception*, together with the boundary condition, prove that $\varphi = V$ on $(-\infty, T] \times \mathbb{R}^n$.

8 Feedback Synthesis from Semisolutions

The Hamilton–Jacobi inequality

$$\bar{h}(x, \partial_P \varphi(t, x)) \leq 0,$$

together with the boundary condition

$$\varphi(T, \cdot) \geq \ell(\cdot)$$

defines what is called a *semisolution*. Such a function can be useful in producing upper bounds for the value V of our optimal control problem; as we saw, these conditions imply $V \leq \varphi$ (Corollary 7.6). Thus for each $(\tau, \alpha) \in (-\infty, T] \times \mathbb{R}^n$ there is a trajectory $\bar{x}$ with $x(\tau) = \alpha$ and $\ell(\bar{x}(T)) \leq \varphi(\tau, \alpha)$. We address now the issue of actually constructing such a trajectory. In the special case in which $\varphi = V$, this becomes the problem of finding optimal trajectories.

It is interesting to recall the classical approach to this issue. Assuming that φ is smooth, this would direct us to select for each (t, x) a point $\bar{v}(t, x)$ in $F(x)$ at which the minimum defining $\bar{h}(x, \nabla \varphi(t, x))$ is attained; i.e., such that

$$\varphi_t(t, x) + \langle \varphi_x(t, x), \bar{v} \rangle = \bar{h}(x, \nabla \varphi(t, x)).$$

Then, we proceed to define a trajectory $\bar{x}$ via

$$\dot{\bar{x}}(t) = \bar{v}(t, \bar{x}(t)), \quad \bar{x}(\tau) = \alpha.$$

If all this is possible, we derive

$$\ell(\bar{x}(T)) \leq \varphi(\tau, \alpha)$$

as follows:
$$\ell(\bar{x}(T)) - \varphi(\tau,\alpha) \leq \varphi(T,\bar{x}(T)) - \varphi(\tau,\alpha)$$
$$= \int_0^T \frac{d}{dt}\varphi(t,\bar{x}(t))\,dt$$
$$= \int_0^T \{\varphi_t(t,\bar{x}(t)) + \langle\varphi_x(t,\bar{x}(t)),\dot{\bar{x}}(t)\rangle\}\,dt$$
$$= \int_0^T \bar{h}(\bar{x}(t),\nabla\varphi(t,\bar{x}(t)))\,dt \leq 0.$$

The difficulties with this "dynamic programming" approach are intrinsic (smoothness of φ, regularity of $\bar{v}$, existence of $\bar{x}$), but it is of note that it attempts to construct a *feedback* giving rise to the required trajectory.

Proximal aiming allows us to rescue the approach in essentially these terms, and for merely *lower semicontinuous* semisolutions. The integration step above is still not possible, but proximal methods produce the required system monotonicity. The following result is obtained. It is not just an abstract existence theorem: we construct a feedback $\bar{v}$ quite explicitly.

8.1. Theorem. *Let F be locally Lipschitz, and let $\varphi \in \mathcal{F}((-\infty,T)\times\mathbb{R}^n)$ satisfy*
$$\bar{h}(x,\partial_P\varphi(t,x)) \leq 0 \;\forall(t,x) \in (-\infty,T)\times\mathbb{R}^n$$
and
$$\ell(x) \leq \liminf_{\substack{t'\uparrow T \\ x'\to x}} \varphi(t',x') \;\forall x \in \mathbb{R}^n.$$
Then for given $(\tau,\alpha) \in (-\infty,T)\times\mathbb{R}^n$, there exists a feedback selection $\bar{v}$ of F with the property that every Euler solution $\bar{x}$ of the initial-value problem
$$\dot{x} = \bar{v}(t,x), \quad x(\tau) = \alpha,$$
satisfies $\ell(\bar{x}(T)) \leq \varphi(\tau,\alpha)$.

Proof. Let us consider the set
$$S := \{(t,x) \in (-\infty,T)\times\mathbb{R}^n : \varphi(t,x) \leq \varphi(\tau,\alpha)\} \cup \{(t,x): t \geq T, x \in \mathbb{R}^n\},$$
and note that S is closed, since φ is lower semicontinuous, and also that the system $(S,\bar{F})$ is weakly invariant, where $\bar{F}(x) := \{1\}\times F(x)$. This is essentially a restatement of the fact that (φ,F) is weakly decreasing on $t < T$, which follows from Exercise 6.4.

We now take the feedback selection g_P for $\bar{F}$ defined in Theorem 3.4; necessarily, it is of the form $(1,\bar{v}(t,x))$, where $\bar{v}$ is a feedback selection for F. It follows that any Euler solution of $\dot{x} = \bar{v}(t,x)$, $x(\tau) = \alpha$ is such that $(t,x(t)) \in S \;\forall t \geq \tau$. Thus
$$\varphi(\tau,\alpha) \geq \liminf_{\varepsilon\downarrow 0}\varphi(T-\varepsilon,x(T-\varepsilon)) \geq \ell(x(T)). \qquad\square$$

230   4. A Short Course in Control Theory

We recall that (as shown in the proof of Theorem 3.4), the feedback $\bar{v}(t,x)$ is constructed as follows: for given (t,x), we first find any point $(t',x') \in \text{proj}_S(t,x)$; next, we locate $v \in F(x')$ minimizing the function $v \mapsto \langle v, x - x' \rangle$ over $F(x')$. Finally, we take $\bar{v}(t,x)$ in $F(x)$ closest to v. This *same* $\bar{v}$ will provide trajectories x satisfying $\ell(x(T)) \leq \varphi(\tau,\alpha)$ from any initial data (τ',α') for which $\varphi(\tau',\alpha') \leq \varphi(\tau,\alpha)$. In general $\bar{v}$ is discontinuous, and $\bar{v}(t,x)$ is an arbitrary element of $F(t,x)$ when $x' = x$. We remark that a refinement of this construction can be made so as to define, for any compact subset C and $\varepsilon > 0$, a feedback selection generating trajectories x that satisfy $\ell(x(T)) \leq \varphi(\tau,\alpha) + \varepsilon$ whenever (τ,α) lies in C, though we shall omit this topic.

8.2. Exercise. For $\varphi = V$ in Exercise 7.11(b), and for $(\tau,\alpha) = (0,0)$, describe the set S and the optimal feedback $\bar{v}$. Sketch typical directions of $\bar{v}$ in the (t,x) plane. Note that $\bar{v}$ cannot be defined so as to be continuous on $(-\infty, 1] \times \mathbb{R}$. Prove that no continuous feedback can be optimal.

8.3. Exercise. With $n = 1$, set $h(x,p) := -|xp|$, and let

$$\varphi(t,x) = \begin{cases} xe^{t-1} & \text{if } x \geq 0, \\ xe^{1-t} & \text{if } x \leq 0. \end{cases}$$

Prove that φ is the unique proximal/minimax/viscosity solution on $(-\infty, 1]$ of the Hamilton–Jacobi boundary-value problem

$$\varphi_t + h(x, \varphi_x) = 0, \quad \varphi(1,y) = y \; \forall y.$$

(Which of the three solution concepts is easiest to use in order to do this?) Why does this immediately imply that $\varphi = V$, where V is the value function of Exercise 7.4?

9 Necessary Conditions for Optimal Control

We have studied in §7 a verification method that can confirm the optimality of a suspect, and we have shown how to calculate optimal arcs by feedback in §8, if the value function is known (or suboptimal arcs, if a semisolution is at hand). But we still lack the necessary conditions that can be used to *identify* potential optimal solutions. That is the subject of this section, in which is considered the following optimal control problem (P):

$$\text{minimize } \{\ell(x(b)) : \dot{x} \in F(x) \text{ a.e., } x(a) = x_0\}.$$

That is, we wish as before to minimize $\ell(x(b))$ over the trajectories x on $[a,b]$ with initial-value x_0. The interval $[a,b]$ is given, together with $x_0 \in \mathbb{R}^n$. We assume that ℓ is a locally Lipschitz function, and that the autonomous multifunction F is locally Lipschitz. We will say that x is a *local solution* to (P) if, for some $\varepsilon_0 > 0$, we have

$$\ell\big(x(b)\big) \leq \ell\big(y(b)\big)$$

whenever y is a trajectory on $[a,b]$ satisfying $y(a) = x_0$ as well as

$$\|y - x\|_\infty \leq \varepsilon_0.$$

Recall that the upper Hamiltonian H corresponding to F is the function H given by
$$H(x,p) := \max\{\langle p, v \rangle : v \in F(x)\}.$$

In the following set of necessary conditions for a local minimum, the first conclusion is known as the *Hamiltonian inclusion*; the condition at (b) is called the *transversality condition*.

9.1. Theorem. *Let x be a local solution to the optimal control problem (P). Then there exists an arc p on $[a,b]$ which, together with x, satisfies*

$$\big(-\dot{p}(t), \dot{x}(t)\big) \in \partial_C H\big(x(t), p(t)\big) \text{ a.e.,} \quad a \leq t \leq b,$$
$$-p(b) \in \partial_L \ell\big(x(b)\big).$$

Proof. This is a long and fairly involved proof, in which nonsmooth calculus plays the major role. We will require the following facts regarding $\partial_C H$; the Generalized Gradient Formula 2.8.1 is helpful in proving them:

9.2. Exercise. *H is locally Lipschitz, and if (q,v) belongs to $\partial_C H(x,p)$ then:*

(a) *$v \in F(x)$ and $\langle p, v \rangle = H(x,p)$.*

(b) *$\|q\| \leq K\|p\|$, where K is a local Lipschitz constant for F.*

(c) *For any $\lambda \geq 0$, $(\lambda q, v) \in \partial_C H(x, \lambda p)$.*

(d) *For any $w \in F(x)$, we have $(0, w) \in \partial_C H(x, 0)$.*

There is no loss of generality, we claim, in assuming that F is globally Lipschitz and bounded. For there is an a priori bound M on $\|x\|_\infty$ for any trajectory x on $[a,b]$ with $x(a) = x_0$ (see Exercise 1.5); let us redefine $F(x)$ for $\|x\| > 2M$ (only) as follows:

$$\widetilde{F}(x) = F\left(\frac{2Mx}{\|x\|}\right).$$

Then $\widetilde{F}$ satisfies the Standing Hypotheses, and can be shown to be globally Lipschitz and bounded. Furthermore, the arc that solved the initial problem continues to solve the problem in which $\widetilde{F}$ replaces F, since the trajectories from x_0 are the same. Finally, observe that H and $\widetilde{H}$ coincide

232   4. A Short Course in Control Theory

in a neighborhood of the optimal arc, and therefore so do $\partial_C H$ and $\partial_C \widetilde{H}$. Thus the conclusion of the theorem is indifferent to whether F or $\widetilde{F}$ is employed. To summarize, we assume that for some $K > 0$, $F(x) \subset KB$ for all x, and F is globally Lipschitz of rank K.

We will simplify notation by taking $[a,b] = [0,1]$, and $x_0 = 0$, and we denote the optimal trajectory by $\bar{x}$, and its derivative by $\bar{v}$.

Let us now proceed to define two subsets of $X := L_n^2[a,b]$:

$$S := \left\{ v \in X : v(t) \in F\left(\int_0^t v(s)\,ds\right) \text{ a.e.} \right\},$$

$$\Sigma := \left\{ v \in X : \left\| \bar{x} - \int_0^t v(s)\,ds \right\|_\infty \leq \varepsilon_0 \right\}.$$

We denote by $\widetilde{\ell}$ the following function on X:

$$\widetilde{\ell}(v) := \ell\left(\int_0^1 v(s)\,ds\right).$$

We ask the reader to verify that $\bar{v}$ minimizes $\widetilde{\ell}(v)$ over $v \in S \cap \Sigma$, and to provide the following facts:

9.3. Exercise.

(a) S is a closed subset of X.

(b) $\bar{v} \in \text{int}\,\Sigma$.

(c) $\widetilde{\ell}$ is locally Lipschitz.

It follows from the above that we have

$$0 \in \partial_P \{\widetilde{\ell} + I_S\}(\bar{v}),$$

and from the Proximal Sum Rule 1.8.2 that, for any $\varepsilon > 0$, for some v_1 and v_2 within ε of $\bar{v}$, we have

$$0 \in \partial_P \widetilde{\ell}(v_1) + N_S^P(v_2) + \varepsilon B. \tag{1}$$

The bulk of our effort will go toward the calculation of N_S^P. To that end, it is useful to introduce the following subset C of $X \times X$:

$$C := \{(u,v) \in X \times X : (u(t), v(t)) \in \text{graph}(F) \text{ a.e.}\}.$$

Before going on to the study of N_S^P, we will need:

9.4. Lemma. *Let $(\zeta, \xi) \in N_C^P(u,v)$. Then*

$$(-\zeta(t), v(t)) \in \partial_C H(u(t), \xi(t)) \text{ a.e.}, \quad 0 \leq t \leq 1.$$

9 Necessary Conditions for Optimal Control

Proof. To begin, observe that by Proposition 3.5.7, there exists $\sigma > 0$ such that for almost all $t \in [0, T]$, we have

$$\langle \zeta(t), u' - u(t) \rangle + \langle \xi(t), v' - v(t) \rangle$$
$$\leq \sigma \|(u' - u(t), v' - v(t))\|^2 \quad \forall (u', v') \in \text{graph}(F). \quad (2)$$

Let us fix a value of t for which (2) holds.

By setting $u' = u(t)$ in (2), we derive that for any $v' \in F(u(t))$,

$$\langle \xi(t), v' \rangle \leq \langle \xi(t), v(t) \rangle + \sigma \|v' - v(t)\|^2.$$

In other terms, $\xi(t)$ lies in $N^P_{F(u(t))}(v(t))$. Since $F(u(t))$ is a convex set, this is equivalent to

$$\langle \xi(t), v' \rangle \leq \langle \xi(t), v(t) \rangle \quad \forall v' \in F(u(t)).$$

It follows that we have

$$H(u(t), \xi(t)) = \langle \xi(t), v(t) \rangle. \quad (3)$$

We proceed to define a useful function $g \colon \mathbb{R}^n \times \mathbb{R}^n \to \mathbb{R}$ as follows:

$$g(x, p) := \langle \xi(t) - p, v(t) \rangle + \frac{\|\xi(t) - p\|^2}{4\sigma}$$
$$+ \langle \zeta(t), x - u(t) \rangle - \sigma \|x - u(t)\|^2 + H(x, p).$$

Note that g is strictly convex in p for each x. Since $|H(x, p)| \leq K\|p\|$, where K is a uniform bound for $F(x)$, it is easy to see that for all x in some neighborhood of $u(t)$, the function $p \mapsto g(x, p)$ attains a unique minimum at $p = p(x)$, and that we have (for all x near $u(t)$) $\|p(x)\| \leq c$ for some constant c.

We claim further that when $x = u(t)$, the p at which said minimum occurs is $\xi(t)$. Since

$$g(u(t), \xi(t)) = H(u(t), \xi(t)) = \langle \xi(t), v(t) \rangle$$

by (3), this claim follows from the following calculation:

$$\min_p \{g(u(t), p) - \langle \xi(t), v(t) \rangle\}$$

$$= \min_p \left\{ H(u(t), p) - \langle p, v(t) \rangle + \frac{\|\xi(t) - p\|^2}{4\sigma} \right\}$$

$$= \min_p \max_{v' \in F(u(t))} \left\{ \langle p, v' - v(t) \rangle + \frac{\|\xi(t) - p\|^2}{4\sigma} \right\}$$

$$= \max_{v' \in F(u(t))} \min_p \left\{ \langle p, v' - v(t) \rangle + \frac{\|\xi(t) - p\|^2}{4\sigma} \right\}$$

(by the Minimax Theorem)

$$= \max_{v' \in F(u(t))} \{\langle \xi(t), v' - v(t)\rangle - \sigma\|v' - v(t)\|^2\} = 0 \quad \text{(by (2))}.$$

The preceding facts concerning g will allow us to complete the proof of Lemma 9.4; we record here in generic terms the max-min principle involved for its independent interest.

9.5. Lemma. *Let $g(x,p)$ be a locally Lipschitz function such that for each x in a neighborhood of $\bar{x}$, the function $p \mapsto g(x,p)$ has a unique minimum at $p = p(x)$, where for some c we have $\|p(x)\| \leq c$ for all x near $\bar{x}$. Let $p(\bar{x}) = \bar{p}$, and suppose that the function $x \mapsto \min_p g(x,p)$ has a local maximum at $x = \bar{x}$. Then $(0,0) \in \partial_C g(\bar{x},\bar{p})$.*

Proof. Take any (y,q), together with a positive sequence λ_i decreasing to 0. Let $p_i = p(\bar{x} - \lambda_i y)$, the minimizer of $g(\bar{x} - \lambda_i y, \cdot)$. Then $\|p_i\| \leq c$ for all i sufficiently large, and by passing to a subsequence we can suppose that $p_i \to p_0$ for some p_0. We have, for any p,

$$g(\bar{x} - \lambda_i y, p) \geq g(\bar{x} - \lambda_i y, p_i),$$

whence, in the limit, $g(\bar{x}, p) \geq g(\bar{x}, p_0)$. By uniqueness of the minimizer, we deduce $p_0 = p(\bar{x}) = \bar{p}$. Now consider

$$g(\bar{x} - \lambda_i y, p_i) = \min_p g(\bar{x} - \lambda_i y, p)$$
$$\leq \min_p g(\bar{x}, \cdot) \quad \text{(by the maximizing property of } \bar{x}\text{)}$$
$$\leq g(\bar{x}, p_i + \lambda_i q).$$

Hence

$$g(\bar{x}, p_i + \lambda_i q) - g(\bar{x} - \lambda_i y, p_i) \geq 0.$$

Dividing this by λ_i and taking upper limits gives:

$$g^\circ(\bar{x}, \bar{p}; y, q) \geq \limsup_{i \to \infty} \frac{g(\bar{x}, p_i + \lambda_i q) - g(\bar{x} - \lambda_i y, p_i)}{\lambda_i} \geq 0.$$

Since (y,q) is arbitrary, we obtain $(0,0) \in \partial_C g(\bar{x},\bar{p})$ as claimed, proving Lemma 9.5.

We wish to apply Lemma 9.5 to our particular function g, at the point $\bar{x} = u(t)$, $\bar{p} = \xi(t)$. To do so, it suffices to check that the function $x \mapsto \min_p g(x,p)$ has a maximum at $x = u(t)$.

We calculate

$$\min_p g(x,p) \leq \langle \zeta(t), x - u(t)\rangle - \sigma\|x - u(t)\|^2 + H(x, \xi(t))$$
$$= \langle \zeta(t), x - u(t)\rangle - \sigma\|x - u(t)\|^2 + \langle \xi(t), v'\rangle$$

9 Necessary Conditions for Optimal Control   235

(for some $v' \in F(x)$)

$$\leq \langle \xi(t), v(t) \rangle = g(u(t), \xi(t)) = \min_p g(u(t), p),$$

which gives the required conclusion.

The application of Lemma 9.5 gives $(0,0) \in \partial_C g(u(t), \xi(t))$, which reduces precisely to the conclusion of Lemma 9.4, which is therefore proven.

We return now to characterizing N_S^P.

9.6. Lemma. *Let $\zeta \in N_S^P(v_0)$. Then there is an arc q on $[0,1]$ with $q(1) = 0$ such that*

$$(-\dot{q}, v_0) \in \partial_C H(u_0, q + \zeta) \ a.e.,$$

where $u_0(t) := \int_0^t v_0(s)\,ds$.

Proof. Observe first that for some $\sigma_0 > 0$, the following function φ of v is minimized over S at $v = v_0$:

$$\varphi(v) := \langle -\zeta, v \rangle + \sigma_0 \|v - v_0\|^2.$$

We can assume (by increasing σ_0 if necessary) that v_0 is the unique minimizer of φ on S. We introduce the value function $V \colon X \to (-\infty, \infty)$ of a certain minimization problem on $X \times X$:

$$V(\alpha) := \inf \left\{ \varphi(v) \colon (u, v) \in C, u(t) - \int_0^t v(s)\,ds = \alpha(t) \text{ a.e.} \right\}.$$

9.7. Exercise.

(a) $V(0) = \varphi(v_0)$.

(b) The infimum defining $V(\alpha)$ is attained.

(c) If $\alpha_i \to 0$ in X and $V(\alpha_i) \to V(0)$, and if (u_i, v_i) is a solution of the problem defining $V(\alpha_i)$, then there is a subsequence of $\{(u_i, v_i)\}$ converging in $X \times X$ to (u_0, v_0), where

$$u_0(t) := \int_0^t v_0(s)\,ds.$$

We now take a subsequence $\alpha_i \to 0$ for which $(u_i, v_i) \to (u_0, v_0)$ as in the exercise above, and for which $\partial_P V(\alpha_i)$ contains an element ζ_i (we have invoked the Proximal Density Theorem).

As in §3.1, the proximal subgradient inequality for V at α_i translates immediately to the following assertion: the function $f_i(u, v)$ has a local minimum over C at $(u, v) = (u_i, v_i)$, where

$$f_i(u, v) := \langle -\zeta, v \rangle + \sigma_0 \|v - v_0\|^2 - \left\langle \zeta_i, u - \int_0^t v \right\rangle + \sigma_i \left\| u - \int_0^t v - u_i + \int_0^t v_i \right\|^2.$$

236   4. A Short Course in Control Theory

9.8. Exercise. f_i is C^2, with

$$f_i'(u_i, v_i) = \left(-\zeta_i, -\zeta + 2\sigma_0(v_i - v_0) + \int_t^1 \zeta_i\right).$$

In light of this exercise, the necessary condition

$$-f_i'(u_i, v_i) \in N_C^P(u_i, v_i)$$

becomes

$$\left(\zeta_i, \zeta - 2\sigma_0(v_i - v_0) - \int_t^1 \zeta_i\right) \in N_C^P(u_i, v_i).$$

Calling upon Lemma 9.4, this gives

$$(-\zeta_i, v_i) \in \partial_C H\left(u_i, \zeta - 2\sigma_0(v_i - v_0) - \int_t^1 \zeta_i\right) \text{ a.e.}$$

Let us relabel as follows:

$$q_i(t) = -\int_t^1 \zeta_i.$$

Then

$$(-\dot{q}_i, v_i) \in \partial_C H\left(u_i, \zeta + q_i - 2\sigma_0(v_i - v_0)\right) \text{ a.e.} \qquad (4)$$

Since F is globally Lipschitz of rank K, this implies

$$\|\dot{q}_i(t)\| \leq K\|\zeta + q_i(t) - 2\sigma_0(v_i(t) - v_0(t))\| \text{ a.e.,}$$

and Gronwall's Lemma leads to a uniform bound on $\|q_i\|_\infty$, and so to a uniform bound on $\|\dot{q}_i\|_2$. Taking a subsequence to arrange the weak convergence of $\dot{q}_i$, we are then able to pass to the limit in (4) (with the help of Theorem 3.5.24) to obtain the conclusion of Lemma 9.6.

We are ready now to return to the proof of Theorem 9.1, picking up the thread of the argument at the point (1).

We recall from Exercise 3.5.20 that an element ζ of $\partial_P \tilde{\ell}(v_1)$ is of the form

$$\zeta(t) = \zeta_0 \in \partial_L \ell\left(\int_0^1 v_1(s)\,ds\right).$$

Combining this with Lemma 9.6, we deduce from (1) that for some such ζ_0, for some element w of X having $\|w\| \leq \varepsilon$, and for some arc q having $q(1) = 0$, we have

$$(-\dot{q}, v_2) \in \partial_C H(u_2, -\zeta_0 + q + w) \text{ a.e.,}$$

where $u_2(t) := \int_0^t v_2(s)\,ds$. Let us now observe this conclusion for a sequence $\varepsilon_i \downarrow 0$, and label the corresponding $(u_2, v_2, q, \zeta_0, w)$ in the form $(u_i, v_i, q_i, \zeta_i, w_i)$. We have

$$(-\dot{q}_i, v_i) \in \partial_C H(u_i, -\zeta_i + q_i + w_i) \text{ a.e.,}$$

where $w_i \to 0$, $v_i \to \bar{v}$, $u_i \to \bar{x}$, and where

$$\zeta_i \in \partial_L \ell\left(\int_0^1 v_i(s)\,ds\right).$$

Let us set $p_i(t) := q_i(t) - \zeta_i$. Then another straightforward use of Gronwall's Lemma and the Sequential Compactness Theorem 3.5.24 shows that a subsequence of $\{p_i\}$ converges uniformly to an arc p satisfying the conclusions of Theorem 9.1, whose proof is now complete. □

Remark. The Hamiltonian inclusion incorporates the equality, or *maximum principle*

$$H(x(t), p(t)) = \langle p(t), \dot{x}(t) \rangle \text{ a.e.,} \qquad (5)$$

as a consequence of Exercise 9.2.

9.9. Exercise. Let $F(x) = \{Ax + Bu: u \in C\}$, where C is a compact convex subset of $\mathbb{R}^m$, and where A, B are $n \times n$ and $n \times m$ matrices, respectively.

(a) Prove that F satisfies the Standing Hypotheses and is globally Lipschitz.

(b) Let $\bar{x}$ be a trajectory of F on $[0, 1]$, and let $\bar{u}$ be a control function realizing $\bar{x}$; i.e., such that $\bar{u}(t) \in C$ and $\dot{\bar{x}}(t) = A\bar{x}(t) + B\bar{u}(t)$ a.e. Show that the Hamiltonian inclusion of Theorem 9.1 is equivalent to

$$-\dot{p}(t) = A^* p(t), \quad \max_{u \in C}\langle p(t), Bu \rangle = \langle p(t), B\bar{u}(t) \rangle \text{ a.e.}$$

(This is the linear case of *Pontryagin's Maximum Principle*.)

(c) Observe that H is convex, and so admits one-sided directional derivatives.

(d) Prove that any pair of arcs (x, p) satisfying the Hamiltonian inclusion is such that $t \mapsto H(x(t), p(t))$ is constant on the underlying interval.

The following technical extension of the theorem will be of use to us later.

9.10. Exercise. Consider again the context of Theorem 9.1, but extended as follows: for some $r \geq 0$, for some $z(\cdot) \in L_n^2[a, b]$, the trajectory $\bar{x}$ solves locally the problem of minimizing

$$\ell(x(b)) + r\int_a^b \|x(t) - z(t)\|^2\,dt$$

relative to the trajectories of F originating at x_0. Adapt the proof of Theorem 9.1 to deduce the same conclusions with one change: the Hamiltonian inclusion becomes

$$(-\dot{p}(t), \dot{\bar{x}}(t)) \in \partial_C H(\bar{x}(t), p(t)) - 2r(\bar{x}(t) - z(t), 0) \text{ a.e.}$$

(*Hint.* The derivative of the functional

$$x \mapsto \int_a^b \|x(t) - z(t)\|^2 \, dt = \|x - z\|_2^2$$

is apparent.)

The Case of Terminal Constraints

We consider now the optimal control problem in the presence of explicit constraints on $x(b)$. Specifically, we examine the problem of minimizing $\ell(x(b))$ over the trajectories x of F on $[a, b]$ which satisfy

$$x(a) = x_0, \quad x(b) \in S.$$

Here, S is a given closed subset of $\mathbb{R}^n$, and the hypotheses on the rest of the data are unchanged. The proof of the following result uses *value function analysis* to derive the necessary conditions for the terminally constrained problem from those previously derived for the free endpoint case.

9.11. Theorem. *Let x solve locally the optimal control problem described above. Then there exists an arc p on $[a, b]$ and a scalar $\lambda_0 = 0$ or 1 such that $\lambda_0 + \|p(t)\| \neq 0 \ \forall t \in [a, b]$, and such that*

$$(-\dot{p}(t), \dot{x}(t)) \in \partial_C H(x(t), p(t)) \text{ a.e.}, \quad a \leq t \leq b,$$
$$-p(b) \in \lambda_0 \partial_L \ell(x(b)) + N_S^L(x(b)).$$

Proof. We take $[a, b] = [0, 1]$, $x_0 = 0$, and denote the solution by $\bar{x}$. For $\alpha \in \mathbb{R}^n$, consider the problem $P(\alpha)$ of minimizing

$$\ell(x(1)) + \int_0^1 \|x(t) - \bar{x}(t)\|^2 \, dt$$

over the trajectories of F satisfying

$$x(0) = 0, \quad x(1) \in S + \alpha, \quad \|x - \bar{x}\|_\infty \leq \varepsilon_0.$$

Evidently, $\bar{x}$ is the *unique* solution of $P(0)$. We designate by $V(\alpha) \in (-\infty, \infty]$ the value of the problem $P(\alpha)$. Sequential compactness of trajectories provides a few salient properties of V:

9.12. Exercise.

(a) Whenever $V(\alpha) < \infty$, the problem $P(\alpha)$ admits a solution.

(b) If x_i solves $P(\alpha_i)$, where $\alpha_i \to 0$ and x_i converges uniformly to an arc x, then $x = \bar{x}$.

(c) V is lower semicontinuous.

Suppose that V admits a proximal subgradient ζ_α at $\alpha \in \operatorname{dom} V$; then, for all α' sufficiently near α, we have

$$V(\alpha') - V(\alpha) + \sigma \|\alpha' - \alpha\|^2 \geq \langle \zeta_\alpha, \alpha' - \alpha \rangle.$$

Let x_α solve $P(\alpha)$, and let $c_\alpha \in S$ satisfy

$$x_\alpha(1) = c_\alpha + \alpha.$$

In light of Exercise 9.12, we can suppose that $\|x_\alpha - \bar{x}\|_\infty < \varepsilon_0$ when α is small.

For any trajectory x originating at 0 and for any $c \in S$, we have $x(1) \in S + (x(1) - c)$, whence, if $\|x - \bar{x}\|_\infty < \varepsilon_0$,

$$\ell(x(1)) + \int_0^1 \|x(t) - \bar{x}(t)\|^2 \, dt \geq V(x(1) - c).$$

If $\|x - x_\alpha\|_\infty$ and $\|c - c_\alpha\|$ are small enough, this allows us to set $\alpha' = x(1) - c$ in the preceding inequality and combine it with the last one to deduce that the functional

$$\ell(x(1)) + \int_0^1 \|x(t) - \bar{x}(t)\|^2 \, dt - \langle \zeta_\alpha, x(1) - c \rangle + \sigma \|x(1) - x_\alpha(1) - c + c_\alpha\|^2$$

is minimized locally by $x = x_\alpha$, $c = c_\alpha$ (relative to trajectories x of F originating at 0 and points $c \in S$).

Setting $c = c_\alpha$, we observe that x_α solves locally the problem of minimizing over trajectories x originating at 0 the cost functional

$$\ell(x(1)) - \langle \zeta_\alpha, x(1) \rangle + \sigma \|x(1) - x_\alpha(1)\|^2 + \int_0^1 \|x(t) - \bar{x}(t)\|^2 \, dt$$

subject to no terminal constraint.

Setting $x = x_\alpha$ leads to the conclusion that for all $c \in S$ sufficiently near c_α we have

$$\langle \zeta_\alpha, c \rangle + \sigma \|c - c_\alpha\|^2 \geq \langle \zeta_\alpha, c_\alpha \rangle;$$

this implies $-\zeta_\alpha \in N_S^P(c_\alpha)$.

Applying Exercise 9.10 to the free endpoint problem solved by x_α, we deduce the existence of an arc p_α such that

$$(-\dot{p}_\alpha, \dot{x}_\alpha) \in \partial_C H(x_\alpha, p_\alpha) - 2(x_\alpha(t) - \bar{x}(t), 0) \text{ a.e.,} \tag{6}$$

$$-p_\alpha(1) + \zeta_\alpha \in \partial_L \ell(x_\alpha(1)), \tag{7}$$

$$-\zeta_\alpha \in N_S^P(c_\alpha). \tag{8}$$

240   4. A Short Course in Control Theory

We now consider a sequence $\alpha_i \to 0$ admitting $\zeta_i \in \partial_P V(\alpha_i)$; this is possible by the Proximal Density Theorem 1.3.1. Denote the x_{α_i}, p_{α_i}, c_{α_i} above by the simpler notation x_i, p_i, c_i. We may suppose in light of Exercise 9.12 that the corresponding solutions x_i converge uniformly to $\bar{x}$, by passing to a subsequence. Note that $c_i = x_i(1) - \alpha_i$ converges to $\bar{x}(1)$.

The end of the proof consists of passing to the limit in (6), (7), and (8), and there are two cases depending on whether or not the sequence ζ_i is bounded.

Consider first the case in which ζ_i is bounded. Then we may suppose that $\zeta_i \to \zeta \in N_S^L(\bar{x}(1))$, and an application of the Sequential Compactness Theorem 3.5.24 leads to an arc p satisfying

$$(-\dot{p}, \dot{\bar{x}}) \in \partial_C H(\bar{x}, p), \quad -p(1) \in \partial_L \ell(\bar{x}(1)) + N_S^L(\bar{x}(1)),$$

that is, the conclusions of the theorem with $\lambda_0 = 1$.

Now suppose that ζ_i is unbounded, and pass to a subsequence to arrange $\|\zeta_i\| \to \infty$. Dividing by $\|\zeta_i\|$ in (6) and (7) and setting $q_i := p_i/\|\zeta_i\|$, we obtain (Exercise 9.2):

$$(-\dot{q}_i, \dot{x}_i) \in \partial_C H(x_i, q_i) - 2\|\zeta_i\|^{-1}(x_\alpha(t) - \bar{x}(t), 0),$$
$$-q_i(1) \in \|\zeta_i\|^{-1}\partial_L \ell(x_i(1)) - \zeta_i/\|\zeta_i\|,$$

where $-\zeta_i/\|\zeta_i\| \in N_S^P(c_i)$. We use Gronwall's Lemma to deduce that $\|q_i\|_\infty$ is bounded, following which Theorem 3.5.24 once more applies to give an arc $p = \lim q_i$ such that

$$(-\dot{p}, \dot{\bar{x}}) \in \partial_C H(\bar{x}, p), \quad 0 \neq -p(1) \in N_S^L(\bar{x}(1)).$$

These are the required conclusions with $\lambda_0 = 0$, there remaining only to check that $p(t) \neq 0$ $\forall t \in [0, 1]$. But we have $\|\dot{p}(t)\| \leq K\|p(t)\|$, by Exercise 9.2, so that if $p(t) = 0$ for some t, then $p(t) = 0$ everywhere (by Gronwall's Lemma). Since $p(1) \neq 0$, this does not occur. □

Remark. If the solution x of Theorem 9.11 is such that $x(b) \in \text{int } S$, then $\lambda_0 = 1$ necessarily and the necessary conditions become those of Theorem 9.1 for the free endpoint problem. The condition that $\lambda_0 + \|p(t)\|$ be different from 0 is required in Theorem 9.11 to avoid triviality of the necessary conditions, for they always hold with $\lambda_0 = 0$, $p \equiv 0$, when x is a trajectory. The case $\lambda_0 = 0$ is termed *abnormal*, and can be shown to arise when the constraints are so tight as to make the cost functional irrelevant.

9.13. Exercise.

(a) Show that $\lambda_0 = 1$ in Theorem 9.11 if $x(b) \in \text{int } S$.

(b) Show that $\lambda_0 = 0$, $p \equiv 0$ satisfies the Hamiltonian inclusion when x is any trajectory.

(c) Let $n = 2$, $[a,b] = [0,1]$, $x_0 = (0,0)$, $F(x,y) \equiv \overline{B}$, $S = \{1\} \times \mathbb{R}$. Let $\ell(x,y) = \ell(y)$ be any function such that $0 \notin \partial_L \ell(0)$. What is the unique trajectory solving the problem of Theorem 9.11 in this case? Show that $\lambda_0 = 0$ necessarily.

(d) Prove that in the context of Theorem 9.11, when $\lambda_0 = 0$, then $p(t) = 0$ for some t iff p is identically zero.

Constancy of the Hamiltonian

In the general context of Theorem 9.11, in contrast to the setting of Exercise 9.9, it does *not* follow from the Hamiltonian inclusion that $H(x(t), p(t))$ is constant. But this condition *can* be obtained for at least one arc p, as we now show.

9.14. Theorem. *In Theorem 9.11, we may add to the conditions satisfied by (x,p) the constancy of the function $t \mapsto H(x(t), p(t))$ on $[a,b]$.*

Proof. The proof will consist of "bootstrapping" from Theorem 9.11, by means of a device known as the Erdmann transform. Let us consider $(n+1)$-dimensional arcs which are trajectories of the multifunction $\widetilde{F}$ defined by

$$\widetilde{F}(x^0, x) := \{(u,v) \in \mathbb{R} \times \mathbb{R}^n : |u| \leq \alpha, v \in (1+u)F(x)\},$$

where α is a fixed parameter in $(0,1)$. We will take $[a,b] = [0,1]$, and we consider problem (E) of minimizing $\ell(x(1))$ over the trajectories (x^0, x) of $\widetilde{F}$ on $[0,1]$ satisfying

$$(x^0, x)(0) = (1, x_0), \quad x^0(1) = 1, \quad x(1) \in S.$$

We denote by $\bar{x}$ the solution of the original problem (P).

9.15. Lemma. *For α sufficiently small, the arc $(1, \bar{x})$ is an admissible trajectory of $\widetilde{F}$ which solves locally problem (E).*

That $(1, \bar{x})$ is an admissible trajectory for (E) is easily observed. Suppose that (x^0, x) is another such, and that $\ell(x(1)) < \ell(\bar{x}(1))$; we will manufacture a contradiction by exhibiting an arc y feasible for (P) with $\ell(y(1)) < \ell(\bar{x}(1))$.

For each $t \in [0,1]$, a unique $\tau(t)$ in $[0,1]$ is defined by the equation

$$\tau + x^0(\tau) - 1 = t,$$

since the function $\tau \mapsto \tau + x^0(\tau) - 1$ is strictly monotone increasing, with value 0 at 0 and 1 at 1. The function $\tau(t)$ is Lipschitz by the Inverse Function Theorem 3.3.12. Thus the relation $y(t) := x(\tau(t))$ defines y as an arc on $[0,1]$ with $y(0) = x_0$ and $y(1) = x(1) \in S$. We calculate

$$\frac{d}{dt}y(t) = \frac{\dot{x}(\tau)}{1 + \dot{x}^0(\tau)},$$

from which it follows that y is a trajectory of F. Thus y is feasible for (P); yet, $\ell(y(1)) = \ell(x(1)) < \ell(\bar{x}(1))$. This contradiction completes the proof, except that we have ignored the fact that $\bar{x}$ may be only a *local* solution of (P) relative to $\|x - \bar{x}\|_\infty \leq \varepsilon_0$. To take account of this, it suffices to observe that taking both (x^0, x) uniformly close enough to $(1, \bar{x})$ and α sufficiently small guarantees that y satisfies $\|y - \bar{x}\|_\infty < \varepsilon_0$.

The next step in the proof is to apply the necessary conditions of Theorem 9.11 to the solution $(1, \bar{x})$ of problem (E). This will involve the Hamiltonian

$$\widetilde{H}(x^0, x, p^0, p) := \max\{p^0 u + \langle p, v \rangle : |u| \leq \alpha, v \in (1+u)F(x)\}$$
$$= \max\{p^0 u + (1+u)H(x, p) : |u| \leq \alpha\}$$
$$= H(x, p) + \alpha|p^0 + H(x, p)|.$$

Since this is independent of x^0, it follows from the Hamiltonian inclusion that p^0 is constant, and that we have

$$(-\dot{p}, \dot{\bar{x}}) \in \partial_C H(\dot{\bar{x}}, p) + (\alpha K \|p\| \bar{B}) \times (\alpha K \bar{B}), \tag{9}$$

where (as in the proof of Theorem 9.1) the global bound and Lipschitz constant K for F can be assumed to exist.

The Hamiltonian inclusion for $\widetilde{H}$ also yields a.e. (see (5)):

$$\widetilde{H}(1, \bar{x}, p^0, p) = \langle (0, \dot{\bar{x}}), (p^0, p) \rangle = \langle \dot{\bar{x}}, p \rangle$$
$$\leq H(\bar{x}, p) \leq H(\bar{x}, p) + \alpha|p^0 + H(\bar{x}, p)| = \widetilde{H}(1, \bar{x}, p^0, p).$$

It follows from this that

$$H(\bar{x}(t), p(t)) \equiv -p^0 \ \forall t \in [0, 1]. \tag{10}$$

Note also that the transversality conditions applied to (E) are the following:

$$-(p^0(1), p(1)) \in \lambda_0(0, \partial_L \ell(\bar{x}(1))) + N^L_{\{1\} \times S}(1, \bar{x}(1))$$

which implies

$$-p(1) \in \lambda_0 \partial_L \ell(\bar{x}(1)) + N^L_S(\bar{x}(1)). \tag{11}$$

Finally, we have the nontriviality condition $\lambda_0 + \|(p^0, p)\| \neq 0$. If it were the case that λ_0 and $\|p(t)\|$ were both 0 for some t, then $p^0 = 0$ also (in light of (10)), which cannot be. Thus we conclude

$$\lambda_0 + \|p(t)\| \neq 0 \ \forall t \in [0, 1], \tag{12}$$

as required.

9 Necessary Conditions for Optimal Control

If we examine (9)–(12) we see exactly the set of conditions that we are attempting to obtain, except for the superfluous term involving α in (9). The next step is evident: we consider having done all the above for a sequence $\alpha_i \downarrow 0$, and pass to the limit. The corresponding arcs p_i satisfy

$$\|\dot{p}_i\| \leq K(1 + \alpha_i)\|p_i\|$$

as a consequence of (9), and so $\|p_i\|_\infty$ is bounded iff $p_i(1)$ is bounded (by Gronwall's Lemma). In the unbounded case, we arrange $\|p_i(1)\| \to \infty$, and then normalize p_i in (9)–(11) by passing to $p_i/\|p_i(1)\|$, much as in the last step of the proof of Theorem 9.11. The required conclusions then emerge in the limit via sequential compactness (and with $\lambda_0 = 0$). The resulting arc p has $\|p(1)\| = 1$, so p is never zero (Exercise 9.13(d)) and we have nontriviality.

When $p_i(1)$, and hence $\|p_i\|_\infty$ is bounded, sequential compactness can be applied directly to (9)–(11) without renormalizing to yield the required Hamiltonian inclusion, constancy, and transversality, but a danger lurks: triviality; perhaps the sequences p_i and λ_{0_i} both converge to 0. We deal with this by considering two subcases. In the first, $\lambda_{0_i} = 1$ infinitely often; in that event the danger does not arise. In the limit, we get $\lambda_0 = 1$. In the second, all $\lambda_{0_i} = 0$ beyond a certain point. Then we have $p_i(1) \neq 0$ necessarily, in view of (12), and we can renormalize (9)–(11) by dividing by $\|p_i(1)\|$. Then sequential compactness applies to give the required (nontrivial) conclusions in the limit, with $\lambda_0 = 0$. □

Free Time Problems

The transformation device introduced in the proof of the theorem can be used to treat *free time* problems, in which the underlying interval $[a, b]$ is itself a variable. We illustrate this now in a simple setting.

9.16. Exercise. Let $\bar{x}$ and $\overline{T} > 0$ solve the following problem: to minimize $\ell(T, x(T))$ over those trajectories x on $[0, T]$ originating at x which satisfy $(t, x(t)) \in \Omega \ \forall t \in [0, T]$, where Ω is a given open set in $\mathbb{R} \times \mathbb{R}^n$. (Thus a form of merely local optimality is involved; we stress that T is a choice variable in this problem.) The function ℓ is locally Lipschitz.

(a) Referring to the proof of Theorem 9.14, show that the arc $(\overline{T}, \bar{x})$ minimizes locally, over trajectories (x^0, x) of $\widetilde{F}$ on $[0, \overline{T}]$ originating at $(\overline{T}, x_0)$, the function $\ell(x^0(\overline{T}), x(\overline{T}))$, subject now to *no* explicit constraint on the value of $(x^0(\overline{T}), x(\overline{T}))$. Observe that this is a situation to which Theorem 9.1 applies. Express the resulting necessary conditions.

(b) Show that the necessary conditions of (a) translate directly as follows: for some arc p on $[0, \overline{T}]$ we have

$$(-\dot{p}(t), \dot{\bar{x}}(t)) \in \partial_C H(\bar{x}(t), p(t)) \text{ a.e.,} \quad 0 \leq t \leq \overline{T},$$
$$H(\bar{x}(t), p(t)) = h \ (= \text{constant}), \quad 0 \leq t \leq \overline{T},$$
$$(h, -p(\overline{T})) \in \partial_L \ell(\overline{T}, \bar{x}(\overline{T})).$$

10 Normality and Controllability

We now take up the issue of whether it is possible to reach a given equilibrium point x^* in finite time (from a neighborhood of the point, or globally). Without loss of generality, we take $x^* = 0$, so we are dealing with *null controllability*, where it is the case that 0 belongs to $F(0)$. We assume throughout this section that F is locally Lipschitz.

Recall that if $F(x) = \{f(x)\}$, then no point different from 0 can reach the equilibrium 0 in finite time. The opposite extreme occurs when $0 \in \text{int } F(0)$.

10.1. Exercise. Let $0 \in \text{int } F(0)$. Prove that every x_0 sufficiently near 0 can be steered to the origin in finite time; i.e., admits a trajectory x on some interval $[0, T]$ with $x(0) = x_0$, $x(T) = 0$.

We are interested in situations *intermediate* to the extremes above. It turns out that the Hamiltonian inclusion provides a criterion for local null controllability. We will say that *the origin is normal* provided that for some $T > 0$, the only arc p on $[0, T]$ which satisfies the two conditions

$$(-\dot{p}(t), 0) \in \partial_C H(0, p(t)) \text{ a.e.,} \quad 0 \leq t \leq T,$$
$$H(0, p(t)) = 0 \ \forall t \in [0, T],$$

is $p(t) \equiv 0$. (That $p \equiv 0$ *does* satisfy them is apparent from Exercise 9.2.)

10.2. Exercise.

(a) If $0 \in \text{int } F(0)$, then the origin is normal.

(b) Let $n = 2$ and $F(x, y) := \{(y, u): -1 \leq u \leq 1\}$. Then $0 \notin \text{int } F(0)$. Show that the origin is normal.

(c) Let $F(x) = Ax + BC$ as in Exercise 9.9, where now $0 \in \text{int } C$. Show that the origin is normal iff the following $n \times nm$ matrix is of maximal rank: $[B \ AB \ A^2 B \ldots A^{n-1} B]$.

10.3. Theorem. *If the origin is normal, then every point x_0 sufficiently near it can be steered to it in finite time.*

Proof. Let $b > 0$ be such that no nontrivial arc p on $[0, b]$ satisfies $(-\dot{p}, 0) \in \partial_C H(0, p)$, $H(0, p(t)) = 0$ on $[0, b]$. For $\alpha \in \mathbb{R}^n$, we define $V(\alpha) \in [0, \infty]$ to be the infimum of

$$(T - b)^2 + \int_0^T \|x(t)\| \, dt$$

over all $T > 0$ and arcs x for $-F$ on $[0, T]$ originating at 0 which satisfy $x(T) = \alpha$. (Note that $-F$, not F, is used here.)

10.4. Exercise.

(a) $V(0) = 0$, and for $\alpha = 0$ the unique solution to the problem defining $V(0)$ is the arc identically 0 on $[0, b]$.

(b) If $V(\alpha) < \infty$, then the infimum defining $V(\alpha)$ is attained. (Note: Exercise 1.13 helps in dealing with variable intervals.)

(c) V is lower semicontinuous.

(d) If $\alpha_i \to 0$, and if (T_i, x_i) is a solution corresponding to α_i, then $T_i \to b$ and $\max\{\|x_i(t)\| : 0 \le t \le T_i\} \to 0$.

We will show that V is Lipschitz (and hence finite) in a neighborhood of 0. This implies that for every x_0 near 0, there is a trajectory x of $-F$ on an interval $[0, T]$ satisfying $x(0) = 0$, $x(T) = x_0$. Then, reversing time, we see that x_0 can be steered to 0 in finite time.

To show that V is Lipschitz near 0, we will prove that its proximal subgradients are locally bounded. We proceed by supposing the contrary: for some sequence $\alpha_i \to 0$, and sequence $\alpha_i \in \partial_P V(\alpha_i)$, we have $\|\zeta_i\| \to \infty$. We will derive a contradiction to the normality hypothesis.

Arguing as in the proof of Theorem 9.11, we let (T_i, x_i) be a solution corresponding to α_i, and from $\zeta_i \in \partial_P V(\alpha_i)$ we deduce that for some $\sigma_i > 0$, (T_i, x_i) furnishes a minimum for the problem of minimizing

$$\langle -\zeta_i, x(T) \rangle + (T - b)^2 + \int_0^T \|x(t)\| \, dt + \sigma_i \|x(T) - x_i(T_i)\|^2$$

over trajectories x of $-F$ on $[0, T]$ which originate at 0. (The minimum is local in that $x(T)$ must be near $x_i(T_i)$.) Because of the integral term in the cost, this is not the type of problem to which our previous results apply. We will fix this by redefining the dynamics so as to "absorb" the integral.

We set

$$\widetilde{F}(x, y) := -F(x) \times \{\|x\|\}, \quad y_i(t) := \int_0^T \|x_i(\tau)\| \, dt.$$

Then (x_i, y_i) on $[0, T_i]$ locally minimizes the free-time cost functional

$$\langle -\zeta_i, x(T) \rangle + (T - b)^2 + y(T) + \sigma_i \|x(T) - x_i(T_i)\|^2$$

246   4. A Short Course in Control Theory

over the trajectories for $\widetilde{F}$ originating at $(0,0)$.

The necessary conditions of Exercise 9.16 are available now. After the dust settles, here is what they give in terms of the original variables:

$$(-\dot{p}_i, \dot{x}_i) \in \partial_C \tilde{h}(x_i, p_i) + \overline{B} \times \{0\} \text{ a.e.}, \quad 0 \leq t \leq T_i,$$
$$\tilde{h}(x_i, p_i) - \|x_i(t)\| = 2(T_i - b) \ \forall t \in [0, T_i],$$
$$p_i(T_i) = \zeta_i,$$

where $\tilde{h}(x, p) := H(x, -p)$ appears here because $-F$ is involved rather than F. We now replace p_i by $p_i/\|\zeta_i\|$ in these conditions, and implement the now-familiar passage to the limit.

We arrive at a nontrivial arc $\bar{p}$ satisfying

$$(-\dot{\bar{p}}, 0) \in \partial_C \tilde{h}(0, \bar{p}) \text{ a.e.}, \quad 0 \leq t \leq b,$$
$$\tilde{h}(0, \bar{p}(t)) = 0 \ \forall t \in [0, b].$$

In terms of H, $\bar{p}$ satisfies

$$(-\dot{\bar{p}}, 0) \in \partial_C H(0, -\bar{p}) \text{ a.e.},$$
$$H(0, -\bar{p}(t)) = 0.$$

Now define $p(t) := -\bar{p}(b - t)$; then on $[0, b]$ we have

$$(-\dot{p}, 0) \in \partial_C H(0, p) \text{ a.e.},$$
$$H(0, p(t)) = 0, \quad p \neq 0.$$

This contradicts the normality of the origin.   □

The usual route to a conclusion of global null controllability is to combine global asymptotic controllability with local null controllability, as in the following result that builds upon Theorem 5.5.

10.5. Corollary. *Let 0 be a normal equilibrium of F admitting a Lyapounov pair (Q, W). Then any point in $\operatorname{dom} Q$ can be steered to 0 in finite time. If $\operatorname{dom} Q = \mathbb{R}^n$, then all points can be steered to the origin in finite time.*

10.6. Exercise. Prove Corollary 10.5.

If a system is globally null controllable, then in fact a Lyapounov pair (Q, W) does exist. This provides a partial converse to Corollary 10.5:

10.7. Exercise. We define the well-known *minimal time function*:

$$Q(\alpha) := \inf\{T \geq 0 : \text{some trajectory } x \text{ has } x(0) = \alpha, \ x(T) = 0\}.$$

(As usual, we set $Q(\alpha) = \infty$ when no trajectory joins α to 0 in finite time.)

(a) Prove that the infimum is attained if $Q(\alpha) < \infty$.
(b) Prove that the system $(Q(x)+t, \{1\} \times F(x))$ is weakly decreasing on the set $\mathbb{R}^n \setminus \{0\}$. Call upon Theorem 6.1 to deduce
$$h(x, \partial_P Q(x)) \leq -1 \; \forall x \in \mathbb{R}^n \setminus \{0\}.$$
Conclude that setting $W(x) = 1$ for $x \neq 0$ and $W(0) = 0$ leads to a Lyapounov pair (Q, W) for the origin. Observe that $\operatorname{dom} Q = \mathbb{R}^n$ iff the control system is globally null controllable.
(c) Prove that in fact Q satisfies
$$h(x, \partial_P Q(x)) = -1 \; \forall x \in \mathbb{R}^n \setminus \{0\}.$$
(A uniqueness theorem for Q appears among the problems in the next section.)

11 Problems on Chapter 4

11.1. Let $v \colon [a, b] \to \mathbb{R}^n$ be a bounded measurable function such that $v(t) \in K$ a.e., where $K \subseteq \mathbb{R}^n$ is a given closed convex set. Prove that
$$\frac{1}{b-a} \int_a^b v(t) \, dt \in K.$$

11.2. Consider the ordinary differential system
$$\dot{x} = -x + 2y^3, \quad \dot{y} = -x.$$
(a) Show that the set $E_r := \{(x, y) \colon x^2 + y^4 \leq r^2\}$ is invariant under its flow for all $r \geq 0$.
(b) Show that $r\overline{B}$ is invariant iff $r = 0$ or $1/\sqrt{2}$.
(c) Show that $S_r := \{(x, y) \colon \max[|x|, |y|] \leq r\}$ is invariant only for $r = 0$.

11.3. We now consider a controlled version of the preceding exercise. Specifically we take
$$\dot{x} = -x + 2y^3 + u, \quad \dot{y} = -x + v,$$
where u and v are measurable control functions satisfying
$$|u(t)| \leq \delta \text{ and } |v(t)| \leq \Delta \text{ a.e.}$$
Given $r \geq 0$, find values of δ and Δ which ensure that S_r is weakly invariant (where S_r is as defined in Problem 11.2).

11.4. A multifunction F on $\mathbb{R}^n$ has lower Hamiltonian
$$h(x,p) = \begin{cases} -\|x\|\,\|p\| & \text{if } \|x\| < 1, \\ -2\|p\| & \text{if } \|x\| \geq 1. \end{cases}$$
What is F?

11.5. Let $\{S_\alpha\}$ be a family of subsets of $\mathbb{R}^n$, each weakly invariant with respect to F. Prove that $\operatorname{cl}\bigcup_\alpha S_\alpha$ is weakly invariant with respect to F.

11.6. Consider
$$F(x,y) = \begin{cases} \left(1, \dfrac{1}{\sqrt{2}}\right) & \text{if } y > \sqrt{1+x^2}, \\ (1,0) & \text{if } y < \sqrt{1+x^2}, \\ \{1\} \times \left[0, \dfrac{1}{\sqrt{2}}\right] & \text{if } y = \sqrt{1+x^2}. \end{cases}$$

(a) Show that F satisfies the Standing Hypotheses.

(b) Show that for any selection f for F, the unique Euler solution on $[0,1]$ to $(\dot{x}, \dot{y}) = f(x,y)$ starting at $(0,1)$ is one of two linear functions, depending entirely upon $f(0,1)$.

(c) Show that the curve $(t, \sqrt{1+t^2})$ does not arise as an Euler solution, but that it is nonetheless a trajectory on $[0,1]$ of the differential inclusion $(\dot{x}, \dot{y}) \in F(x,y)$.

(d) Find all trajectories of the differential inclusion on $[0,1]$ beginning at $(0,1)$; there are infinitely many.

11.7. Consider
$$F(x,y) = \{[xu, yv] : u \geq 0, v \geq 0, u+v = 1\}.$$
Confirm that the Standing Hypotheses hold, and prove that the attainable set $\mathcal{A}((1,1);1)$ is not convex. (*Hint.* Show that while both $(e,1)$ and $(1,e)$ are attainable from $(1,1)$ in time 1, their midpoint is not.)

11.8. Let F satisfy the Standing Hypotheses, and let $S \subseteq \mathbb{R}^n$ be a nonempty, compact, wedged subset of $\mathbb{R}^n$ which is *weakly avoidable* with respect to F; that is, $\operatorname{cl}[\operatorname{comp}(S)]$ is weakly invariant with respect to F. Assume also that $\operatorname{cl}[\operatorname{comp}(S)]$ is regular. Then S contains a zero of F.

11.9. Let $g : \mathbb{R}^n \to \mathbb{R}^n$ be a continuous function satisfying the tangency condition
$$\lim_{\substack{x' \to x \\ x' \in S \\ \lambda \downarrow 0}} \frac{d_S[\lambda g(x') + (1-\lambda)x]}{\lambda} = 0,$$
where $S \subseteq \mathbb{R}^n$ is a compact, wedged, and homeomorphically convex set.

(a) Prove that g has a fixed point in S.

(b) Show that the tangency condition always holds if S is convex and g maps S to itself (the case of Brouwer's Theorem).

11.10. Prove the following Fixed Point Theorem of Kakutani: let C be a compact convex subset of $\mathbb{R}^n$, and let the multifunction G satisfy the Standing Hypotheses. Suppose that

$$G(x) \cap C \neq \emptyset \ \forall x \in S.$$

Then there exists $\hat{x} \in C$ such that $\hat{x} \in G(\hat{x})$.

11.11. Let F be an upper semicontinuous multifunction on $\mathbb{R}^n$, with images that are nonempty compact convex subsets of $\mathbb{R}^n$. Prove that for any given $\varepsilon > 0$ there exists a locally Lipschitz selection

$$f_\varepsilon(x) \in F(x) + \varepsilon B \ \forall x \in \mathbb{R}^n.$$

11.12. Consider the linear differential system

$$\dot{x}(t) = Ax(t),$$

where A is an $n \times n$ constant real matrix. Prove that the following are equivalent:

(a) the nonnegative orthant $\mathbb{R}^n_+$ is invariant;

(b) $e^{tA}\mathbb{R}^n_+ \subseteq \mathbb{R}^n_+$ for all $t \geq 0$; and

(c) A is a *Metzler matrix*; that is, its off-diagonal entries are nonnegative.

11.13. Let S be a nonempty compact subset of $\mathbb{R}^n$, and let $f \colon \mathbb{R}^n \to \mathbb{R}^n$ be continuous. Prove that $(S, \{f\})$ is weakly invariant iff $f(x) \in T^C_S(x)$ $\forall x \in \mathbb{R}^n$.

11.14.

(a) Suppose that S is a compact subset of $\mathbb{R}^n$, and that

$$h(x, \zeta) < 0 \ \forall \zeta \in N^P_S(x), \quad \zeta \neq 0, \ \forall x \in S.$$

Show by example that S may nonetheless fail to be locally attainable.

(b) Let the compact set S in $\mathbb{R}^n$ be described in the form

$$\{x \in \mathbb{R}^n : f(x) \leq 0\},$$

where the locally Lipschitz function $f \colon \mathbb{R}^n \to \mathbb{R}$ satisfies

$$f(x) = 0 \implies 0 \notin \partial_L f(x).$$

Show that if

$$h(x,\zeta) < 0 \;\forall \zeta \in \partial_L f(x),\; \forall x \in S \text{ such that } f(x) = 0,$$

then S is locally attainable.

11.15. Lyapounov theory can be developed for *sets* as well as for equilibrium points. Let S be a nonempty compact subset of $\mathbb{R}^n$. (The case of an equilibrium x^* corresponds to taking $S = \{x^*\}$.) A Lyapounov pair (Q, W) for S is defined to be a pair of strictly positive functions $Q, W \in \mathcal{F}(\text{comp } S)$ such that the sublevel sets $\{x \colon Q(x) \le q\}$ are compact and the following condition holds:

$$h\bigl(x, \partial_P Q(x)\bigr) \le -W(x) \;\forall x \in \text{comp } S.$$

(a) Show that if such a pair (Q, W) exists, then another pair $(Q, \widetilde{W})$ exists for which $\widetilde{W}$ is globally defined and locally Lipschitz on $\mathbb{R}^n$, is equal to zero precisely on S, and satisfies a linear growth condition.

(b) Show that if a Lyapounov pair (Q, W) exists for S, then any $\alpha \in \text{dom}\, Q$ can be steered at least asymptotically to S (i.e., there is a trajectory $x(\cdot)$ which begins at α and either reaches S in finite time, or else satisfies $\lim_{t\to\infty} d_S(x(t)) = 0$).

(c) A partial converse to (b) is available. Suppose that S has the property that any $\alpha \in \mathbb{R}^n$ can be steered at least asymptotically to S. Let $\varepsilon > 0$. Then the set $S + \varepsilon \overline{B}$ admits a Lyapounov pair (Q, W) for which Q is finite. (*Hint.* Let

$$Q(\alpha) := \min\{T \ge 0 \colon \text{some trajectory } x \text{ has } x(0) = \alpha, x(T) \in S + \varepsilon \overline{B}\}.)$$

11.16. Let (φ, F) be weakly decreasing, where $\varphi \colon \mathbb{R}^n \to \mathbb{R}$ is continuous. Prove that given $\alpha \in \mathbb{R}^n$, there is a trajectory x of F with $x(0) = \alpha$ such that the function $t \mapsto \varphi(x(t))$ is decreasing. (It is not known whether this is true when φ is merely assumed to belong to $\mathcal{F}(\mathbb{R}^n)$.)

11.17. For $n = 1$, let $f(x) = 0$ if $x \in (0,1)$ and x is rational, $f(x) = -1$ otherwise. Show that the following function $\bar{x}$ is an Euler solution of the initial-value problem $\dot{x} = f(x),\; x(0) = 1$:

$$\bar{x}(t) = \begin{cases} 1 - t & \text{if } 0 \le t \le 1, \\ 0 & \text{if } t \ge 1. \end{cases}$$

Nonetheless, $\bar{x}$ restricted to $[1, \infty)$ is *not* an Euler solution of the initial-value problem $\dot{x} = f(x),\; x(1) = 0$. (The moral is that Euler solutions cannot be truncated to give Euler solutions.)

11.18. Let φ be continuous and let F be locally Lipschitz. Show by examples that the following monotonicity properties of the system (φ, F) are all distinct, and then go on to prove that each is characterized by its corresponding proximal Hamiltonian inequality holding for all x:

(a) *weakly decreasing*: $h(x, \partial_P \varphi(x)) \leq 0$;

(b) *weakly increasing*: $H(x, \partial^P \varphi(x)) \geq 0$;

(c) *strongly decreasing*: $H(x, \partial_P \varphi(x)) \leq 0$ or $H(x, \partial^P \varphi(x)) \leq 0$;

(d) *strongly increasing*: $h(x, \partial_P \varphi(x)) \geq 0$ or $h(x, \partial^P \varphi(x)) \geq 0$;

(e) *weakly predecreasing* (that is, given any x_0, there exists a trajectory x of F on $(-\infty, 0]$ such that $x(0) = x_0$ and $\varphi(x(t)) \geq \varphi(x_0)$, $t \leq 0$): $h(x, \partial^P \varphi(x)) \leq 0$; and

(f) *weakly preincreasing* (the definition is analogous to (e)): $H(x, \partial_P \varphi(x)) \geq 0$.

In which of these cases is the Lipschitz hypothesis on F actually required? (*Hint.* Exercise 2.11 (c), (d) is relevant to (e) above.)

11.19. For a given initial time t_0 and compact subset A of $\mathbb{R}^n$, the *reachable set* $\mathcal{R}$ from A is defined as follows:

$$\mathcal{R} := \{(t, x(t)) : t \geq t_0, x \text{ is a trajectory for } F \text{ on } [t_0, t], x(t_0) \in A\}.$$

Here we consider F depending on both t and x. We denote the "slice" at time T of $\mathcal{R}$ by $\mathcal{R}_T$; that is,

$$\mathcal{R}_T := \{x : (T, x) \in \mathcal{R}\}.$$

Then clearly

$$\mathcal{R}_T = \bigcup \{\mathcal{A}(t_0, x_0; T) : x_0 \in A\}.$$

We will see that the intrinsic invariance properties of $\mathcal{R}$ allow us to characterize it via a certain Hamilton–Jacobi relationship satisfied by its proximal normal vectors.

(a) Prove that $\mathcal{R}$ is closed and that each $\mathcal{R}_T$ is compact. Show that $\mathcal{R}_{t_0} = A$.

(b) Prove that $\mathcal{R}$ is *uniformly bounded near* t_0; i.e., there exists $\varepsilon > 0$ and a compact set C such that

$$\mathcal{R}_T \subseteq C \ \forall T \in [t_0, t_0 + \varepsilon].$$

252   4. A Short Course in Control Theory

(c) Let S be a closed subset of $[t_0, \infty) \times \mathbb{R}^n$ which is uniformly bounded near t_0, and for which $S_{t_0} = A$. Prove that for any $\varepsilon > 0$ there exists $\delta > 0$ such that $S_T \subseteq A + \varepsilon B$ for all $T \in [t_0, t_0 + \delta]$.

(d) Prove the following:

Theorem. *Let F be locally Lipschitz. Then $\mathcal{R}$ is the unique closed subset S of $[t_0, \infty) \times \mathbb{R}^n$ which is uniformly bounded near t_0 and satisfies*:

(i) $0 + H(t, x, \zeta) = 0 \; \forall (\theta, \zeta) \in N_S^P(t, x), \; \forall (t, x) \in (t_0, \infty) \times \mathbb{R}^n$; *and*

(ii) $S_{t_0} = A$.

11.20. Let (Q_1, W_1) and (Q_2, W_2) be Lyapounov pairs for x^*. Prove that $\bigl(\min\{Q_1, Q_2\}, \min\{W_1, W_2\}\bigr)$ is also a Lyapounov pair for x^*.

11.21. For $n = 2$, F induces the lower Hamiltonian

$$h(x, y, p, q) = -\bigl|(x - y)(p - 1)\bigr|.$$

(a) Find F.

(b) Let

$$\varphi_1(\tau, \alpha, \beta) := e^{2\tau - 2}|\alpha - \beta|, \quad \varphi_2(\tau, \alpha, \beta) := -e^{-2\tau + 2}|\alpha - \beta|.$$

Which of φ_1, φ_2 is the value function of a problem of the form

$$\text{minimize } \{\ell(x(1)) : \dot{x} \in F(x), x(\tau) = (\alpha, \beta)\}?$$

11.22. In the context of Theorem 9.1, suppose that $F(x)$ has the form $f(x, U)$, where U is a compact subset of $\mathbb{R}^m$ and f is continuously differentiable. Suppose in addition that $F(x)$ is *strictly convex* for each x, in the following sense:

$$v, w \in F(x), \quad v \neq w, \quad \lambda \in (0, 1) \implies \lambda v + (1 - \lambda)w \in \text{int } F(x).$$

Derive from the Hamiltonian inclusion of Theorem 9.1 the following conclusion: there exists a measurable function $u(\cdot)$ on $[a, b]$ with values in U such that for almost every t in $[a, b]$ we have

$$\dot{x}(t) = f(x(t), u(t)), \quad -\dot{p}(t) = f'_x(x(t), u(t))^* p(t),$$
$$\max_{u' \in U} \langle p(t), f(x(t), u') \rangle = \langle p(t), \dot{x}(t) \rangle.$$

These conclusions (together with the transversality condition) constitute the *Pontryagin Maximum Principle*; they can be obtained under weaker assumptions than those in force here. (*Hint.* Problem 2.9.13.)

11.23. Consider the problem of minimizing

$$\ell(x(b)) + \int_a^b \varphi(x(t))\,dt$$

over those trajectories of F on $[a,b]$ satisfying $x(a) = x_0$. Here, x_0 is given, and ℓ and φ are given locally Lipschitz functions. Let $x(\cdot)$ be a solution. Use the device introduced in the proof of Theorem 10.3 (absorbing φ into the dynamics) to prove the following necessary conditions: there exists an arc p such that

$$(-\dot{p}(t), \dot{x}(t)) \in \partial_C H(x(t), p(t)) - \partial_C \varphi(x(t)) \times \{0\}, \quad t \in [a,b] \text{ a.e.},$$
$$H(x(t), p(t)) - \varphi(x(t)) = \text{constant}, \quad t \in [a,b],$$
$$-p(b) \in \partial_L \ell(x(b)).$$

(*Hint.* Without loss of generality, we can assume that φ satisfies a linear growth condition.)

11.24. With $n = 1$, let $x(\cdot)$ solve the problem of minimizing

$$-\beta x(1) + \int_0^1 |x(t)|\,dt$$

over those arcs x on $[0,1]$ satisfying $x(0) = \alpha$ and $|\dot{x}(t)| \leq 1$, $t \in [0,1]$ a.e., where α and β are given constants.

(a) Prove that x is piecewise affine, and that only three certain values are possible for the slopes of the affine portions. (*Hint.* Study the curves $H(x,p) - |x| = $ constant.)

(b) Find the unique solution x when α and β are positive.

11.25. Let F be locally Lipschitz, and assume that the system is globally null controllable. Then the minimal time function Q (see Exercise 10.7) is everywhere finite. Prove that Q is the unique function $\varphi \colon \mathbb{R}^n \to \mathbb{R}$ satisfying the following conditions:

(i) $\varphi \in \mathcal{F}(\mathbb{R}^n)$, $\varphi(0) = 0$, $\varphi(x) > 0$ for $x \neq 0$;

(ii) $\liminf_{\substack{x' \to x \\ x' \neq 0}} \varphi(x') = 0$; and

(iii) $h(x, \partial_P \varphi(x)) = -1$ on $\mathbb{R}^n \setminus \{0\}$.

11.26. We will prove that when the origin is normal, the minimal-time function is continuous at 0. The context is that of Theorem 10.3, in which it was shown that for certain $b > 0$, the value function V_b used in the proof is Lipschitz near 0: for some $\delta_b > 0$, we have

$$0 \leq V_b(\alpha) \leq K_b \|\alpha\| \quad \text{when } \|\alpha\| \leq \delta_b.$$

254   4. A Short Course in Control Theory

(a) Prove that this conclusion holds for all $b > 0$ sufficiently small.

(b) If Q is the minimal-time function, then show that
$$Q(\alpha) \leq \sqrt{K_b \|\alpha\|} + b \quad \text{when } \|\alpha\| < \delta_b.$$

(c) Given $\varepsilon > 0$, pick $b > 0$ so that $\sqrt{b} + b < \varepsilon$. Then, for $\|\alpha\| < \min(\delta_b, b/K_b)$, we have $Q(\alpha) < \varepsilon$. Deduce that Q is continuous at 0.

11.27. Let $\mathcal{R}_0$ be the reachable set from the origin:
$$\mathcal{R}_0 := \{(t, x(t)): t \geq 0, x \text{ is a trajectory}, x(0) = 0\}.$$

We assume that F is locally Lipschitz and autonomous. Suppose that (T, β) is a *boundary point* of $\mathcal{R}_0$, $T > 0$.

(a) Prove the existence of an arc p on $[0, T]$, and a trajectory x of F on $[0, T]$ with $x(0) = 0$, $x(T) = \beta$ such that for some constant h we have $|h| + \|p\|_\infty \neq 0$ and
$$(-\dot{p}(t), \dot{x}(t)) \in \partial_C H(x(t), p(t)), \quad t \in [0, T] \text{ a.e.,}$$
$$H(x(t), p(t)) = h, \quad t \in [0, T],$$
$$(h, -p(T)) \in N^L_{\mathcal{R}_0}(T, \beta).$$

(*Hint.* Consider first the case in which $N^P_{\mathcal{R}_0}(T, \beta)$ is nontrivial; then some point $(\tau, \gamma) \in \text{comp}\,\mathcal{R}_0$ admits (T, β) as the closest point in $\mathcal{R}_0$; apply Exercise 9.16.)

(b) Deduce that some solution x of the minimal-time problem (Exercise 10.7) admits an arc p and a constant h with $|h| + \|p\|_\infty \neq 0$ such that
$$(-\dot{p}(t), \dot{x}(t)) \in \partial_C H(x(t), p(t)), \quad t \in [0, T] \text{ a.e.,}$$
$$H(x(t), p(t)) = h, \quad t \in [0, T].$$

(c) Consider the problem of finding the function $x \colon [0, T] \to \mathbb{R}$ which steers an initial position $\alpha = x(0)$ and an initial velocity $v = \dot{x}(0)$ to the origin and at rest in minimal time T under the constraint $|\ddot{x}(t)| \leq 1$. Interpret it as a special case of the minimal-time problem, with $n = 2$ and $F(x, y) := \{(y, u): |u| \leq 1\}$.

(d) Interpret the necessary conditions of part (b) for this case. (*Hint.* Exercise 9.9.) Proceed to find the unique solution of the problem, as a function of the initial condition (α, v).

(e) Show that the origin is normal.

(f) By calculating the minimal-time function explicitly, verify that it is continuous, but not Lipschitz, at zero.

11.28. Let $f\colon \mathbb{R}^m \times \mathbb{R}^n \to \mathbb{R}$ be locally Lipschitz, and suppose that $f(x,y)$ is concave as a function of x and convex as a function of y. Prove that

$$\partial_C f(x,y) = \partial_C f(\cdot, y)(x) \times \partial_C f(x, \cdot)(y).$$

(*Hint.* See Problem 2.9.15, as well as the proof of Lemma 9.5.)

11.29. We consider the context of Theorem 9.2, under the additional assumption that the set $G := \operatorname{graph} F$ is convex.

(a) Prove that $H(x,p)$ is concave as a function of x.

(b) Show that the Hamiltonian inclusion is equivalent to the relations

$$-\dot{p}(t) \in \partial_C H\bigl(\cdot, p(t)\bigr)\bigl(x(t)\bigr), \quad \dot{x}(t) \in \partial_C H\bigl(x(t), \cdot\bigr)\bigl(p(t)\bigr).$$

(c) Show that the Hamiltonian inclusion is also equivalent to

$$\bigl(\dot{p}(t), p(t)\bigr) \in N_G^P\bigl(x(t), \dot{x}(t)\bigr).$$

(d) If in addition it is assumed that $\ell(\cdot)$ is convex, then prove that any arc x admissible for (P) which satisfies (together with some arc p) the Hamiltonian inclusion and the transversality condition is a solution of (P) (i.e., the necessary conditions of Theorem 9.1 are also *sufficient* for optimality).

11.30. Let V be the value function of §7:

$$V(\tau, \alpha) := \min\{\ell(x(T)) : \dot{x} \in F(x), x(\tau) = \alpha\},$$

where ℓ and F are locally Lipschitz. Let $M(\tau, \alpha)$ denote the set of arcs x on $[\tau, T]$ satisfying $x(\tau) = \alpha$ and admitting an arc p such that

$$\bigl(-\dot{p}(t), \dot{x}(t)\bigr) \in \partial_C H\bigl(x(t), p(t)\bigr), \quad t \in [\tau, T],$$
$$H\bigl(x(t), p(t)\bigr) = \text{constant}, \quad t \in [\tau, T],$$
$$-p(T) \in \partial_L \ell\bigl(x(T)\bigr).$$

Prove that
$$V(\tau, \alpha) = \min\{\ell(x(T)) : x \in M(\tau, \alpha)\}.$$

(The fact that V, the solution of the Hamilton–Jacobi equation, is generated by solutions of a Hamiltonian system is known in a classical setting as the *method of characteristics*.)

11.31. In the context of Theorem 5.5, suppose that Q and W are continuously differentiable. Prove that the system (F, x^*) admits a *globally stabilizing feedback* $\hat{v}(\cdot)$; i.e., a selection $\hat{v}$ for F all of whose Euler solutions on $[0, \infty)$, for any initial condition, tend to x^* as $t \to \infty$. (*Hint.* Consider the multifunction

$$\widehat{F}(x) := \{v \in F(x) \colon \langle \nabla Q(x), v \rangle \leq -W(x)\}.)$$

We remark that a "nice" system which is globally asymptotically controllable to the origin need not admit a *smooth* Lyapounov pair. An example is the *non-holonomic integrator*, in which $n = 3$ and

$$F(x) := \{(u_1, u_2, x_1 u_2 - u_1 x_2) \colon \|(u_1, u_2)\| \leq 1\}.$$

Notes and Comments

To be neglected before one's time, must be very vexatious.

—Jane Austen, *Mansfield Park*

The theory of generalized gradients, proposed by Clarke in 1973, demonstrated the possibility of developing a useful calculus in a fully nonsmooth, nonconvex setting. This theory has served as the pattern for subsequent ones, by indicating how and why to treat functions and sets on an equal footing, how to pass back and forth from one to the other, and in so doing obtain a coherent, complete calculus. Certain elements of this design were foreshadowed by convex analysis, in the work of Moreau and Rockafellar. But the opportunity to shed convexity hypotheses greatly expanded the potential applications of the subject.

The basic construct of proximal analysis, the proximal normal vector, also appears in Clarke's early work, under the name "perpendicular." It was used there to generate the normal cone (and hence, via epigraphs, subgradients). In that sense, proximal subgradients were implicit from the beginning. However, the development of proximal calculus per se came later, and it gradually became clear that proximal analysis should be viewed as a distinct branch of the subject in its own right. The work of Rockafellar, Ioffe, Mordukhovich, and Borwein is to be mentioned in this regard.

We will not attempt to list, in an introductory text, the hundreds of references, nor the dozens of other names, that could justifiably be mentioned. However, the Bibliography contains a selection of recent articles and books on both theory and applications that can serve as a starting point for the interested reader. In the notes below, when no citation is given, the result (or a variant of it) is likely to be found in one or both of the two basic sources for this work, [C4] and [C5], which also contain many other references and related topics.

Other general works that we recommend to the reader include, for nonsmooth and nonlinear analysis, Aubin [A2] (which emphasizes games and economics), Demyanov and Rubinov [DR] and Loewen [L1] (emphasis on optimal control), and Mäkelä and Neittaanmäki [MN] (emphasizes applications). Convex analysis is treated, for example, by Hiriart-Urruty and Lemaréchal in [HUL]. Phelps [Ph] is an excellent source for differentiability. As standard references on control theory, we suggest Fleming and Soner [FS], Roxin [Ro], Sontag [So], and Zabczyk [Z]. The books [HL], [PBGM], and [Y] are old favorites of ours.

Chapter 0

The elementary discussion of Dini derivates in Boas [Bo] is a nice one; the definitive references in this area remain the classic books of Saks [Sa] and Bruckner [Bru]. The argument used to prove Theorem 1.1 is motivated by that in [CR]. The existence theorem in the calculus of variations that is alluded to appears in [C6]. The tangential characterization of flow-invariance given by Theorem 2.3 has a history of rediscovery, beginning with Nagumo in 1942. The normality characterization of Theorem 2.4 goes back to Bony. Optimization of eigenvalues is discussed in [BO], [HY], and [O]; see Cox and Overton [CO] and [Cx] for the strongest column. Torricelli's table was inspired by an example in Lyusternik [Ly]. Tikhomirov's book [T1] is a charming, elementary introduction to optimization.

Chapter 1

The proof technique of Theorem 3.1 is now a standard one (see [LS]) in the subject of minimization principles, for which a general reference is [DGZ]. The best known such principle is Ekeland's [E]; however, it is of limited interest in proximal analysis since it introduces a term with a "corner," in contrast to the "smooth variational principle" of Borwein and Preiss [BP]. We follow [CLW] in deriving this result from proximal density. This article contains an example of a C^1 function on the line whose proximal subdifferential is empty except on a set which is small, both in the sense of measure and of category.

Theorem 7.3 has many predecessors; the approach used here appears in [CSW], [RW], and [W]. The result also follows immediately from the Mean

Value Inequality of §3.2. The fuzzy Sum Rule Theorem 8.3 goes back to Ioffe in its essentials, while the limiting Sum Rule 10.1 is due to Mordukhovich. Problem 11.23 is a result due to Rockafellar in finite dimensions.

Chapter 2

The results of this chapter follow closely those in Clarke [C4], which has significantly more material, however. The Chain Rule 2.5 is an improvement on the original treatment. Because the notation ∂, N_S, and T_S is now so widely used, we have opted to retain it, while simultaneously introducing the alternate notation ∂_C, N_S^C, and T_S^C for cases in which other constructs also intervene (as in Chapter 4).

Chapter 3

Constrained optimization is treated at greater length in Chapter 6 of [C4]. The Mean Value Inequality of Theorem 2.6 is due to Clarke and Ledyaev [CL2]. Surprisingly perhaps, the case in which the values at two different *sets* are compared appears to be quite different; it is treated in [CL1]. The results of §3 have innumerable antecedents, though the specific approach seems to be new. The Lipschitz Inverse Function Theorem is due to Clarke. For Theorem 4.2, see, for example, Subbotin [Su]. Theorem 4.16, and the resulting proof of Rademacher's Theorem, are drawn from Bessis and Clarke [BC]. Section 5 follows [C4] for the most part. We refer to [C5] for further results in the calculus of variations. It was Rockafellar who first pointed out the importance of the tangent cone T_S^C having a nonempty interior, and the properties that result from this. What we have called "wedged" is a new term that we are proposing here instead of "epi-Lipschitz." Theorem 6.12 is due to Aubin and Clarke [AC1], and the function in Problem 7.30 is due to Rockafellar.

Chapter 4

A great deal of work on what we have called weak invariance has been done under the label *viability*, see Aubin [A1] and Aubin and Cellina [AC]. The book by Deimling [D] is another valuable reference for differential inclusions (note: neither "invariance" nor "viability" is employed here; the author simply speaks of existence). The concept of Euler solutions is borrowed from Krasovskii and Subbotin [KS], where it plays an important role in differential games; our proximal aiming is inspired by their "extremal aiming method." Theorem 2.10 is an amalgam of several results, including certain ones for the case of a function (see the comments for Chapter 0), as well as theorems of Haddad and of Veliov [V] (to whom is first due the proximal criterion). See [CLSW] for further comment. The Strong Invariance Theorem 3.8 has its antecedents in [C2]. The results of §4 are taken from [CLS1]. We remark that all the results of this chapter can be extended to the case in which the linear growth hypothesis is dropped, by taking account of possible finite "blow-up" times.

Several schools have participated over the past decades in developing the nonsmooth theory of the Hamilton–Jacobi equation. The state of the art circa 1976 is described in [B] where the "almost everywhere" type of (Lipschitz) solution dominates, and where Fleming's "artificial viscosity" approximation method is outlined. To our knowledge, the first truly pointwise (subdifferential) definition of the generalized solution (proposed by Clarke, again for Lipschitz functions) appeared in 1977 [H] in connection with the verification method. These turn out to be semisolutions in the viscosity sense. The approach is synthesized by Clarke and Vinter in [CV]. In 1980 Subbotin (see [Su]) inaugurated the two-sided approach to defining a nonsmooth (Lipschitz) solution, with Dini derivates and in the context of differential games. Subsequently, Crandall and Lions (see [CIL]) developed further Fleming's approach; the (eventual) two-sided, subdifferential approach to defining viscosity solutions, carried out for continuous functions and with the attendant uniqueness theorem, constituted a breakthrough that vindicated the nonsmooth analysis approach to the issue. It appears to be Barron and Jensen [BJ] who first demonstrated the (surprising) possibility of giving a single subdifferential characterization in certain cases, and for merely lower semicontinuous solutions. The results of §7 owe an intellectual debt to them, and also to Subbotin, who first stressed the relevance of invariance. See also [AF2], [Ba], [BaS], [CaF], [CD], and [He] for other examples of current research directions.

Theorem 9.1 is due to Clarke; Theorem 10.3 is taken from [CL]. A number of results relating the adjoint variable to the sensitivity of the problem are due to Clarke and Vinter. We refer to [C5] for these and other refinements of the necessary conditions in optimal control, and for examples of their use, under less stringent hypotheses; in this connection, see [C3] for the maximum principle. In particular, it is not assumed that the velocity set is convex. To put this another way, the current chapter deals with the "relaxed problem" only. The forthcoming book of Vinter [Vi] surveys the question of necessary conditions in optimal control, including results of Ioffe, Loewen, Mordukhovich, and Rockafellar aimed at refining the Hamiltonian inclusion.

It is a celebrated theorem of Brockett that (smooth) systems which are globally asymptotically controllable to the origin need not admit a continuous feedback control which stabilizes the system. In [CLSS], it is shown that time-independent stabilizing feedbacks *do* exist, if they are allowed to be discontinuous and if the corresponding solution concept is defined as in Chapter 4. This is achieved essentially as follows: a result of Sontag is invoked to deduce the existence of a *continuous* Lyapounov pair, and then an analogue of Problem 11.31 is carried out. Insight on the existence or otherwise of *smooth* Lyapounov pairs is provided in [CLS2]. See also [Ber], [CPT], [Kry], [KS], and [RolV] for other approaches to feedback construction.

List of Notation

We list here the principal constructs that appear in the book.

$\text{proj}_S(x)$	projection of x onto S.
$d_S(x)$ or $d(x;S)$	distance from x to S.
$N_S^P(x)$ or $N^P(x;S)$	proximal normal cone to S at x.
$\partial_P f(x)$	proximal subdifferential of f at x.
$\partial^P f(x)$	proximal superdifferential of f at x.
$\text{dom } f$	(effective) domain of f.
$\text{gr } f$	graph of f.
$\text{epi } f$	epigraph of f.
$\mathcal{F}(U)$	all $f: U \to (-\infty, \infty]$ that are lower semicontinuous and not identically $+\infty$.
$\partial_L f(x)$	limiting subdifferential of f at x.
$N_S^L(x)$	limiting normal cone to S at x
$f'(x;v)$	directional derivative of f at x in direction v.
$f'_G(x)$	Gâteaux derivative of f at x.
$f'(x)$	Fréchet derivative of f at x.
$I_S(\cdot)$ or $I(\cdot;S)$	indicator function of a set S.
$f^\circ(x;v)$	generalized directional derivative of f at x in direction v.
$h_S(\cdot), H_S(\cdot)$	lower and upper support functions of a set S.
$\partial f(x)$ or $\partial_C f(x)$	generalized gradient of f at x.
$T_S(x)$ or $T_S^C(x)$	(generalized) tangent cone to S at x.
$T_S^B(x)$	Bouligand tangent cone to S at x.
$N_S(x)$ or $N_S^C(x)$	(generalized) normal cone to S at x.

$Df(x;v)$	subderivate of f at x in direction v.
$\partial_D f(x)$	directional subdifferential of f at x.
$\nabla f(x)$	gradient vector of f at x.
h, H	lower and upper Hamiltonians.

We remark that at least four people in the field agree on these notational choices, which is certainly a record.

Bibliography

[Ar] Artstein, Z., Stabilization with relaxed controls, *Nonlinear Anal.* **7** (1983), 1163–1173.

[At] Attouch, H., Viscosity solutions of minimization problems, *SIAM J. Optim.* **6** (1996), 769–806.

[A1] Aubin, J.-P., *Viability Theory*, Birkhäuser, Boston, 1991.

[A2] Aubin, J. P., *Optima and Equilibria—An Introduction to Nonlinear Analysis*, Graduate Texts in Mathematics, vol. 140, Springer-Verlag, New York, 1993.

[AC] Aubin, J.-P. and Cellina, A., *Differential Inclusions*, Springer-Verlag, Berlin, 1994.

[ACl] Aubin, J.-P. and Clarke, F. H., Monotone invariant solutions to differential inclusions, *J. London Math. Soc.* **16** (2) (1977), 357–366.

[AF1] Aubin, J.-P. and Frankowska, H., *Set-Valued Analysis*, Birkhäuser, Boston, 1990.

[AF2] Aubin, J.-P. and Frankowska, H., Partial differential inclusions governing feedback controls, *J. Convex Anal.* **2** (1995), 19–40.

[BaS] Bardi, M. and Soravia, P., A comparison result for Hamilton–Jacobi equations and applications to some differential games lacking controllability, *Funkcial. Ekvac.* **37** (1994), 19–43.

[Ba]   Barles, G., Discontinuous viscosity solutions of first-order Hamilton–Jacobi equations: A guided visit, *Nonlinear Anal.* **20** (1993), 1123–1134.

[BJ]   Barron, E. N. and Jensen, R., Optimal control and semicontinuous viscosity solutions, *Proc. Amer. Math. Soc.* **113** (1991), 397–402.

[BB]   Basar, T. and Bernhard, P., H^∞-*Optimal Control and Related Minimax Design Problems*, Birkhäuser, Berlin, 1991.

[B]   Benton, S. H., *The Hamilton–Jacobi Equation: A Global Approach*, Academic Press, New York, 1977.

[Ber] Berkovitz, L. D., Optimal feedback controls, *SIAM J. Control Optim.* **27** (1989), 991–1006.

[BC]   Bessis, D. and Clarke, F. H., Partial subdifferentials, derivates, and Rademacher's Theorem, *Trans. Amer. Math. Soc.*, in press.

[BEFB] Boyd, S., El Ghaoui, L., Feron, E. and Balakrishnan, V., *Linear Matrix Inequalities in System and Control Theory*, SIAM, Philadelphia, PA, 1994.

[Bo]   Boas, R. P., *A Primer of Real Functions*, Carus Mathematical Monographs, Mathematical Association of America, Washington, DC, 1960.

[BoCo] Bonnisseau, J.-M. and Cornet, B., Existence of marginal cost pricing equilibria: The nonsmooth case, *Internat. Econom. Rev.* **31** (1990), 685–708.

[BI]   Borwein, J. M. and Ioffe, A., Proximal analysis in smooth spaces, *Set-Valued Anal.* **4** (1996), 1–24.

[BP]   Borwein, J. M. and Preiss, D., A smooth variational principle with applications to subdifferentiability and to differentiability of convex functions, *Trans. Amer. Math. Soc.* **303** (1987), 517–527.

[BZ]   Borwein, J. M. and Zhu, Q. J., Variational analysis in nonreflexive spaces and applications to control problems with L^1 perturbations, *Nonlinear Anal.* **28** (1997), 889–915.

[Br]   Brézis, H., *Analyse fonctionnelle: Théorie et applications*, Masson, Paris, 1983.

[Bru] Bruckner, A., *Differentiation of Real Functions*, 2nd ed., CRM Monograph Series 5, American Mathematical Society, Providence, RI, 1994.

[BO] Burke, J. V. and Overton, M. L., Differential properties of the spectral abscissa and the spectral radius for analytic matrix-valued mappings, *Nonlinear Anal.* **23** (1994), 467–488.

[By] Byrnes, C. I., On the control of certain infinite dimensional systems by algebro-geometric techniques, *Amer. J. Math.* **100** (1978), 1333–1381.

[CD] Cannarsa, P. and DaPrato, G., Second-order Hamilton–Jacobi equations in infinite dimensions, *SIAM J. Control Optim.* **29** (1991), 474–492.

[CaF] Cannarsa, P. and Frankowska, H., Value function and optimality condition for semilinear control problems. II. The parabolic case, *Appl. Math. Optim.* **33** (1996), 1–33.

[C1] Clarke, F. H., *Necessary Conditions for Nonsmooth Problems in Optimal Control and the Calculus of Variations*, Doctoral thesis, University of Washington, 1973.

[C2] Clarke, F. H., Generalized gradients and applications, *Trans. Amer. Math. Soc.* **205** (1975), 247–262.

[C3] Clarke, F. H., The maximum principle under minimal hypotheses, *SIAM J. Control Optim.* **14** (1976), 1078–1091.

[C4] Clarke, F. H., *Optimization and Nonsmooth Analysis*, Wiley Interscience, New York, 1983; reprinted as vol. 5 of Classics in Applied Mathematics, SIAM, Philadelphia, PA, 1990; Russian translation, Nauka, Moscow, 1988.

[C5] Clarke, F. H., *Methods of Dynamic and Nonsmooth Optimization*, CBMS/NSF Regional Conf. Ser. in Appl. Math., vol. 57, SIAM, Philadelphia, PA, 1989.

[C6] Clarke, F. H., An indirect method in the calculus of variations, *Trans. Amer. Math. Soc.* **336** (1993), 655–673.

[CL1] Clarke, F. H. and Ledyaev, Yu., Mean value inequalities, *Proc. Amer. Math. Soc.* **122** (1994), 1075–1083.

[CL2] Clarke, F. H. and Ledyaev, Yu., Mean value inequalities in Hilbert space, *Trans. Amer. Math. Soc.* **344** (1994), 307–324.

[CL] Clarke, F. H. and Loewen, P. D., The value function in optimal control: Sensitivity, controllability and time-optimality, *SIAM J. Control Optim.* **24** (1986), 243–263.

[CLS1] Clarke, F. H., Ledyaev, Yu., and Stern, R. J., Fixed points and equilibria in nonconvex sets, *Nonlinear Anal.* **25** (1995), 145–161.

[CLS2] Clarke, F. H., Ledyaev, Yu., and Stern, R. J., Asymptotic stability and smooth Lyapounov functions, preprint, 1997.

[CLSS] Clarke, F. H., Ledyaev, Yu., Sontag, E. D., and Subbotin, A. I., Asymptotic controllability implies feedback stabilization, *IEEE Trans. Automat. Control*, in press (1997).

[CLSW] Clarke, F. H., Ledyaev, Yu., Stern, R. J., and Wolenski, P. R., Qualitative properties of trajectories of control systems: A survey, *J. Dynam. Control Systems* **1** (1995), 1–47.

[CLW] Clarke, F. H., Ledyaev, Yu., and Wolenski, P. R., Proximal analysis and minimization principles, *J. Math. Anal. Appl.* **196** (1995), 722–735.

[CR] Clarke, F. H. and Redheffer, R., The proximal subgradient and constancy, *Canad. Math. Bull.* **36** (1993), 30–32.

[CSW] Clarke, F. H., Stern, R. J., and Wolenski, P. R., Subgradient criteria for Lipschitz behavior, monotonicity and convexity, *Canad. J. Math.* **45** (1993), 1167–1183.

[CV] Clarke, F. H. and Vinter, R. B., Local optimality conditions and Lipschitzian solutions to the Hamilton–Jacobi equations, *SIAM J. Control Optim.* **21** (1983), 856–870.

[CPT] Coron, J.-M., Praly, L., and Teel, A., Feedback stabilization of nonlinear systems: Sufficient conditions and Lyapounov and input–output techniques, In *Trends in Control* (A. Isidori, Ed.), Springer-Verlag, New York, 1995.

[Cx] Cox, S. J., The shape of the ideal column, *Math. Intelligencer* **14** (1) (1992), 16–24.

[CO] Cox, S. J. and Overton, M. L., On the optimal design of columns against buckling, *SIAM J. Math. Anal.* **23** (1992), 287–325.

[CIL] Crandall, M. G., Ishii, H., and Lions, P.-L., User's guide to viscosity solutions of second-order partial differential equations, *Bull. Amer. Math. Soc.* **27** (1992), 1–67.

[D] Deimling, K., *Multivalued Differential Equations*, de Gruyter, Berlin, 1992.

[DR] Demyanov, V. F. and Rubinov, A. M., *Constructive Nonsmooth Analysis*, Peter Lang, Frankfurt, 1995.

[DGZ] Deville, R., Godefroy, G., and Zizler, V., *Smoothness and Renormings in Banach Spaces*, Longman and Wiley, New York, 1993.

[E] Ekeland, I., Nonconvex minimization problems, *Bull. Amer. Math. Soc.* (New Series) **1** (1979), 443–474.

[F] Filippov, A. F., *Differential Equations with Discontinuous Right-Hand Sides*, Kluwer Academic, Dordrecht, 1988.

[FS] Fleming, W. H. and Soner, H. M., *Controlled Markov Processes and Viscosity Solutions*, Springer-Verlag, New York, 1993.

[Fr] Frankowska, H., Lower semicontinuous solutions of the Hamilton–Jacobi equation, *SIAM J. Control Optim.* **31** (1993), 257–272.

[FK] Freeman, R. A. and Kokotovic, P. V., *Robust Nonlinear Control Design. State-Space and Lyapounov Techniques*, Birkhäuser, Boston, 1996.

[H] Havelock, D., *A Generalization of the Hamilton–Jacobi Equation*, Master's thesis, University of British Columbia, Vancouver, Canada, 1977.

[He] Hermes, H., Resonance, stabilizing feedback controls, and regularity of viscosity solutions of Hamilton–Jacobi–Bellman equations, *Math. Control Signals Systems* **9** (1996), 59–72.

[HL] Hermes, H. and Lasalle, J. P., *Functional Analysis and Time Optimal Control*, Academic Press, New York, 1969.

[HS] Heymann, M. and Stern, R. J., Ω rest points in autonomous control systems, *J. Differential Equations* **20** (1976), 389–398.

[HUL] Hiriart-Urruty, J.-B. and Lemaréchal, C., *Convex Analysis and Minimization Algorithms*. I. *Fundamentals*, Springer-Verlag, Berlin, 1993.

[HY] Hiriart-Urruty, J.-B. and Ye, D., Sensitivity analysis of all eigenvalues of a symmetric matrix, *Numer. Math.* **70** (1995), 45-72.

[I] Ioffe, A. D., Proximal analysis and approximate subdifferentials, *J. London Math. Soc.* **41** (1990), 175–192.

[Is] Isidori, A., *Nonlinear Control Systems: An Introduction*, Springer-Verlag, Berlin, 1985.

[JDT] Jofré, A., Dihn, T. L., and Théra, M., ε-Subdifferential calculus for nonconvex functions and ε-monotonicity, *C. R. Acad. Sci. Paris Sér. I Math.* **323** (1996), 735–740.

[KS] Krasovskii, N. N. and Subbotin, A. I., *Game-Theoretical Control Problems*, Springer-Verlag, New York, 1988.

[Kry]   Kryazhimskii, A., Optimization of the ensured result for dynamical systems, *Proceedings of the International Congress of Mathematicians*, vols. 1, 2 (Berkeley, CA, 1986), pp. 1171–1179, American Mathematical Society, Providence, RI, 1987.

[LLM]   Lakshmikantham, V. Leela, S., and Martynyuk, A., *Practical Stability of Nonlinear Systems*, World Scientific, Singapore, 1990.

[Le]   Leitmann, G., One approach to the control of uncertain dynamical systems, *Appl. Math. Comput.* **70** (1995), 261–272.

[LS]   Li, Y. and Shi, S., A generalization of Ekeland's ε-variational principle and of its Borwein–Preiss smooth variant, *J. Math. Anal. Appl.*, to appear.

[L]   Lions, J. L., *Contrôle optimal de systèmes gouvernés par des équations aux dérivées partielles*, Dunod, Paris, 1968.

[L1]   Loewen, P. D., *Optimal Control via Nonsmooth Analysis*, CRM Proc. Lecture Notes, vol. 2, American Mathematical Society, Providence, RI, 1993.

[L2]   Loewen, P. D., A mean value theorem for Fréchet subgradients, *Nonlinear Anal.* **23** (1994), 1365–1381.

[Ly]   Lyusternik, L. A., *The Shortest Lines—Variational Problems*, MIR, Moscow, 1976.

[MN]   Mäkelä, M. M. and Neittaanmäki, P., *Nonsmooth Optimization*, World Scientific, London, 1992.

[MW]   Mawhin, J. and Willem, M., *Critical Point Theory and Hamiltonian Systems*, Springer-Verlag, New York, 1989.

[MS]   Mordukhovich, B. S. and Shao, Y. H., On nonconvex subdifferential calculus in Banach spaces, *J. Convex Anal.* **2** (1995), 211–227.

[O]   Overton, M. L. Large-scale optimization of eigenvalues, *SIAM J. Optim.* **2** (1992), 88–120.

[Pa]   Panagiotopoulos, P. D., *Hemivariational Inequalities. Applications in Mechanics and Engineering*, Springer-Verlag, Berlin, 1993.

[PQ]   Penot, J.-P. and Quang, P. H., Generalized convexity of functions and generalized monotonicity of set-valued maps, *J. Optim. Theory Appl.* **92** (1997), 343–356.

[Ph]   Phelps, R. R., *Convex Functions, Monotone Operators and Differentiability*, 2nd ed., Lecture Notes in Mathematics, vol. 1364, Springer-Verlag, New York, 1993.

[Po]   Polak, E.,, On the mathematical foundations of nondifferentiable optimization in engineering design, *SIAM Rev.* **29** (1987), 21–89.

[PBGM]   Pontryagin, L. S., Boltyanskii, R. V., Gamkrelidze, R. V., and Mischenko, E. F., *The Mathematical Theory of Optimal Processes*, Wiley, New York, 1962.

[RW]   Redheffer, R. and Walter, W., The subgradient in $\mathbb{R}^n$, *Nonlinear Anal.* **20** (1993), 1345–1348.

[R1]   Rockafellar, R. T., *Convex Analysis*, Princeton Mathematical Series, vol. 28, Princeton University Press, Princeton, NJ, 1970.

[R2]   Rockafellar, R. T., *The Theory of Subgradients and Its Applications to Problems of Optimization: Convex and Nonconvex Functions*, Helderman Verlag, Berlin, 1981.

[R3]   Rockafellar, R. T., Lagrange multipliers and optimality, *SIAM Rev.* **35** (1993), 183–238.

[RolV]   Roland, J. D. and Vinter, R. B., Construction of optimal feedback control, *System Control Lett.* **16** (1991), 357–367.

[Ro]   Roxin, E. O., *Control Theory and its Applications*, Gordon and Breach, New York, 1996.

[Ru1]   Rudin, W., *Real and Complex Analysis*, McGraw-Hill, New York, 1966.

[Ru2]   Rudin, W., *Functional Analysis*, McGraw-Hill, New York, 1973.

[Sa]   Saks, S., *Theory of the Integral*, Monografie Matematyczne Ser., no. 7 (1937); 2nd rev. ed., Dover, New York, 1964.

[So]   Sontag, E. D., *Mathematical Control Theory*, Texts in Applied Mathematics vol. 6, Springer-Verlag, New York, 1990.

[Su]   Subbotin, A. I., *Generalized Solutions of First-Order PDEs*, Birkhäuser, Boston, 1995.

[SSY]   Sussmann, H., Sontag, E., and Yang, Y., A general result on the stabilization of linear systems using bounded controls, *IEEE Trans. Automat. Control* **39** (1994), 2411–2425.

[T1]   Tikhomirov, V. M., *Stories about Maxima and Minima*, American Mathematical Society, Providence, RI, 1990.

[T2] Tikhomirov, V. M., *Convex Analysis and Approximation Theory, Analysis* II, Encyclopaedia of Mathematical Sciences (R. V. Gamkrelidze, Ed.), vol. 14, Springer-Verlag, New York, 1990.

[V] Veliov, V., Sufficient conditions for viability under imperfect measurement, *Set-Valued Anal.* **1** (1993), 305–317.

[Vi] Vinter, R. B., forthcoming monograph.

[W] Weckesser, V., The subdifferential in Banach spaces, *Nonlinear Anal.* **20** (1993), 1349–1354.

[WZ] Wolenski, P. R. and Zhuang, Y., Proximal analysis and the minimal time function, *SIAM J. Control Optim.*, to appear.

[Y] Young, L. C., *Lectures on the Calculus of Variations and Optimal Control Theory*, Saunders, Philadelphia, PA, 1969.

[Z] Zabczyk, J., *Mathematical Control Theory: An Introduction*, Birkhäuser, Boston, 1992.

Index

abnormal, 107, 241
absolutely continuous, 162, 177
almost everywhere solution, 228
approximation by P-subgradients, 138
attainability, 215
attainable set, 193
augmented Hamiltonian, 223
Aumann, 157
autonomous, 190

Borwein and Preiss, 43
Bouligand tangent cone, 8, 90
Brouwer's theorem, 203, 249

calculus of variations, 162, 163
calm, 170
Carathéodory, 223
chain rule, 32, 48, 58, 76
closed-valued, 150
closest point, 22
compactness of approximate trajectories, 185
comparison theorems, 225
constancy of the Hamiltonian, 241

constrained optimization, 50, 103
constraint qualification, 128
contingent, 90
continuous function, 28
continuously differentiable, 33
controllability, 244
convex functions, 29, 51, 80
convex set, 26

D-normal cone, 141
D-subdifferential, 138
D-subgradient, 138
D-tangent cone, 141
Danskin's theorem, 99
decoupling, 55
decrease principle, 122, 171
decreasing function, 2
density theorem, 39
dependence on initial conditions, 201
derivate, 2, 136
diameter, 181
differential inclusion, 177
Dini, vii, 2
Dini subderivate, 98

direction of descent, 97
directional calculus, 139
directional derivative, 31
directional subderivate, 4
directional subgradient, 138
distance function, 8, 23
domain, 27
Dubois–Reymond lemma, 162
dynamic feedback, 215

eigenvalue, 10
endpoint cost, 222
epigraph, 28
equilibria, 202
equilibrium, 1, 202
Erdmann condition, 164
Erdmann transform, 242
Euler equation, 163
Euler inclusion, 163
Euler polygonal arc, 181
Euler solution, 181
Euler solutions, 180
exact penalization, 50
extended real-valued, 27

feasible set, 104
feedback, 192, 197, 229
feedback selections, 192
feedback synthesis, 228
Fermat's rule, 2
Filippov's lemma, 174, 178
fixed point theorem, 203
flow-invariant, 7
Fréchet derivative, 31
Fréchet differentiable, 31, 148
free time problems, 244
Fritz John necessary conditions, 100
fuzzy sum rule, 56

Gâteaux differentiable, 31
generalized directional derivative, 70
generalized gradient, 6, 69, 72, 160
generalized gradient formula, 93

generalized Jacobian, 108, 133
global null controllability, 247
globally asymptotically stable, 209
gradient, 104
graph, 28, 150
Grave–Lyusternik, 127
Gronwall's lemma, 179
growth, 209, 210
growth hypothesis, 105
Gâteaux derivative, 78

half-space, 27
Hamilton–Jacobi equation, 17, 224
Hamilton–Jacobi inequality, 223, 228
Hamiltonian function, 17
Hamiltonian inclusion, 231, 255
Hausdorff continuous, 175
Hausdorff distance, 175
Hessian, 33
homeomorphic, 203
horizontal approximation theorem, 67
hyperplane, 27

implicit and inverse functions, 108, 129
indicator function, 28
inf-convolution, 44
infinitesimal decrease, 209, 210
initial-value problem, 180
inner product, 18
integral functionals, 148
invariance, 188
invariant embedding, 223
inverse function theorem, 2
inverse functions, 133

Jacobian, 2, 104, 106

Kakutani, 249

L^1-optimization, 11
Lagrange Multiplier Rule, 15, 65, 104

Lagrangian, 110
Lebourg's Mean Value Theorem, 75
Legendre, 222
limiting calculus, 61
limiting chain rule, 65
limiting normal, 62
limiting subdifferential, 61, 160
Line-Fitting, 98
linear growth, 178
linearization, 1
Lipschitz, 39, 51, 52
Lipschitz Inverse Function Theorem, 135
Lipschitz near x, 39
local attainable, 220
locally attainable, 220, 250
locally Lipschitz, 39, 196
locally Lipschitz selection, 249
lower Hamiltonian, 188
lower semicontinuity, 28, 167
Lyapounov functions, 209
Lyapounov pair, 210

manifold, 26
maximal rank, 245
maximum principle, 237
mean value inequality, 111, 117
Mean Value Theorem, 32, 111
measurable multifunctions, 149
measurable selections, 149, 151
mesh size, 181
method of characteristics, 256
metric projection, 192
Metzler matrix, 250
minimal time function, 16, 247
minimal time problem, 255
minimax problems, 10
minimax solutions, 226
minimax theorem, 234
minimization principles, 21, 43
monotonicity, 65, 209
Moreau–Yosida approximation, 64
multidirectionality, 112
multifunction, 177

multiplier, 15, 110
multiplier rule, 15, 104
multiplier set, 106, 107

necessary conditions, 231
nonautonomous case, 200
normal, 107, 245
normal cone, 85
normality, 244
null controllability, 244

open ball, 18
optimal control problem, 222
orthogonal decomposition, 25

partial proximal subdifferential, 38
partial proximal subgradients, 38
partial subdifferentials, 143
partial subgradients, 66
partition of unity, 205
pointed, 168
polar, 65, 85, 168
Pontryagin's Maximum Principle, 238
Pontryagin's maximum principle, 253
positive definiteness, 209, 210
positively homogeneous, 70
principle of optimality, 16, 224
projection, 9, 22
proximal aiming, 189
proximal Gronwall inequality, 68
proximal Hamilton–Jacobi equation, 224
proximal mean value theorem, 64
proximal normal cone, 22
proximal normal direction, 9
proximal normal inequality, 25
proximal normals, 21
proximal solution, 224
proximal subdifferential, 5, 29
proximal subgradient, 21, 29
proximal subgradient inequality, 33
proximal superdifferential, 37

quadratic inf-convolutions, 44

Rademacher's Theorem, 6, 93
Rayleigh's formula, 11
reachable set, 252
regular, 69, 81, 91
rest point, 202

selection, 180
semigroup property, 193
semisolution, 228
sensitivity, 105
separation theorem, 27
smooth manifold, 7
solution, 177
solvability, 105, 126
stabilizing feedback, 214
standard control system, 178
Standing Hypotheses, 178
state augmentation, 200
static feedback, 215
Stegall's minimization principle, 43
strictly convex, 253
strictly differentiable, 96
strong invariance, 195, 198
strong weak monotonicity, 123
strongly decreasing, 123, 124, 217, 219, 251
strongly increasing, 219, 251
strongly invariant, 198
subadditive, 70
Subbotin, 137
subderivate, 136
sum rule, 38, 54
support function, 71, 124

tangency, 69
tangent, 83
tangent cone, 84
tangent space, 7
terminal constraints, 238
the origin is normal, 245
Torricelli point, 12
Torricelli's Table, 11
trajectory, 178
trajectory continuation, 187
transversality condition, 164, 231

unilateral constraint set, 149
upper envelopes, 65
upper Hamiltonian, 188
upper semicontinuous, 28, 73, 179

value function, 104, 224
value function analysis, 238
variable intervals, 187
variational problem, 162
verification functions, 222
viability, 191
viscosity solutions, 142, 227

weak invariance, 188, 190, 192
weak sequential compactness, 164
weakly avoidable, 249
weakly decreasing, 124, 212, 215, 251
weakly increasing, 251
weakly lower semicontinuous, 55
weakly predecreasing, 251
weakly preincreasing, 251
weakly preinvariant, 194
wedged, 166

zero of a multifunction, 202

Graduate Texts in Mathematics

continued from page ii

61 WHITEHEAD. Elements of Homotopy Theory.
62 KARGAPOLOV/MERLZJAKOV. Fundamentals of the Theory of Groups.
63 BOLLOBAS. Graph Theory.
64 EDWARDS. Fourier Series. Vol. I 2nd ed.
65 WELLS. Differential Analysis on Complex Manifolds. 2nd ed.
66 WATERHOUSE. Introduction to Affine Group Schemes.
67 SERRE. Local Fields.
68 WEIDMANN. Linear Operators in Hilbert Spaces.
69 LANG. Cyclotomic Fields II.
70 MASSEY. Singular Homology Theory.
71 FARKAS/KRA. Riemann Surfaces. 2nd ed.
72 STILLWELL. Classical Topology and Combinatorial Group Theory. 2nd ed.
73 HUNGERFORD. Algebra.
74 DAVENPORT. Multiplicative Number Theory. 2nd ed.
75 HOCHSCHILD. Basic Theory of Algebraic Groups and Lie Algebras.
76 IITAKA. Algebraic Geometry.
77 HECKE. Lectures on the Theory of Algebraic Numbers.
78 BURRIS/SANKAPPANAVAR. A Course in Universal Algebra.
79 WALTERS. An Introduction to Ergodic Theory.
80 ROBINSON. A Course in the Theory of Groups. 2nd ed.
81 FORSTER. Lectures on Riemann Surfaces.
82 BOTT/TU. Differential Forms in Algebraic Topology.
83 WASHINGTON. Introduction to Cyclotomic Fields. 2nd ed.
84 IRELAND/ROSEN. A Classical Introduction to Modern Number Theory. 2nd ed.
85 EDWARDS. Fourier Series. Vol. II. 2nd ed.
86 VAN LINT. Introduction to Coding Theory. 2nd ed.
87 BROWN. Cohomology of Groups.
88 PIERCE. Associative Algebras.
89 LANG. Introduction to Algebraic and Abelian Functions. 2nd ed.
90 BRØNDSTED. An Introduction to Convex Polytopes.
91 BEARDON. On the Geometry of Discrete Groups.
92 DIESTEL. Sequences and Series in Banach Spaces.
93 DUBROVIN/FOMENKO/NOVIKOV. Modern Geometry—Methods and Applications. Part I. 2nd ed.
94 WARNER. Foundations of Differentiable Manifolds and Lie Groups.
95 SHIRYAEV. Probability. 2nd ed.
96 CONWAY. A Course in Functional Analysis. 2nd ed.
97 KOBLITZ. Introduction to Elliptic Curves and Modular Forms. 2nd ed.
98 BRÖCKER/TOM DIECK. Representations of Compact Lie Groups.
99 GROVE/BENSON. Finite Reflection Groups. 2nd ed.
100 BERG/CHRISTENSEN/RESSEL. Harmonic Analysis on Semigroups: Theory of Positive Definite and Related Functions.
101 EDWARDS. Galois Theory.
102 VARADARAJAN. Lie Groups, Lie Algebras and Their Representations.
103 LANG. Complex Analysis. 3rd ed.
104 DUBROVIN/FOMENKO/NOVIKOV. Modern Geometry—Methods and Applications. Part II.
105 LANG. $SL_2(\mathbf{R})$.
106 SILVERMAN. The Arithmetic of Elliptic Curves.
107 OLVER. Applications of Lie Groups to Differential Equations. 2nd ed.
108 RANGE. Holomorphic Functions and Integral Representations in Several Complex Variables.
109 LEHTO. Univalent Functions and Teichmüller Spaces.
110 LANG. Algebraic Number Theory.
111 HUSEMÖLLER. Elliptic Curves.
112 LANG. Elliptic Functions.
113 KARATZAS/SHREVE. Brownian Motion and Stochastic Calculus. 2nd ed.
114 KOBLITZ. A Course in Number Theory and Cryptography. 2nd ed.
115 BERGER/GOSTIAUX. Differential Geometry: Manifolds, Curves, and Surfaces.
116 KELLEY/SRINIVASAN. Measure and Integral. Vol. I.
117 SERRE. Algebraic Groups and Class Fields.
118 PEDERSEN. Analysis Now.

119 ROTMAN. An Introduction to Algebraic Topology.
120 ZIEMER. Weakly Differentiable Functions: Sobolev Spaces and Functions of Bounded Variation.
121 LANG. Cyclotomic Fields I and II. Combined 2nd ed.
122 REMMERT. Theory of Complex Functions. *Readings in Mathematics*
123 EBBINGHAUS/HERMES et al. Numbers. *Readings in Mathematics*
124 DUBROVIN/FOMENKO/NOVIKOV. Modern Geometry—Methods and Applications. Part III.
125 BERENSTEIN/GAY. Complex Variables: An Introduction.
126 BOREL. Linear Algebraic Groups. 2nd ed.
127 MASSEY. A Basic Course in Algebraic Topology.
128 RAUCH. Partial Differential Equations.
129 FULTON/HARRIS. Representation Theory: A First Course. *Readings in Mathematics*
130 DODSON/POSTON. Tensor Geometry.
131 LAM. A First Course in Noncommutative Rings.
132 BEARDON. Iteration of Rational Functions.
133 HARRIS. Algebraic Geometry: A First Course.
134 ROMAN. Coding and Information Theory.
135 ROMAN. Advanced Linear Algebra.
136 ADKINS/WEINTRAUB. Algebra: An Approach via Module Theory.
137 AXLER/BOURDON/RAMEY. Harmonic Function Theory.
138 COHEN. A Course in Computational Algebraic Number Theory.
139 BREDON. Topology and Geometry.
140 AUBIN. Optima and Equilibria. An Introduction to Nonlinear Analysis.
141 BECKER/WEISPFENNING/KREDEL. Gröbner Bases. A Computational Approach to Commutative Algebra.
142 LANG. Real and Functional Analysis. 3rd ed.
143 DOOB. Measure Theory.
144 DENNIS/FARB. Noncommutative Algebra.
145 VICK. Homology Theory. An Introduction to Algebraic Topology. 2nd ed.
146 BRIDGES. Computability: A Mathematical Sketchbook.
147 ROSENBERG. Algebraic K-Theory and Its Applications.
148 ROTMAN. An Introduction to the Theory of Groups. 4th ed.
149 RATCLIFFE. Foundations of Hyperbolic Manifolds.
150 EISENBUD. Commutative Algebra with a View Toward Algebraic Geometry.
151 SILVERMAN. Advanced Topics in the Arithmetic of Elliptic Curves.
152 ZIEGLER. Lectures on Polytopes.
153 FULTON. Algebraic Topology: A First Course.
154 BROWN/PEARCY. An Introduction to Analysis.
155 KASSEL. Quantum Groups.
156 KECHRIS. Classical Descriptive Set Theory.
157 MALLIAVIN. Integration and Probability.
158 ROMAN. Field Theory.
159 CONWAY. Functions of One Complex Variable II.
160 LANG. Differential and Riemannian Manifolds.
161 BORWEIN/ERDÉLYI. Polynomials and Polynomial Inequalities.
162 ALPERIN/BELL. Groups and Representations.
163 DIXON/MORTIMER. Permutation Groups.
164 NATHANSON. Additive Number Theory: The Classical Bases.
165 NATHANSON. Additive Number Theory: Inverse Problems and the Geometry of Sumsets.
166 SHARPE. Differential Geometry: Cartan's Generalization of Klein's Erlangen Program.
167 MORANDI. Field and Galois Theory.
168 EWALD. Combinatorial Convexity and Algebraic Geometry.
169 BHATIA. Matrix Analysis.
170 BREDON. Sheaf Theory. 2nd ed.
171 PETERSEN. Riemannian Geometry.
172 REMMERT. Classical Topics in Complex Function Theory.
173 DIESTEL. Graph Theory.
174 BRIDGES. Foundations of Real and Abstract Analysis.
175 LICKORISH. An Introduction to Knot Theory.
176 LEE. Riemannian Manifolds.
177 NEWMAN. Analytic Number Theory.
178 CLARKE/LEDYAEV/STERN/WOLENSKI. Nonsmooth Analysis and Control Theory.